SUPER YOU

How Technology Is Revolutionizing What It Means to Be Human

ANDY WALKER, KAY WALKER, AND SEAN CARRUTHERS

800 East 96th Street,
Indianapolis, Indiana 46240 USA

Super You

ISBN-13: 978-0-7897-5486-8
ISBN-10: 0-7897-5486-X

Library of Congress Control Number: 2016937879

Printed in the United States of America

First Printing: June 2016

Trademarks

Warning and Disclaimer

Special Sales

For information about buying this title in bulk quantities, or for special sales opportunities (which may include electronic versions; custom cover designs; and content particular to your business, training goals, marketing focus, or branding interests), please contact our corporate sales department at corpsales@pearsoned.com or (800) 382-3419.

For government sales inquiries, please contact governmentsales@pearsoned.com.

For questions about sales outside the U.S., please contact intlcs@pearson.com.

Editor-in-Chief
Greg Wiegand

Executive Editor
Rick Kughen

Development Editor
Rick Kughen

Managing Editor
Sandra Schroeder

Project Editor
Lori Lyons

Project Manager
Ellora Sengupta

Copy Editor
Paul Gottehrer

Indexer
Cheryl Lenser

Proofreader
Srimathy

Publishing Coordinators
Cindy Teeters

Cover Designer
Steve Huntriss

Compositor
codeMantra

CONTENTS AT A GLANCE

Introduction: The Confronting Nature
of Becoming Super 1

1 The Emergence of (You) the Human Machine 5

2 Baby Science: How to Conceive a Tennis Star and
Other Procreative Miracles 25

3 Beauty Hacks: Becoming Barbie, a Lizard, or Whatever
You Want to Be 71

4 Lifesaving Hacks: Whirring Hearts, Printed Organs, and
Miraculous Medicine 125

5 The Human Computer: How to Rewire and Turbo-Boost
Your Ape Brain 163

6 Franken-You: A Better Life Through Cyborg Technology 211

7 In Hacks We Trust? The Political and Religious Backlash
Against the Future 241

8 Hyper Longevity: How to Make Death Obsolete 263

9 Human 2.0: The Future Is You 297

Index 319

TABLE OF CONTENTS

Introduction: The Confronting Nature of
    Becoming Super   **1**

1   **The Emergence of (You) the Human Machine**   **5**

    Thumbs Up to Opposable Thumbs   10

    Is Evolution Obsolete?   11

    Is Technology the New Evolution?   13

        How Logarithmic Improvements Will Change
            Your Life   14

    How the Technology Singularity Will Change Your Life   21

2   **Baby Science: How to Conceive a Tennis Star
and Other Procreative Miracles**   **25**

    Baby Technologies   27

        Baby Food   27

    Ultrasound Technology   31

        3D and 4D Ultrasound   31

        Smartphone-Based Ultrasound   32

    The Future of Ultrasound   32

        High-Intensity Focused Ultrasound   33

        Magneto-Acoustic Imaging   33

    Birth Control Technologies   33

        Types of Birth Control   34

    The Future of Birth Control   37

        Remote Control Birth Control   37

        Male Contraception   38

    Curing Mr. Happy   39

    Baby-Making Technologies   42

        The First Test Tube Baby   42

    The Future of Reproduction Technology   45

        Next Generation DNA Sequencing   45

    Baby-Saving Technologies   47

        Stem Cells from Cord Blood   47

        Genome Sequencing   48

Nanotechnology for Babies 49

3D Printing To Save Babies 51

Designer Babies 52

Gender Selection 54

    Nature's Method 54

    Amniocentesis 55

    The Ericsson Method 56

    Sperm Spinning 56

    Preimplantation Genetic Diagnosis 57

The Fertility Doctor, the Media, and the Vatican 58

Genetically Engineered Babies 61

Testing 1, 2, 3 ... Is Your Baby Genetically Healthy? 62

Daddy, Mommy, and Your Extra Mommy 64

Pregnant Men? 66

    The First "Pregnant Man" 66

    The Future of Male Pregnancy 67

    Lab-Grown Vaginas 67

From Birth to Forever 68

3 Beauty Hacks: Becoming Barbie, a Lizard, or Whatever You Want to Be   71

Cosmetic Surgery Trends in America 73

Lizardman 75

Extreme Cosmetic Hall of Fame 78

    Valeria Lukyanova—"Real Life Barbie" 78

    Justin Jedlica—"The Human Ken Doll" 79

    Maria Jose Cristerna—"Vampire Mom" 79

    Patricia Krentcil—"Tanning Mom" 79

    Lacey Wildd—Largest Breasts 80

    ORLAN—Performance Artist 80

Cosmetic Enhancements, Non-Weird Edition 81

    Cosmeceuticals 81

    Needle-Free Botox 83

    Guyliner: Makeup for Men? 84

Electro Cosmetics 85

Cosmetic Stickers 86

Tattoos 86

    Tattoos Today 87

    Freedom-2-Ink 88

    LED Tattoos 89

    Gadget-Activated Tattoos 89

Hair 91

    Hair Removal 91

    Hair Loss 93

Weight Loss 97

    Soylent 98

    Weight-Loss Surgery 101

    Gastric Bypass 102

    Sleeve Gastrectomy 103

The Future of Thin 103

    Zerona Laser 105

    Liposonix 106

Cosmetic Surgery: A History 106

Cosmetic Surgery Today 108

    Nose Job 108

    Liposuction 108

    Eyelid Surgery 109

    Tummy Tuck 109

    Facelift 109

    Dermabrasion 109

    Cosmetic Dentistry 110

    Vaginoplasty 110

    Breast Surgery 110

    Breast Augmentation 111

    Bigger Boobs ... Today 112

    Fat Transfer Breast Augmentation 115

    Ideal Implants 116

Sex Reassignment 117

Weird and Wonderful Cosmetic Surgery 119

    Palm Alterations 119

    Double Eyelid Surgery 119

    Iris Surgery 119

Foot Surgery 120

Limb Lengthening 120

The Future of Beauty 121

**4 Lifesaving Hacks: Whirring Hearts,
Printed Organs, and Miraculous Medicine   125**

A Brief History of Medicine 126

There's a Future in Plastics 128

Taking a Peek Inside 129

Have a Heart—Pacemakers, Transplants,
and Artificial Hearts 130

Who Needs a Pulse, Anyway? 135

It's Now, and It's New: The Future Frontiers
of Medicine 138

Nano, Nano 138

I Sing the Body Electronic! 141

Up Close and Personal with Your Genome 146

DIY Body Parts 149

Stop the Bleeding 152

A Sensor of Wonder 153

Hi, I'm Your Robot Doctor ... Wait! Come back! ... 156

DIY Health 160

**5 The Human Computer: How to Rewire and
Turbo-Boost Your Ape Brain   163**

A Brief Early History of the Brain 165

Your Brain Is Plastic? What? 169

Do it Yourself (DIY) Brain Technology 173

Go Learn Something 173

Meditate on This 175

Brain Fixers 178

Brain-Zapping Fixes 180

Deep Brain Stimulation 182

Brain Drugs 183

No New Brain Cells: Myth Busted 184

Brain Enhancers 185

Over-the-Counter Drugs to Improve Brain Plasticity ... 187

This Is Your Brain on Video Games 188

More Super Cool Brain Projects 190

    Brain-Controlled Exoskeleton 191

    OpenWorm Project 192

    Human Brains in Bots 193

    Brain-to-Brain Communication 194

    Human-to-Human Brain Communication 195

Unsolved Mysteries of the Brain 196

    What Is Consciousness? 196

    Why Do We Sleep? 197

    Do We Have Free Will? 197

    How Are Memories Processed? 198

Is that It? 199

The Future 200

    Expanding Human Intelligence 200

    Artificial Intelligence (AI) 203

Will the Machines Rise Up? 205

    … And What About Conscious Robots? "GULP" 206

Super Us? Here Come Virtual Helpers Armed
with Strong AI 209

**6   Franken-You: A Better Life Through
Cyborg Technology   211**

Are you a Cyborg? Or just Bionic? 213

What the Heck Is Cybernetics? 215

    Mostly Evil Killing Machines? 216

    This Cyborg Life 217

Let's Start with Bionics 219

I ♥ Technology 220

I See U 221

I Am the Very Model of a Modern Cybernetic Man 221

Super Senses 225

    Mr. Cyborg: Steve Mann 225

    Neil Harbisson: The Man Who Can Hear Red 226

    Jens Naumann: Now You See Me, Now You Don't 227

    Michael Chorost and the 100-Acre Wood 228

    Rob Spence: Life Imitates Art 228

Super Strength 229

Amanda Boxtel: Exoskeleton Pioneer 230

Super Powers 231

Jerry Jalava: To USB or Not to USB 231

Amal Graafstra: Open Sesame 231

Kevin Warwick: Wife-Fi Connector 232

Pranav Mistry: Cruise Control 233

Super Body: Wearables for Amputees 235

Jesse Sullivan: Resume Hugging 235

Nigel Ackland: Give this Guy a Hand 235

DIY Bio-Hacking 236

Gabriel Licina: Cat's Eyes 236

The Grinder Movement: DIY Surgery 236

The Future of Cyborg? 238

7 In Hacks We Trust? The Political and Religious Backlash Against the Future   241

The Future Is ... Now? 244

Body Modification 245

Judaism 246

Christianity 247

Islam 248

Hinduism 248

Genetic Engineering 250

Cloning 252

Stem Cell Research and Genetic Therapy 254

Political Views on Stem Cell Research 256

Religious Views on Stem Cell Research 257

Emerging Technologies 260

8 Hyper Longevity: How to Make Death Obsolete   263

The History of Aging 264

The Methuselah Award Goes to 266

What We Know About Super Agers 267

Longevity Research Is Still Young 268

Lifestyle Secrets: Live Long and Prosper 270

Centenarian Studies 271

    The Blue Zones 271

    Power 9: The Nine Lifestyle Choices that Promote
        Longer Life Spans 272

    The New England Centenarian Study 275

The Longevity Genes Project 277

Strategies for a Longer Life 278

    Calorie Restriction 278

    Red Wine and Resveratrol 281

Current Bodies of Research in Longevity 283

    Sirtuin Studies 283

    Amazing mTOR 286

    Insulin Signaling (Long Live the Worms) 287

    AMPK, the Cellular Housekeeper 289

Living Forever: The Research of Dr. Aubrey de Grey 289

Cryonics: Freeze Me When I Die So I Can Live Forever 292

Reports of Your Death Are Greatly Exaggerated 294

Extendgame, Not the Endgame 295

9 Human 2.0: The Future Is You   297

Look Like Who You Want to Be and Be Who You
    Want to Be 299

Customize Your Children So They Live a Life
    Free of Disease 301

Live Your Life Disease Free 303

Be Superhuman 306

Live Forever, If You Choose 308

Robots Replacing Jobs—Less Work and More Fun 310

    Stage 1: The Bots You Know 311

    Stage 2: The Bots Are Coming 311

    Stage 3: Smarter, Strategic Bots 312

    Stage 4: Here Come the Lawyer Bots, et al 312

Restructured Society, Economy and Political System 314

A Final Word 318

Index   319

Foreword

Let's face it, the world is moving quickly into a future full of uncertainty. One where for the first time in history, human beings may no longer be classified as a mammal. One where technology will naturally be a part of man's evolutionary footprint.

There are many people today that have done the research. They understand the future we are inevitably walking into. Technology is growing faster than it ever has before. There will come a day where man will build machines that match and then supersede his intelligence.

These groups use different words to describe the new era: Transhumanist, posthuman, techno-optimism, cyborgism, humanity+, immortalist, machine intelligence, robotopia, life extension, or Singularity.

While there is overlap, each name represents a unique camp of thought, strategy, and possible historical outcome for the people promoting their vision of the future. Collectively, they all believe in the same experience of life, something the authors have captured so expertly in this book. We are walking into the "Age of Super You."

This book is a guide to that future. It explores some of the major themes we face in this new era. How will science and technology impact our humanity? What will we look like? What will our children look like and become? Will we be healthy? Will we live for hundreds of years? Can death be cured? Who will be in the way of all this? Can our leaders, government, and clerics keep up with it all?

As this book goes to press, I am in the final months of running for President of the United States. I'm running as the leader of the Transhumanist Party, a political organization I founded seeking to use science and technology to radically improve the human being and the society we live in.

In addition to upholding American values, prosperity, and security, many of these issues are on my mind.

The three primary goals of my political agenda are as follows:

1. Attempt to do everything possible to make it so this country's amazing scientists and technologists have resources to overcome human death and aging within 15 to 20 years—a goal an increasing number of leading scientists think is reachable.

2. Create a cultural mindset in America that will embrace and produce a radical technology and science that is in the best interest of our nation and species.

3. Create national and global safeguards and programs that protect people against abusive technology and other possible planetary perils we might face as we transition into the transhumanist era.

There is a burgeoning movement in the United States, and as well as many other countries, that we need to prepare for a robust technology-enhanced future. The authors have crisply defined the trend in their book and have cleverly looked back to see how it happened and look forward to see where it is going. Even though it wasn't intended to be a guide to the future and how we got there, in many ways it is.

The book examines how humans are transforming themselves through science and technology to become better versions of themselves, to live healthier and more fulfilling lives, and to hyperextend their human capabilities, including their longevity.

There's increasing evidence everyday from the frontiers of science—and already in the warehouses of ecommerce companies—that much of what this book examines and predicts is, or soon will be, a near-term reality.

In 1969, as man first walked on the moon, we lived in a largely analog society. There was no sign of the digital economy that was about to sweep the world in the next three to four decades, and yet, life has since radically changed. First, with the personal computer in the 1980s and the birth of the consumer Internet revolution in the 1990s and the mobile revolution in the 2000s and 2010s. Information that was once limited to black and white television screens on the evening news and in newsprint each morning now flows freely and ubiquitously on-demand, and with no regard for your location.

All that was around the corner back then, but only a few could see it. The Jetson future was about the raw power of rockets, a push-button work week, and flying cars, rather than the simplicity and elegance of the Internet, the transformational nature of a digital economy, and the radical hybridization of machine and human that's about to arrive.

The future is hard to predict with any accuracy. Still, here in 2016, we know a major shift is about to happen again. We can see the macro-trends of the technology Singularity coming, and fast. It will be tectonic in how it shifts society.

You'd think every politician in the twenty first century would be publicly and passionately pursuing and preparing for the future. But they're not. They're more interested in landing your votes, making you slave away at low-paying jobs, keeping you addicted to shopping, forcing you to accept bandage medicine and its death culture, and getting you to pay as much tax as possible to fund far-off wars. There's no regard for the future.

And if transhumanists—a growing group consisting of futurists, life extensionists, biohackers, technologists, singularitarians, cryonicists, techno-optimists, a few authors, and many other scientific-minded people—are serious about the pending future, then it's time to get involved. If it's new territory for you, start by reading this book.

Zoltan Istvan
Presidential candidate and founder of The Transhumanist Party
April 2016

About the Authors

Andy Walker has had a long career as one of North America's top technology journalists. In the last two decades, he has written about consumer technology for dozens of national newspapers, magazines, and websites. His personal technology advice column was syndicated across Canada and today his body of work is published at technologytips.com where more than 50 million unique visitors have read the advice over the last decade.

Andy was also a cohost on the internationally syndicated TV show *Call for Help* with Leo Laporte on G4TechTV as well as writer and host of several spinoff shows.

Super You is his fifth book (he has written four with Pearson Education).

He has also worked with some of the top luminaries in technology publishing. Between 2002 and 2004, Andy was the executive editor of Berkeley-based *Dig_iT* magazine, a publication focused on the digital lifestyle. It was founded by David Bunnell and Fred Davis, the publishing pioneers behind *PC* magazine, *PC World*, *MacUser*, and *MacWorld*.

Walker has a passion for technology literacy. He created the Canadian charity Little Geeks, which gives computers to children and families in need. He is also a recipient of the Queen Elizabeth II's Diamond Jubilee Medal for his work in technology literacy and digital publishing.

Andy was a pioneer in video podcasting with the hit Internet show *Lab Rats*, which he cocreated with *Super You* coauthor Sean Carruthers, and can also be seen and heard regularly across the dial on national radio and television commenting on emerging technology trends.

He has had consulting roles on content and business development projects for Microsoft, Yahoo!, and Canadian Press Enterprises.

Andy was born in the UK, educated and raised in Canada, and now lives in Tampa, Florida, with Kay, his wife and coauthor, and Carter, their first child.

They also run Cyberwalker Digital (Cyberwalker.com), an online marketing agency.

You can reach Andy, and learn more about him and his coauthors at readsuperyou.com.

Kay Walker is a life hacker. She teaches people noninvasive tools—neuroplasticity exercises, emotional IQ skills, and personal development tactics—they can use to access their full potential and overcome their biological limitations that hold them back from living a life they love. She's the creator of AwesomeLifeClub.com, an exclusive club for individuals who want to learn tangible tools they can use to become super performers in all areas of life.

Walker is well known for her advocacy work in the mental health field. She runs a resource site Depression Zone (http://depression.zone) where she provides online support, books, courses, and private coaching services for people suffering from depression.

She's also married to coauthor Andy Walker. It's not the first project the two have collaborated on. They run a digital marketing agency, Cyberwalker Digital (based in Tampa, Florida) where they teach businesses and entrepreneurs how to strategically market their businesses on the Internet.

Though she's well-versed in digital marketing, Kay is the least "techie" of the three authors. She helped refine *Super You* into a book for a mainstream audience. She also brings a female perspective to some of the more gender-specific topics covered in the book, such as designer babies and cosmetic surgery.

Sean Carruthers has been writing, podcasting, and broadcasting about technology for nearly two decades. Sean was a content producer on the G4TechTV programs *The Lab with Leo Laporte, Gadgets and Gizmos, Torrent,* and *Call For Help.* He was also one of the early pioneers in the world of video podcasting, where he cocreated, cohosted, and edited the long-running technology program *Lab Rats* with coauthor Andy Walker.

He served as the Test Lab Editor for *The Computer Paper* and *HUB: Digital Living* magazine, where he was always up to his eyebrows in the hottest new technology. His writing has been featured in *The Globe and Mail,* the *Village Voice,* allmusic.com, *ITWorld Canada,* and various other technology and music publications. He has also produced audio and video content for various outlets, including CBC Radio's program *Spark.*

Currently, Sean is the manager of the custom video department of The Canadian Press Enterprises Pagemasters North America subsidiary, where he has overseen the production of nearly ten thousand videos.

Dedication

The authors dedicate this book to Carter Devon Walker, the son of authors Andy and Kay, who was birthed, learned to walk, talk, and work a touch screen, in the time it took to write this book. His parents and "Uncle Sean" believe his generation will be the greatest yet.

Acknowledgments

The authors would like to acknowledge the amazing contributions of the following people, without whom this book would not have been possible.

First, a warm thank you to everyone who agreed to speak to us during our research. We were just thrilled to talk to so many dedicated and talented scientists and experts. They were generous with their time and expertise. So thank you to Ray Kurzweil, Zoltan Istvan, David Sinclair, Jeffrey Steinberg, Eric Sprague, Bertelan Mesko, Aubrey de Grey, David Bunnell, Amber Case, Billy Cohn, Kevin Warwick, Robert Murphy, and all the amazing geniuses we spoke to while researching this book.

This book was also a true family affair:

Cornelia Svela: Cornelia is Kay's mom, an amazing artist and illustrator who agreed to provide the amazing illustrations you see in this book. That, plus undying love, support, and enthusiasm. Thanks Cornelia, we love you!

Kris Svela: Kay's talented and hardworking journalist father provided indispensable research for several chapters in this book. He was a champion of the project all the way through. We love you, Kris!

Ian Svela: Ian is Kay's talented brother. He not only provided volumes of research, but is a talented writer, who has an incredible future as as an author himself. He was an irreplaceable resource, who threw himself into weeks of research for this book and helped us through some of the most intensive chapters, even in the lead up to the birth of his second son. Thank you, Ian. You are incredible!

Then there are our publishing mentors:

Rick Kughen: This book came about when our Executive Editor Rick was watching a documentary on body hackers, who put magnets under the flesh of their finger tips. He has an innate ability to spot a great idea and make it into something awesome. He is funny, supportive, and inspired, and we are lucky to count him as both a friend and colleague. Thank you for your hard work, dedication, and inspiration Rick! You made this book what it is, and it wouldn't have been possible without you.

Sam Hiyate: Our literary agent Sam is an extraordinary champion of great writing, great authors, and great books. He is a fierce advocate for book publishing and perhaps the most generous person around with his time in developing young talent. We love and admire you Sam. Thanks for everything you do.

We would be remiss, if we didn't acknowledge the awesome team at Pearson Education including Laura Norman, Lori Lyons, and Ellora Sengupta. They are a talented bunch we are extremely fortunate to have behind us. Our names may be on the front of this book, but there would be no book without their commitment and dedication. Thank you so much team.

And a very special thanks to Daniel Beylerian, an inspired computer sciences teacher, and his awesome colleagues, teacher Rita de Melo and curriculum coordinator Randy Votary from Trenton High School in Ontario, Canada. They put together an extraordinary group of students who helped with core research for this book. Once a week for several months, teachers and students got together to research questions we posted and provided us with reams of handy information, which we included in this book. Special thanks to the hard-working students: Michelle Abbott, Kayla Egas, Alex Graham, Ryan Montminy, and Morgan Walker. Additional thanks to the students' parents and the leadership at Prince Edward District School Board for supporting the project with their enthusiasm.

Sean would also especially like to thank his wife Eileen, who put up with his disappearing into this book for extended periods, which seriously cut into hiking time.

The authors would also like to thank each other for being so awesome. And to you the reader for getting this far down the page. Without you, we're not much use at all.

We Want to Hear from You!

As the reader of this book, *you* are our most important critic and commentator. We value your opinion and want to know what we're doing right, what we could do better, what areas you'd like to see us publish in, and any other words of wisdom you're willing to pass our way.

We welcome your comments. You can email or write to let us know what you did or didn't like about this book—as well as what we can do to make our books better.

Please note that we cannot help you with technical problems related to the topic of this book.

When you write, please be sure to include this book's title and author as well as your name and email address. We will carefully review your comments and share them with the author and editors who worked on the book.

Email: feedback@quepublishing.com

Mail: Que Publishing
          ATTN: Reader Feedback
          800 East 96th Street
          Indianapolis, IN 46240 USA

Reader Services

Register your copy of *Super You: How Technology Is Revolutionizing What It Means to Be Human* at quepublishing.com for convenient access to downloads, updates, and corrections as they become available. To start the registration process, go to quepublishing.com/register and log in or create an account.* Enter the product ISBN, 9780789754868, and click Submit. Once the process is complete, you will find any available bonus content under Registered Products.

*Be sure to check the box that you would like to hear from us in order to receive exclusive discounts on future editions of this product.

Introduction

The Confronting Nature of Becoming Super

People talk about all kinds things when they eat dinner together: The weather. The family. What's new at work. How Aunt Miriam fell in a fjord on her trip to Norway. You know, mostly humdrum things.

We don't. During the writing of this book, we talked about weirder topics with our friends and family over dinner such as:

- How you are probably going to live forever, if you are under 60 today.

- How you can genetically engineer your baby to be a tennis star.

- How Viagra cures jetlag in hamsters.

- How a scientist is putting rat brains in robots.

- How a replacement "jet engine" heart works without beating.

- How the United States is going to be economically bulldozed by more progressive and less tech-resistant nations such as China if it doesn't fix its resistance to technological progress.

- Who in your neighborhood is a cyborg.

- Will there be nanobots in your bloodstream soon.

- When will you be able to Google song lyrics with your thoughts.

These topics amazed, upset, and generally freaked out many of our dinner companions. And to be honest, we quite enjoyed the mayhem.

You see, the discovery process of writing this book challenged us to think about humans and our humanity in new ways. It was confronting for us. And when we shared it, it was very much so for all those that passed the potatoes. (Note to aspiring authors: Dinner tables are really great focus groups.)

When we encountered anyone with strong religious beliefs (especially those who expect to meet their maker one day), the conversation got quite heated. The prospect of living forever is quite confronting, as is the idea that one day soon (we're talking decades here) we will all be as much machine as human.

We cover all these topics in this book and dozens more. We answered questions such as:

- Why is technology improvement speeding up?
- Why is death being treated like a curable disease?
- Will I live long enough to cure my eventual death?
- Can I eliminate disease in my children?
- Will technology eliminate my job?
- Can I modify my looks to look like anyone (or anything!) I want?

Read this book and each one of these zingers—and many more wacky concepts—can be a topic of fascinating conversation at your own dinner table. In fact, reading this book will make you the most interesting person your friends, family, and colleagues know.

You will be armed with an access to the future that few people have been privy to ever before.

How can this be possible? Because we interviewed some of the most interesting people on the planet for this book. The list of scientists and technologists we spoke to reads like a who's who of science royalty, at least as far as longevity, genetics, nanotechnology, and robotics is concerned.

- Yes, we talked to artificial brain designer and futurist Ray Kurzweil.
- Yes we talked to Zoltan Istvan, the transhumanist presidential candidate who should get your vote.
- Yes, we talked to Jeffrey Steinberg, the fertility doctor who knows how to make your baby's hair curly and her (or his, you choose) eyes blue.
- Yes, we talked to Kevin Warwick, who wirelessly connected his brain to his wife's forearm so he knew when she was thinking of him.
- Yes, we interviewed Erik Sprague, aka Lizardman, a surprisingly intelligent and philosophical man who looks like a green reptile with a forked tongue.

And there are dozens more wonderful and inspiring geniuses in these pages working for a super you.

That said, it's important to know that this book is not some idealized manual for a utopian future. Nor is the book overly dystopian.

We did discover one core macro-theme, however. Humans and their ingenuity are accelerating all humanity toward a fantastic and wondrous future that will inevitably be rife with socially driven detours, political potholes, and philosophical traffic jams.

We discovered that what's in the way of that inevitable future...is humanity itself. Funny, that! If human creativity, inspiration, and perseverance are the engine that will get us there, then ignorance, bureaucracy, and fear are the brakes. Expect a few nasty skid marks along the way.

What is guaranteed is, we will all become ever more super with every tick of the clock. And this book is the guide you'll need to understand what is about to happen to us, as we all become—let's call it—SUPERYOUtopian.

Andy Walker, Kay Walker, and Sean Carruthers
February, 2016

1

The Emergence of (You) the Human Machine

An Extremely Brief History of How Apes Got Smart and Replaced Evolution with Technology

Once upon a time, 3.5 billion years ago, there was this group of amino acids— a bunch of molecules made of carbon, oxygen, and hydrogen—that assembled themselves into DNA—deoxyribonucleic acid.

That ancient chemical miracle was the beginning of you. And me. And all living things.

For the unacquainted, DNA is found in every living cell and is the blueprint for life. That elemental fabric of life allows for simple cells to form. And in turn they have gotten more complex over generations—from a few simple-celled critters to various generations of complex beasties—sponges, fish, lizards, and furry things that looked like aquatic hamsters. Or so we imagine, because we weren't there.

(Proper scientists will tell you something way more accurate. But they will also go on about the pre-Paleozoic Era for hours and point to a lot of dead things imprinted in rocks.)

Eventually this proto-critter would accidently leap out of the water, find lunch on a beach, and decide not to go back to the ocean. One day a long time later—perhaps 70 million years or so—it evolved into something resembling all us modern-day humans: An ape of sorts.

It wasn't very smart. It had no aspirations to invent the Internet, or an electron microscope, or had any illusions that it would one day eat reheated chicken on a tray on a trans-Atlantic flight at 37,000 feet on its way to a nice holiday in Paris.

Somewhere along the way, the ape man was chilly, and discovered that burning stuff made it warmer. So came fire and then a stick with meat on it. And BBQ was born.

This was awesome because it allowed humans to diversify their food sources, which increased their ability to survive, as anyone with a good cookbook knows. After a ponderous epoch or two, it developed into an incrementally improved version of proto-human.

Now it was making progress. But this progress was very, very slow.

And arguably, that same archaic process that biologically nudged humans from their beginnings as furry chittering mammals to Justin Bieber-like Homo sapiens, continues to this day.

It's called natural selection.

In 1831, naturalist and geologist Charles Darwin boarded the good ship HMS *Beagle* and set sail for South America. On the five-year voyage around the world, he encountered birds and animals that got him thinking. They all had commonalities that suggested they evolved from a common ancestor. When he got home, he wrote a page-turner called *The Origin of Species by Means of Natural Selection*.

The work postulated that an animal's survival came from its ability to adapt to its circumstances and the world around it, and critically, to avoid being killed by extreme weather, hunger, larger animals, or random volcano eruptions. Those that survived reproduced. Those that didn't, failed to make babies, thus ending their ability to forward their bloodline. This evolution allowed for strong species to get even stronger as each new generation arrived.

These days, most of us on this planet no longer have to struggle to survive. Outside of accidents or diseases that are related to aging, *most* healthy humans—at least those that live in developed nations—are not likely to die from their environment or circumstances.

How do we know?

One indicator is infant mortality rates.

In Western countries—which possess the most advanced technologies on the planet—infant mortality rates have fallen drastically in the last 50 years as science and technology improvements have accelerated (see Table 1.1).

Table 1.1 Infant Mortality Rates for Technology Using Nations (1950 to 2010)

Country	Mortality rate 1950–1955 (deaths/ 1,000 births)	Mortality rate 1980–1985 (deaths/ 1,000 births)	Mortality rate 2005–2010 (deaths/ 1,000 births)
USA	30.46	11.60	6.81
Canada	38.39	9.31	5.22
UK	28.67	10.83	4.91
Australia	24.05	9.93	4.66
Japan	50.07	6.63	2.62
South Korea	137.95	24.61	3.76

Source: Wikipedia, attributing UN data.

When we speak of technology here, we're not just talking about computers. The generic use of the word "technology" presupposes that science has developed the thinking and has done the research that leads to the application of science so that useable tools can be developed to improve human life. Technology improves and makes affordable food supplies. It provides low cost and efficient fuel sources. It provides clean water. It helps cure disease, improves general human wellness, and optimizes quality of life.

A country that has robust programs for science and technology and has the capacity to apply the resulting innovation to make life easier for its population will see drastic improvements in infant mortality.

 What Is Infant Mortality?

The number of deaths of infants under the age of one year per 1,000 live births. This term is often used as an indicator of the level health in a country.

Technology is the ***relevant*** evolutionary process that impacts our existence.

This process is as true for humans as it is for lizards that can camouflage themselves as rocks, and for camels that can cross a desert and not succumb to thirst.

So it's fair to say that a human's biggest asset in this process is his or her brain. You would think that a big, strong, furry bear with knife-like claws would be much better at getting along on an icy planet compared to us skinny, furless, drippy-nosed ape men. Yet, today we are not living in a world where bears run the show. We humans do.

But why?

Because that gray throbbing goo in our human heads is the most magnificent thing that has ever happened to a living creature. This advanced magic organ gives us the capacity to invent fun and useful ideas that help us survive and ultimately thrive. It is better than bear brains and better than the brains in any other critters. And it gives us the edge on all other living things on the planet, despite evidence to the contrary some days—such as the way people drive on most public highways, or ill-conceived public statements made by some politicians on network television.

About 2.5 million years ago, early humans were carting around a brain weighing about 1 pound; but 2 million years or so later (around 200,000 to 400,000 years ago), the human brain became much bigger than those of competing species. Today humans walk around with brains tipping the scales at about 3 pounds.

The bio-upgrade is something called the neocortex. The "neo" means "new." It is a relatively recent biological development in mammals dating back 200,000 years along the evolutionary timeline. Other nonmammalian animals, such as reptiles and birds, have smaller brains with limited abilities to process sensory information and control behavior.

If you remove a human brain from its skull, 85 percent of what you'd see is neocortex (see Figure 1.1). The neocortex is comprised of four specific lumps of brain (or lobes). Each area has a hard-to-remember name and is capable of a variety of unique functions. A general overview of the four lobes are:

> **Frontal lobe**—This lobe processes reward, attention, memory, planning, and motivation—so the next time you indulge in a chocolate bar or have sex with your partner, you are activating this part of your brain.

> **Occipital lobe**—The part of the brain that enables you to see. If you are out bird-watching on Sunday afternoon, the occipital lobe is at work.

> **Parietal lobe**—This center lobe deals with somatosensory processing, which is really just a fancy word for the way your body reacts to your environment. When you're chopping onions at a high speed in your kitchen, you are using this area of the brain.

> **Temporal lobe**—This area involves a complex structure that looks like a sea-horse. The hippocampus, as it's called, helps you process long-term memory. So, when your boss's criticism triggers flashbacks of that F your 9th grade teacher gave you on your English paper, this area of the brain is at play.

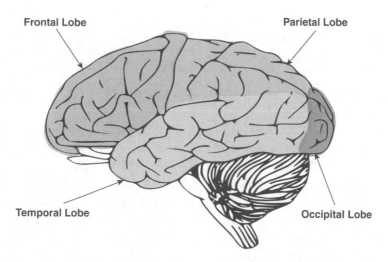

Figure 1.1 *The four lobes of brain you can see here make up the neocortex which gives humans the capacity for language and motor skills such as tool making.* (Illustration by Cornelia Svela)

What is really important is that the neocortex allows for more advanced behaviors particular to humans—things such as high-level consciousness (*deep thinking and decision making*) and the capacity for language and motor skills such as tool making.

This enhanced gray organ helped early humans figure out that fire was a handy tool that made food delicious and storable; that the wheel was a much better method to transport snacks from, say, where they were foraged, to the lunch table; and that a sharp-pointed rock on a long stick could ward off a ferocious bear or another human who had his beady-eyes on your hard-foraged snacks.

DOES (BRAIN) SIZE MATTER?

Who has the largest brain on the planet? No, it's not physicist Stephen Hawking or the guys who invented Ben & Jerry's Ice Cream (though c'mon, they are pretty smart). It's actually the sperm whale. In terms of size, their brains are more than five times the size of humans, weighing in at 17 pounds.

So why aren't sperm whales ruling the world? As it turns out, when it comes to brains, size doesn't matter.

So let's look at the cow. If we compare the cow to a monkey, the theory of size is easily debunked. The monkey has a smaller brain but a higher cognitive ability—that is, the ability to think, judge, and remember things. Which could be why more monkeys follow a career in showbiz or at NASA.

When it came to brain-to-body ratios, we used to think that there was a link in terms of human intelligence. But, the human brain/body ratio is smaller than many birds (which are 1/12) and about the same as a mouse (1/40).

Still, most of us (barring a few politicians and more than a few reality stars) are pretty smart.

So maybe it's brain density that matters. If we flip back to the elephants again, it appears that they have more neurons per square inch than humans, which makes their brains far denser. And, still, we are smarter.

It's a good thing a clever neuroscientist by the name of Suzana Herculano-Houzel took it upon herself to count neurons in the human brain in 2009. She found that the human brain contained 86 billion neurons, compared to 23 billion neurons in elephants, 56 billion in rats, and 75 billion in mice.

It appears that regardless of size, weight, or mass, neurons seem to be the common denominator in terms of smartness.

Thumbs Up to Opposable Thumbs

The other evolutionary innovation that gives humans the edge is the opposable thumb, which allows us the ability to grasp an object with dexterity. Without this thumb, it's very hard to be an everyday human or do anything that requires dexterity.

Try steering your car sometime with four fingers or even pouring a stiff drink (not at the same time, please). Your thumb is critical for so many human tasks. And it's why your dog relies on you to open the door to go outside. Doorknobs are extremely difficult to use without a thumb and a capable, grasping hand.

 The Demise of Doorknobs

Doorknobs are set to become obsolete, at least in Vancouver, Canada. In 2014, all new housing built in the city had to have new accessibility features, including the replacement of doorknobs with easier-to-turn lever handles. So said a city council bylaw passed in late 2013.

This physical adaptation, unique only to humans, helped our species develop fine motor skills. Apes and some monkeys also have opposable thumbs, but humans have longer thumbs and can touch the thumb to the fingers with great strength and more contact area.

This gives us a grip featuring both power and precision, neither of which are found in our hairier primate cousins.

So tool usage in primates is minimal. Chimps are known to use sticks to pull bugs from a hole. Bonobos use water-soaked leaves for grooming. Few other creatures have the cerebral or physical capacity to invent tools that make life better or easier for themselves.

 Here Birdie, Birdie

Recent evidence shows crocodiles are known to use sticks to bait nest-building birds. Birds also use bits of bread, feathers, and leaves to bait fish. None of them, however, have invented anything to rival the Swingline stapler, the Roomba, or duct tape because that's not possible without dexterity given by opposable thumbs.

In the space of 3,000 years or so—a mere blink of an eye in terms of the history of the planet—modern man went from chasing hooved prey on the plains of Africa to driving dad's Chevy Impala to the prom.

All this thanks to this simple formula:

Opposable thumb + Complex brain = Technology innovation

And that's where the backstory ends and this book begins.

Is Evolution Obsolete?

So with all this technology progress, are we still evolving? Most argue that humans are. Some say we aren't.

Perhaps the most public dissenter to ongoing human evolution is British naturalist and broadcaster Sir David Attenborough. In an interview with *Radio Times*, a British TV and radio listings magazine similar to North America's *TV Guide*, Attenborough said: "I think that we've stopped evolving. Because if natural selection, as proposed by Darwin, is the main mechanism of evolution—there may be other things, but it does look as though that's the case—then we've stopped natural selection. We stopped natural selection as soon as we started being able to rear 95–99 percent of our babies that are born. We are the only species to have put a halt to natural selection, of its own free will, as it were."

> "We are the only species to have put a halt to natural selection, of its own free will, as it were."

Is he right?

Sir Attenborough argues our evolutionary process is now cultural. We inherit knowledge from previous generations and build on it. If he is right, the formula for planetary domination by humans needs a tweak.

Opposable thumb + Complex brain + Culture = Technology innovation

Perhaps this is the triple threat that allowed humans to become the rock stars of the Stone Age.

The capacity to create, use, and improve technology has taken over where evolution left off. It's clear that technology has greater impact on day-to-day survival than eye color, height, sensory advantages, or our ability to digest new sources of food, such as cow's milk.

 Milk Mutation

By the way, the ability for milk digestion by humans was allowed by a genetic mutation in Europeans as recently as 7,000 years ago, which in the history of mankind is very recent.

So is technology innovation more important than biological improvement through evolution? It is the most relevant process at play in the human race today with regard to survivability of the species.

And, is evolution obsolete?

Well hang on. Not necessarily.

Some researchers think that we're not quite done. There is still capacity for biological improvement. Natural selection occurs when a mutation gives a species an advantage.

 What Is a Mutation?

A mutation occurs when a segment of DNA in a cell is damaged and then replicates when the cell divides. The damage can be caused by chemicals or radiation, an error in the cell duplication process, or outside forces such as viruses. This can cause disease. It can cause a physical change. Or the error can cause nothing at all. Sometimes this mistake results in an improvement that gives the organism a slight advantage. And if that advantage is great enough, the next generation inherits the mutation and does better than its counterparts and reproduces, passing the mutation on to its offspring.

While our brains and bodies will still improve or adapt over time, helpful mutations will not happen fast enough to make a difference to anyone alive today.

Here's a good example: As many as 35 percent of people are born today without wisdom teeth. Some evolutionary biologists speculate that these back molars will eventually disappear from the human mouth altogether, even though once upon a time all humans were born with them.

Teeth were early man's built-in mode of catching, dismembering, and eating prey, as they are for predatory mammals today. Our distant ancestors survived on a diet of chewy plants as well as raw meat.

However, it doesn't take much to hunt down and disembowel a Big Mac. Industrialized food production and agriculture has eliminated the need for most survival hunting, so the extra molars are not so necessary, and it seems that fewer people are born with them in the twenty first century.

The ability to digest lactose, a sugar component found in milk, is also a helpful human mutation.

Today 95 percent of all North Europeans have the ability to digest lactose. However, before the domestication of livestock and dairy production, that genetic advantage didn't exist.

"Among populations with a long history of cattle herding and milk consumption, the ability to metabolize lactose is maintained into adulthood. These are clear examples that natural selection has recently acted upon our species after the origin of agriculture and the domestication of animals, and independently among different populations," wrote Jay T. Stock, author of "Are Humans Still Evolving?" (EMBO Reports, July 2008, Nature.com.)

Evolution and natural selection are slow processes that impact species over generations. Sad to say, we are mostly stuck with what we were born with.

That's not to say that sudden mutations are impossible.

As UC Berkeley's web page "Understanding Evolution" reports: "Over the past 50 years, we've observed squirrels evolve new breeding times in response to climate change, a fish species evolved resistance to toxins dumped into the Hudson River (in New York state), and a host of microbes evolved resistance to new drugs we've developed."

Is Technology the New Evolution?

If evolution has largely been muted or at least relegated to something that might help our species, but not single individuals in a lifetime, then are we humans doomed to live with our inefficient biology for the duration of our lives?

Not at all.

We can be super human. In fact, we already are thanks to technology.

Chip Walter perhaps said it best in his thought-provoking book entitled *Last Ape Standing: The Seven-Million-Year Story of How and Why We Survived*: "Today we are even manipulating the DNA that makes us possible in the first place—a case of evolution evolving new ways to evolve."

Perhaps technology, as Walter points out so eloquently, is an evolution of evolution—call it Evolution 2.0. A faster, better evolution that impacts and improves single lives as opposed to generations.

Certainly, technology is the new evolution when it starts to integrate directly with our own flesh. This process has already started. The technology of surgery already extends lives. Cosmetic surgery changes the physical and enhances beauty. It even delays the symptoms of aging. Bariatric bands help people lose weight beyond what they can do with their own willpower (we talk about author Andy's personal experience in Chapter 3, "Beauty Hacks: Becoming Barbie, a Lizard, or Whatever You Want to Be"). Cochlear implants help the deaf hear. Google Glass and related wearable computers provide us with distant information on demand that enables us to make informed decisions that affect our well-being, and even opportunities to compete with our human rivals.

Technology is most certainly evolution on speed dial. It can improve very quickly, sometimes seemingly jumping generations in a matter of a year, and sometimes just months.

How Logarithmic Improvements Will Change Your Life

The remarkable twist, however, is the speed at which technology is improving and accelerating. That is to say, the time between innovations is getting shorter. Computers are a good example of this.

Early computers went from being room-sized in the 1950s and 1960s, to cabinet-sized in the 1970s. Then in the late 1970s, companies such as Apple and IBM came out with computers that would fit onto a desktop, helping to spark the personal computer revolution.

They also got faster, and this is where it gets interesting.

In the 1960s, Gordon Moore, one of the cofounders of Intel Corp., noticed that the number of transistors in computer chips was doubling more or less every 1 to 2 years.

What Is a Transistor?

A transistor is a tiny switch in a computer chip. Each transistor can switch between two states: either on or off. A computer counts using these on/off switches. The more transistors there are, the more calculations that are possible, and the faster the processing speed.

This doubling made computers massively faster very quickly. And even if you examine the data zoomed in to the 70s, 80s, and 90s, the trend line looks like a nicely angled graph that shows growth over those decades as a fairly clean 45-degree angle (see Figure 1.2).

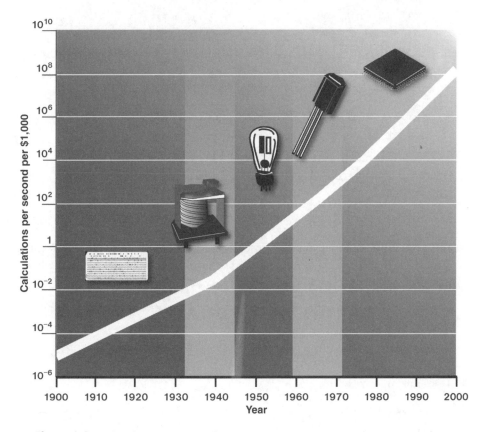

Figure 1.2 *In the short time span of 100 years, computer improvements appear linear.*
(Illustration by Cornelia Svela)

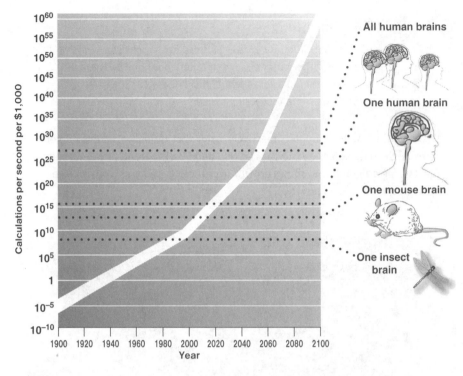

Figure 1.3 *Examined over a long time span, computers are actually improving logarithmically.*
(Illustration by Cornelia Svela)

What you don't see when you look closely at recent data is that chip improvements are just a subset of the era of information processing, or more prosaically, counting machines.

Consider this list of counting machines and their associated dates of invention:

- Sumerian abacus: 2700 BC
- Persian abacus: 600 BC
- Roman wax abacus: 1 AD
- Slide rule: 1625
- Punch card counting loom: 1801
- Difference engine: 1834
- 1/4 abacus for decimal calculations: 1930
- Vacuum tube computer: 1934
- Transistor-based computer: 1947

Notice the time between innovations in the list above. The abacus improved over a long period of time and, as you can see, it was still being improved in the early twentieth century (1930).

As we approach the present day, the technology innovation gap between major iterations of the device goes from 1,000 years to 500 years to a century or so, and then down to decades. By the time we hit the second half of the twentieth century and catch up to transistor-powered processors (and later Intel's microprocessor), the innovation time scale is shortened to months.

So what's happening today?

Space between improvements in most technologies has been reduced to less than a year. In some cases, improvements are seasonal and improvement iteration to iteration is drastically large. For example, the new tablet or smartphone you bought in the spring is a generation behind, by the time fall arrives.

Let's look at the iPhone, from the first generation iPhone to the iPhone 6S. In a little over six years, processor performance improvement went from a 620 MHz processor to a dual 1.8 GHz processor. Memory and storage on the devices also quadrupled in that time. And in Fall 2016, the iPhone 7 will be available, sporting an Apple A10 processor.

If you look at the graph in Figure 1.3, you'll see what most technology watchers in the last three decades have come to understand.

Up-close, technology seems to be improving on a linear scale; however, if we take a longer view, it's actually improving on a logarithmic scale.

To understand this, you have to zoom out from a timeline of a few decades to a timeline of a century or more. The rate of change is improving exponentially with each successive generation. That is to say, the rate of change is accelerating.

Still don't get it? Let's think about toast.

Say you own a high-tech toaster that can toast two pieces of bread in one minute. Using a linear improvement setting (see Figure 1.4) it would produce two pieces of toast in the first minute. In minute two, you get four pieces of toast. In minute three you get six pieces of toast. It is reliable that as each minute passes, more toast is produced on a fixed scale of improvement. (The ability to add two pieces of toast per minute is the improvement on this scale and in this example.)

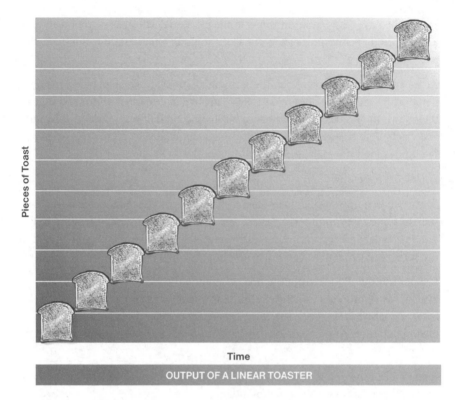

Figure 1.4 *A toaster with a linear setting would produce increasing amounts of toast consistently over time.*

(Illustration by Cornelia Svela)

Compare this to a toaster that has a logarithmic improvement setting (see Figure 1.5), which doubles its output every minute.

So in the first minute, you get two pieces of toast. In the second, it doubles to four pieces. In the third minute you get eight, then sixteen, then thirty two, and so on.

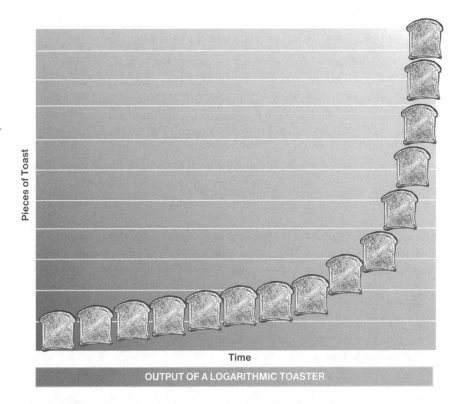

Figure 1.5 *A toaster with a logarithmic setting would start making toast at a slow rate but quickly scales as its output doubles between each segment of time.* (Illustration by Cornelia Svela)

Compare that with the linear toaster over a period of 10 minutes, and you'll see that the difference adds up pretty fast:

Linear toaster:

2 + 4 + 6 + 8 + 10 + 12 + 14 + 16 + 18 + 20 = 110 pieces of toast

Logarithmic toaster:

2 + 4 + 8 + 16 + 32 + 64 + 128 + 256 + 512 + 1,024 = 2,046 pieces of toast

It's important to remember that this is a silly example that is oversimplified for the sake of understanding.

Dear Authors: What is the Meaning of Toast?

After reading this part of the book, some pundits may say: Why did the authors add the results together for each time unit? That's not how the real world works. (True!). Is each piece of toast representative of a particular piece of knowledge? (No.) Or technology? (No.) Or discipline? (No.) So why did you use toast? (We had it for breakfast that day and it was what we were thinking at the time. And, c'mon, toast is funnier than, say, candles.) All we are showing you here is that logarithmic improvements over time are far superior to linear improvements. Next time you look at a piece of toast, you won't see it quite the same way as everyone else at the breakfast table.

It doesn't matter if you followed any of this logic all the way through or not. Here's all you need to remember: Technology is improving logarithmically. And that is good for you and for all humans alive today.

Why?

Logarithmic improvement gets you better results much, much faster over time. Improvements come faster, and their magnitude between each generation is typically larger as time goes by.

Now let's get practical. If that logarithmic improvement is applied to cancer medicine or organ replacement or computers that can think like humans, imagine what's possible in our lifetimes. What's coming in the next decade or so will surely make us superhuman. However, what's coming to humans by midcentury is even more fantastic. So fantastic that is has a term attached to it that suggests that what will happen to human beings will almost be unimaginable.

It's called the *technology Singularity*.

That's a term you might want to know because it is the underpinnings of this book and a lot of the thinking that is behind why you have already and will continue to experience massive and extraordinary personal benefits from today's and future technology.

When we get there, the superhumans we are becoming today will look rather horse and buggyish.

 What Is a Singularity?

In mathematics, the term "singularity" is the point at which a function takes an infinite value, especially in space-time when matter is infinitely dense, such as the center of a black hole.

How the Technology Singularity Will Change Your Life

The Singularity, as it is known (or the technology Singularity, to be more precise), is a point in time in a coming decade when human life changes so drastically that it will be unimaginable to those who came before. In the coming revolution, machine intelligence (or artificial intelligence) will supersede human intelligence.

Yes, your computer will be smarter than you.

But that statement makes it seem like the machines will run the world separately from human beings. That's not the case.

What's more likely to happen is that the machines will be integrated with human flesh and blood. We will be indistinguishable from the machines and they will be indistinguishable from us. And if you are now conjuring some image of your future self as a 1980s movie cyborg, you're on the right track. However, chances are you'll be a less dystopian Terminator and a more Walmart-shopping soccer mom who can access the fridge inventory with her brain. (Although the idea of needing to go to Walmart in the future is kind of dumb because Walmart will come to you on demand and as dictated by the butler-style artificial intelligence routines built into your smart home.)

This process will take us into the posthuman era.

The four major technologies that need to scale logarithmically include:

- **Computers**—Man-made machines that are programmed to carry out arithmetic or logical operations.

- **Nanotechnology**—The engineering of functional systems on a molecular level. That is, technology scaled down to its smallest possible components.

- **Robotics**—The design, construction, operation, and application of robots. In other words, the use of man-made autonomous or semi-autonomous machines.

- **Genetics**—The biological field of study that deals with heredity and DNA makeup in living organisms.

This knowledge and the associated tools will give our organic brains and bodies new abilities. Our flesh and blood will no longer function unaided. The biology that is human will be integrated with machine intelligence and network connections.

Today you look at your smartphone to find the nearest sushi restaurant. After the Singularity, it would be reasonable to imagine that you'll be connected to the Internet at all times. If the information you need about a nearby sushi restaurant is not already locally stored in your gray matter from personal experience, you'll be able to access it from the Internet through thought alone.

It can be difficult to wrap one's (simple and unaided) brain around the concept. But here's the best way to understand why it's difficult to understand:

Imagine if Christopher Columbus stepped off his ship on the beach at the edge of the New World and you jogged up to him in your Lululemons and showed him your iPhone. Imagine how difficult it would be for him to get his fifteenth century head around the little rectangular technology miracle in your hand.

Where would you start in your explanation?

"Hi Chris. I'm from the future. Wanna play Angry Birds?"

His experience would be so far removed from the phenomenon in your hand that it would be difficult to understand where to start your explanation.

Do you explain what an app is? Or explain what a phone is first before you get to the app? Or a maybe explain the idea of a computer and how it developed from a room-sized machine to a pocket-sized wonder that makes, among other things, telephone calls?

Telephone calls?

Then again, maybe you'd start with electricity and alternating versus direct current.

There would be layer upon layer upon layer of concepts to explain before you could even begin to show him how to virtually fling birds at pigs.

The scientific thinkers out there suggest that technology is going to change so quickly and vastly between now and the Technology Singularity's arrival that it will be difficult to comprehend the human experience from our perspective here in 2016. (We are going to try in this book, of course.)

So when is the Singularity likely to arrive? There are conflicting opinions. The betting pool contains some pretty heavy hitters.

Inventor, author, and futurist Ray Kurzweil (see Figure 1.6) thinks he has a handle on it. He points to 2045.

Figure 1.6 *Futurist Ray Kurzweil says 2045 is the year we'll see the Singularity.* *(Photo: Weinberg-Clark Photography.)*

Kurzweil is known as a Singulartarian (defined as someone who espouses the theory of the Technology Singularity).

On the other hand, author and mathematician Vernor Vinge, who originated the term, has predicted 2030 or before. Many others have placed their bets, and the consensus seems to be after 2030 and before 2050.

It's tough to name a date with any certainty.

Kurzweil, by the way, is a director of engineering at Google Inc. He's also the author of five books about the future. He invented text-to-speech technology, optical character recognition, and answered a challenge by Stevie Wonder to create a music keyboard that would mimic the sounds of musical instruments in a way that was indistinguishable from the originals.

We caught up with Kurzweil via a scratchy cell phone connection one chilly evening from our kitchen to ask him about this.

"How long before we're connected directly to the network so our brains can access the Internet and its accumulated knowledge, simply by thinking about it, Ray?"

Kurzweil immediately jumped into a rapid-fire, enlightened sermon about how we're most of the way there already.

"We are already expanding our brains into the Cloud. We can collaborate with other people and access the wisdom of crowds, in ways which we never could do before. We are routinely solving problems with groups very quickly even if the groups

> "We are already expanding our brains into the Cloud."

haven't physically met each other. We have the ability to access all of human knowledge with a few keystrokes—and that is greatly enhancing our intelligence.

"With a group of few people, I can solve problems in a matter of weeks that used to take hundreds of people years. My latest book *How to Create a Mind* was written in one-tenth the time that it took to write my first book, just because of the fantastic ease with which you can access just about any kind of information. So even though these devices are not wired directly into our brains, our mobile devices and all the other devices are extensions of our brain into the Cloud."

It won't take long before that connection makes the short leap from our gray matter to the nearest network connector. He pointed to the exponential rate of improvement that technology is undergoing.

"It's likely to be 1,000 times more powerful within 10 years," he explained. "In 20 years (information) technology will be a million times more than today. Actually more than that by the mid-2030s because there's exponential growth in the rate of exponential growth."

Then he pauses.

"Well there already are computers put inside the body connected into the brain, for example, the Parkinson's implant, which itself is advancing rapidly. The first ones were fairly primitive. They had automatic controllers now you can download software to the computer inside your body connected into your brain from outside the patient."

So there you have it. You will be an extension of the Internet. Or it will be an extension of you. At least that's the thinking.

> "You will be an extension of the Internet. Or it will be an extension of you."

Since publication of his first two books, *The Age of Intelligent Machines* (1990) and *The Age of Spiritual Machines* (1999), Kurzweil has been correct about most of his predictions.

And wrong about some of them, too.

As famed physicist and Pulitzer Prize winner Niels Bohr once said, "Prediction is very difficult, especially about the future."

For the time being, there's a lot to look forward to in this decade, as you'll soon discover in the pages of this book.

It's going to be amazing. We promise.

2

Baby Science: How to Conceive a Tennis Star and Other Procreative Miracles

We can't help ourselves but to get a bit personal with the baby chapter. After all, as two of your authors (Andy and Kay) began writing this book, they had conceived a baby. And during the writing, Kay carried and gave birth to him. And he was raised through his first two and half years of life.

When it came time to write this chapter, the new parents were already heavily invested in the topic, and the research was exciting and highly relevant to the little farty beast in the bassinet next to their bed.

With that, dear reader, please meet Carter Devon Walker (see Figure 2.1).

Figure 2.1 *Carter Devon Walker, born February 15, 2014. Photo taken at six months old.*

If you want to be a parent or have actually become a parent, you know that a longing to know what your baby looks like starts from the moment he or she is a twinkle in your eye.

As with any pregnant woman, Kay had emotional swings throughout her pregnancy. In the last two months or so she'd have tearful moments where she would say: "I can't wait to meet our little man." Andy would get a little steamy, too.

Who doesn't want to meet their baby?

It's natural for parents to wonder what their child will look like. Which parent or grandparent he might resemble? And whether he will have (in our case) Sasquatch feet like his father.

When asked in a 2007 Gallup poll of 1,000 American adults, respondents were split on if, after discovering they were pregnant, they would want to learn their baby's sex. Here are the numbers:

- 51 percent said they would wait until the baby was born

- 47 percent said they would want to know before the birth

- 2 percent had no opinion

Curiously, 58 percent of those who already had a child wanted to know their baby's gender before birth, as did two-thirds of younger adults aged 18 to 34.

As the great American baby room painter Benjamin Moore, once said, "Knowledge is power." Ok, maybe it was philosopher Francis Bacon who said that. (And ok, Benjamin Moore is a company, not a person.) Anyway, you get the idea.

But what if you don't just want to know what they look like? What if you can design how they look so when they come out of mommy's belly, there are no surprises?

Sounds weird and creepy? Perhaps for some. For others it is a choice they'd make. How do we know? Because we talked to a doctor in Los Angeles who has a waiting list full of people who want to design their baby's looks.

In this chapter we explore the history of baby hacking. Test tube babies became possible in the 1980s. And in recent decades, gender selection has become possible. New DNA procedures will soon allow for disease prevention at a genetic level.

We'll tell you how to use the wonders of medical science to choose a boy or a girl and how designer babies can be made. How problems with making babies have been fixed. What is possible today, and what's coming soon. Really soon.

Baby Technologies

Making and birthing babies, raising children, keeping them thriving as teens, and getting them safely into adulthood, where they can ride motorcycles and eat red meat, among other perilous activities, is a lot easier than it used to be thanks to progress in baby technologies.

As we mentioned in the previous chapter, infant mortality has dropped massively in the past century as a direct correlation to accelerating technology improvement. That said, let's look at some of the trends.

Baby Food

Aside from baby poo, there are few things that new parents obsess over more than feeding their child. Parents are crazy. Andy and Kay included. (Sean is exempt here.) They obsess about pretty much anything relating to their baby. But what goes into their baby (or toddler, or child) and what comes out of the little offspring typically dominates parental brain cycles.

"Oh honey, it was very seedy this morning. Not like the runny stuff the other day."

Dinner at Nobu will never be the same again.

As you probably know, a baby's development is highly reliant on what you feed it. In the first five to six months or so, it is breastmilk or infant formula. Then it's baby food. Until you go into full-on kid meals and the ensuing "eat your broccoli and then you can have dessert" battles.

The breast milk versus infant formula decision can be an emotionally charged one, especially for mom. Some mothers might encounter a lot of pressure from family, peers, and medical professionals to breastfeed. And a decision not to breastfeed can often result in feelings of guilt, anxiety, and uncertainty.

If you are raising a super baby, you'll want to know the facts, and what is coming next.

The Invention of Infant Formula

Generations of nonbreastfeeding mothers can thank a man named Justus von Liebig for the miracle that is infant formula. In 1897, he invented Liebig's Soluble Food for Babies, and was the first person to understand the nutritional makeup of mother's milk.

Early infant formula was welcomed by mothers who couldn't afford to hire a wet nurse, a lactating woman who can feed a baby when their mothers can't. Formula feeding at the time was called dry nursing. It raised concerns about proper nutrition: Could a man-made scientifically derived infant formula really be okay for babies?

Thomas Morgan Rotch designed a more natural solution, which he called the "percentage method." It combined cow's milk, cream, milk, water, and honey. By 1907, his homemade formula was all the rage.

But there were problems. Babies who fed on the original recipe were prone to diseases such as scurvy, rickets, and bacterial infections. Orange juice and cod liver oil were added to the mix to remedy the issues. Then came the formula companies. They produced new evaporated milk versions of infant formula, and by the 1950s more than half of babies in the United States were reared on the powdered products.

Hard science paired with the invention of the icebox, the affordability, and the convenience, made formula feeding a trend. By the 1970s, 75 percent of North American babies were fed infant formula. These days, it's the other way around. So what happened? The Nestlé boycott, that's what.

The Nestlé Boycott

In 1977, a boycott was launched in the United States against the company's products. At issue was Nestlé's advertising of its infant-formula baby food as an alternative to breast-feeding, particularly in less-developed countries.

The argument is that mothers in underdeveloped countries lack the ability to read or understand directions, and this impairs their ability to mix the correct ratio of infant formula powder to water. Lack of clean water in some regions can also put formula fed babies at risk.

The U.S. Senate subsequently held a public hearing investigating the claims. An infant formula marketing code was developed then presented to Nestlé, which accepted it. The boycott was subsequently dropped. Since then, 84 countries have enacted legislation implementing all or many of the provisions of the Code, which makes healthcare providers responsible for promoting breast milk over substitutes. It also sets forth labeling requirements for infant formula brands, such as Similac and Enfamil. Manufacturers are required to say that breastmilk is better on all their baby formula products (see Figure 2.2).

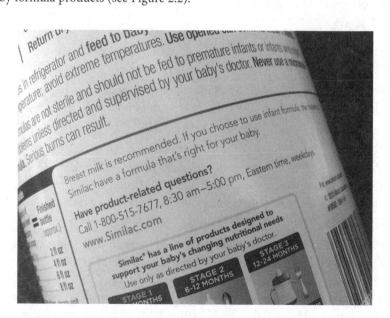

Figure 2.2 *The Similac baby formula label promotes breast milk over formula because of marketing regulations that date back to 1981.*

Study: Breast Might Not Be Best?

In 2014, a controversial study published by Cynthia Colen, Assistant Professor of Sociology, was released by The Ohio State University. It suggests that breastfeeding might not be better than formula feeding, after all.

(That loud series of pops you might have just heard was the breastfeeding lobbyists' brains exploding.)

The research study examined 8,237 children in total, 7,319 siblings and 1,773 pairs of siblings, where children came from the same family but one child was breastfed and one child was formula fed. (These are referred to as "discordant siblings.") The discordant siblings represented 25 percent of the total sibling population tested.

The study included families with different ethnicities and from varying socio-economic situations. All the children were assessed between the ages of 4 and 14". This provided an appropriate amount of time for their bodies to develop and for potential long-term health outcomes to surface.

Colen's team evaluated the children for 11 health issues. When breastfeeding was measured against formula feeding in members of the same family, the criteria collected suggested that breastfeeding was no better than formula feeding.

The one measure that stood out among the 11 examined was asthma. Breastfed babies were affected more negatively by asthma than their formula-fed contemporaries.

Superior Infant Formula

There are a number of reasons why breastfeeding proponents say human milk is the best way to feed your baby. In truth, infant formula is missing a critical ingredient essential for building a strong immune system. It's called 2-fucosyllactose (2FL).

2FL describes one type of human milk oligosaccharides (HMO). HMOs are polymers—large molecules such as protein or DNA—found in breast milk. 2FL essentially protects the body from infection. The bottom line is that it defends babies against pathogens by strengthening their immune systems.

In 2012, food microbiologist Michael Miller and his team at the University of Illinois found a way to synthetically create large amounts of 2FL at a lower cost than was previously possible allowing scientists to study 2FL more readily. Without it, a single study would cost $1million for a supply of 2FL alone. The high manufacturing cost is also why 2FL is missing in current infant formulas.

There's still more work to be done. An improved infant formula, including 2FL, could one day soon appear on the market and rival mother's breast milk.

Ultrasound Technology

A key part of prenatal care is going for a mid-term checkup to get a peek at the baby inside mom. This gives you an opportunity to find out the gender of the child and gives your ob-gyn a tool to ensure the baby is developing well. This is done with a technology called ultrasound. The technology, sometimes referred to as sonography, sends high-frequency sound waves into the body from a probe called a transducer. The sound waves hit tissues and organs in the body. These structures bounce the sound back to the probe in different ways, based on what they are made of and how far away they are from the probe. This information is processed by a computer and a two-dimensional image is generated on an ultrasound screen. It's a handy tool when a doctor needs visual guidance and is otherwise blindly navigating inside the body.

Ultrasound is most commonly used on pregnant women to monitor fetal development. However, it is also used to help diagnose issues that affect organs and tissues. It's ideal for peering at soft tissue in the gut; however, it has some limitations. It's not ideal for examining bones or parts of the body that hold air or gas, such as the bowel.

Ultrasound has improved drastically since its introduction. These days, it produces an image that looks a lot more like baby. If you want to get a sharper look at your baby, you might consider 3D ultrasound.

3D and 4D Ultrasound

Olaf von Ramm and Stephen Smith at Duke University invented the 3D ultrasound in 1987, but it took some time to make its way to the mainstream. Today, 3D ultrasound is available to any parent-to-be who is willing to pay for it. It is similar to a CT scan, which is a form of X-ray that collects virtual slices of a scanned region on the body and builds them into a three-dimensional structure (more on this in Chapter 5 "The Human Computer: How to Rewire and Turbo-Boost Your Ape Brain").

In the case of a 3D ultrasound, sound waves create 2D image segments that are reconstructed with a powerful computer system, which results in a three-dimensional image.

For an even more profound visual experience, 4D ultrasound is the way to go. It captures more detail and does it in real-time. With 4D ultrasound, you can see a live video of the baby moving inside its mother. To see it in action, have a look at this Youtube video: http://superyou.link/4dultrasound.

The FDA hasn't approved 3D and 4D ultrasounds yet. As is the case with any ultrasound, when the emitted sound waves enter the body, they heat the surrounding tissue. Although not illegal, the FDA recommends against them because the long-term effects on the baby and mother are unknown.

Smartphone-Based Ultrasound

MobiUS SP1 Ultrasound Imaging System is the world's first smartphone-sized system (see Figure 2.3).

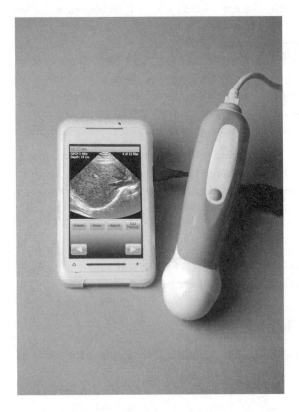

Figure 2.3 *The MobiUS SP1 Ultrasound Imaging System.*
(Permission to use image provided by Mobisante.)

Medical professionals use it to provide remote care. A scan done in one country would be done locally—say on a pregnant mother—and the images would then be sent via the Internet to a medical facility for analysis anywhere in the world. The device has been approved by the FDA in the United States and costs $7,500 to $8,000.

The machine is small enough to fit in a pocket. It can capture eight different types of scans and store them in the device's 8GB of memory. Images are shared via a Wi-Fi connection, a cellular data network, or a wired USB connection.

The Future of Ultrasound

Ultrasound technology is being explored for uses beyond imaging. And you might be able to use devices in your own home sometime in the near future.

High-Intensity Focused Ultrasound

Although ultrasound is traditionally used for imaging, doctors are looking at new ways to use it to destroy tissue in a highly targeted manner. It's called high intensity focused ultrasound (HIFU) and is being tested on patients with various neurological disorders and cancers. However, it's also being used in the baby world.

A 2014 article in *New Scientist* magazine reported the use of HIFU to treat twin fetuses that suffered from a rare disorder known as arterial perfusion. The twins were connected with an umbilical cord that passed through the placenta. This meant that only one of the twins had a heart. He had the big job of supporting the life of both him and his sibling. It's too much work for only one heart, so in many cases both twins die.

For the first time ever, in 2013, doctors used HIFU to sever the connection and save one of the babies. The baby boy was delivered successfully at 37 weeks by Cesarean section.

More work needs to be done to fully understand HIFU, but this is one example of a treatment where it can save lives.

Magneto-Acoustic Imaging

Do-it-yourself ultrasound may be just around the corner. Pregnant women would be able to scan themselves with a home device and send the images to their doctors.

An experimental technology called magneto-acoustic imaging combines ultrasound with a magnetic field to produce resolution up to 1 mm (or 0.04 inches). This technology is exciting because it can be miniaturized. Ultrasound transducers, radiofrequency (RF) electrodes, and related electronics can all be produced on tiny silicon chips.

We can expect a working device in the next three to five years, and possibly a home medical imaging system within a decade.

Birth Control Technologies

Let's turn to another society-bending technology that had a huge impact on parents.

Pop Quiz: Name three things when you think of the 1960s.

Woodstock? JFK assassination? Man on the moon? What else? Maybe you said: the pill. The introduction of oral contraceptives, combined with the women's rights movement, gave women full control over their reproductive cycles for the first time.

Before the pill, various birth control methods had been around a long time, since 3000 B.C.E. to be more specific. The first condoms were made from materials such as fish bladders, linen fabrics, and animal intestines. It was not until 1838 that condoms were more popularly made from vulcanized rubber.

Although condoms were available, there was a lot of religious and political dialogue around birth control. In 1873, the United States government passed the Comstock Act, which was designed for the "suppression of trade in, and circulation of, obscene literature and articles of immoral use." That included birth control technology, and as such, condoms. It also enabled the postal service to confiscate illicit items distributed by mail.

Then Margaret Sanger came along, and, in 1916, she opened the first birth control clinic in the United States. She was subsequently jailed for 30 days as a result.

In 1932, Sanger ordered a diaphragm from Japan to provoke a legal battle. The diaphragm was confiscated by the United States government, and Sanger challenged the action in court. It led to a 1936 court decision overturning an important provision of the Comstock laws, which prevented physicians from obtaining contraceptives. As a result, women gained a bit more freedom.

Sanger continued her work well into her 80s and was involved in some of the research that led to the discovery of the birth control pill.

In 1960, Enovid-10, the first birth control pill, was approved by the FDA. Today, there are many birth control technologies to choose from. And there are some exciting new methods in development that will soon come available.

Types of Birth Control

Birth control methods available today fall within four categories:

- Barrier methods that physically block sperm from contact with an egg.
- Hormone treatments that deliver hormones into a woman's body and cause physiological changes that impair conception.
- Surgical procedures that provide a 100 percent birth control solution.
- The fourth is simple and sometimes overlooked. It's abstinence.

Barrier Methods

- **Condoms**—Sleeves made from latex, polyurethane, or synthetic rubber that block sperm. The condom is the only method of contraception that can also prevent sexually transmitted infections. There are condoms for both men and women.

- **Contraceptive sponge**—This is made of foam and measures about 2 inches in diameter. It works internally in a woman by blocking sperm from entering the cervix and releases spermicide, a toxic solution that destroys sperm.

- **Diaphragm**—A diaphragm is a shallow silicone cup with a steel rim that creates an internal seal against the female cervix, which stops sperm from entering the uterus.

- **Cervical cap**—Similar to the diaphragm, the cervical cap covers the cervix and prevents the entry of sperm. Each cup is tailored to a woman based on whether she has previously given birth.

Hormone Treatment Methods

- **Oral contraception**—Widely referred to as "the pill," there are many brands of oral contraception. Each works in a similar way. Pills must be taken daily so that they release hormones regularly to stop the release of a woman's eggs.

- **Contraceptive patch**—The contraceptive patch is 2 × 2 inch sticker containing hormones that are released into the body through the skin. It prevents a woman from releasing eggs and makes implantation unlikely.

- **Intrauterine device (IUD) or intrauterine system (IUS)**—Small T-shaped devices that are inserted by a physician into a woman's uterus. The IUD is partially made from copper and plastic, whereas the IUS is made of plastic only. The copper found on the IUD inhibits sperm and stops eggs from implanting. The IUS works in a similar way, but releases a synthetic hormone that inhibits pregnancy.

- **Contraceptive implant**—The contraceptive implant is similar to the IUS because it releases a hormone into the body. However, it is a 4-cm (1.6 inches) long plastic tube that is inserted under the skin in a woman's arm. The hormone enters the body through the bloodstream and lasts for up to three years before it needs to be replaced.

- **Vaginal ring**—This is a small ring that is manually inserted and is similar to a diaphragm or cervical cup; however, the ring is not a barrier method. It works by releasing hormones.

Surgical Procedures

- **Female sterilization**—This surgery involves either blocking or sealing off the fallopian tubes so that the ovaries, which produce eggs, are no longer linked to the uterus. Eggs released are reabsorbed and sperm is unable to reach them.

- **Male sterilization**—This procedure is known as a vasectomy. It's a 15-minute procedure where the tubes that carry sperm are blocked or cut by a surgeon. This procedure can usually be reversed.

> "No sex equals no chance of pregnancy. The end."

Abstinence

No sex equals no chance of pregnancy. The end. Unless you're the Virgin Mary. But that puts you in a different bracket.

A BRIEF HISTORY OF THE PREGNANCY TEST

Most women at one time or another have trepidatiously purchased a pregnancy test from their local drug store to see if they were pregnant. The tests available today are a simple 2-minute to 10-minute process. A woman pees on a stick in the privacy of her own home and waits for the results. It usually results in one of two reactions. A sigh of relief or, in author Kay's case, an echoing scream that had Andy come running. And then there was a lot of hugging kissing and bad dancing on Andy's part.

It wasn't always this easy. The home pregnancy test has only been around since 1977. The first one was invented in 1976 by the New Jersey-based company Warner Chilcott. It was approved for sale in the United States in 1977.

They called it e.p.t. or the "early pregnancy test." It was a much more involved process than peeing on a stick. It came with a vial of purified water to drink, an angled mirror, some red blood cells from a sheep, and an eye of newt. All that is true, except for the newt part. It also took 2 hours to provide results.

Before the e.p.t., women had to go to their doctors. And, the early doctor's tests also took 2 hours. In the 1920s and 1930s, the test involved injecting a rabbit's ovaries with a woman's pee. The ovaries were then inspected by a doctor and changes on their surfaces meant the woman was pregnant.

It was always bad news for the bunny. It had to be killed for the test to be conducted.

Further back in time, women had to rely on someone called a Piss Prophet. They were European fortune tellers, during the Middle Ages, who used pee as their medium. Some read the urine's color. Others gave it a little taste. Usually it was the bubbles in the pee that divined the future and the prognosis on a pregnancy.

The ancient Egyptians used a surprisingly reliable method where a woman would pee on wheat and barley seeds. If the seeds grew, she was declared pregnant. It was said to be 70 percent reliable. No one drank the local beer, however.

The most advanced pregnancy tests today are digital urine tests that actually say the words "pregnant" or "not pregnant" for easy interpretation. Some of them also tell you how far along you are.

The Future of Birth Control

As you can see, a lot of technology options have been developed over the years to stop babies from being conceived, and to verify when they have been. Since the back half of the last century, they have been arriving at an accelerated rate, and they are becoming increasingly easier to access and use (and kinder to bunnies). What will be available next, however, is nothing short of extraordinary.

Remote Control Birth Control

Plans for the first remote-controlled contraceptive device for women are currently in development by a company called microCHIPS. Preclinical trials started in 2015 with the ultimate goal to have it hit the United States market by 2018. MicroCHIPS is a Boston-area startup, funded by Microsoft founder Bill Gates and his wife Melinda.

The remote control device is an implant the size of a thick postage stamp. It can be inserted under the skin by a physician in areas such as the buttocks, upper arm, or abdomen. Its lifespan is up to 16 years. Once installed, it administers a hormone called levonorgestrel. The hormone is currently found in many oral contraceptives and it stops ovulation, the process by which a woman's body releases eggs. Once the device is inserted by a healthcare provider it can be toggled on and off with a remote control.

The implant has an air-tight seal made of titanium and platinum on the reservoirs that contain the drug. An electric current is passed through the seal daily by an internal battery which allows a dose of the hormone to be released into the body.

"The idea of using a thin membrane like an electric fuse was the most challenging and the most creative problem we had to solve," said MicroCHIPS president Robert Farra in an interview with *MIT Technology Review*.

Although the prototype works well, researchers still need to figure out how to encrypt the information over a wireless connection. You don't want your neighbor to flip the TV channel and turn off your wife's birth control.

We can see a day when you might be able to manage your birth control with your smartphone. Yes, there'll be an app for that, too.

> "You don't want your neighbor to flip the TV channel and turn off your wife's birth control."

Male Contraception

Most contraceptive devices are designed for women, with the exception of condoms. That could change if an injectable male contraceptive developed in India becomes available. It is currently undergoing trials in the United States and might become available in the next couple of years, likely 2018.

The innovation is called RISUG, which stands for Reversible Inhibition of Sperm Under Guidance. It is an injectable substance composed of two liquids that when combined, form a gel that blocks the passage of sperm in the vas deferens, the anatomical tube that transports them from a man's testicle to his urethra. It's the same conduit that gets tied off during a vasectomy. The compound also kills any sperm that come in contact with it.

The treatment is a 15-minute procedure where the solution is injected directly into the vas deferens. It takes only a few minutes for the entire inside of the tube to be coated.

The procedure was developed by Sujoy K. Guha, a professor of biomedical engineering at the Indian Institute of Technology in Delhi. In 2011, a technology transfer agreement was signed by an American organization, the Parsemus Foundation, for the rights to use the patent. Trials were conducted in the United States in 2014. When the product passes testing, it will be sold under the name VasalGel.

Trials in India have passed Phase I and Phase II trials are underway. In Phase I, volunteers who received the treatment were monitored for 15 years. There were no complications or pregnancies.

Phase II trials have shown to be equally successful. The only side effect so far is a swelling of the testes that can last up to two weeks following the initial procedure. This comes with no pain. (Authors' tip: You might need bigger underpants, for a while.)

It looks like the procedure can be reversed, but this has only been tested in primates. They received the treatment for a year and a half and were given a reversal treatment, either an injectable with baking soda or a low electrical current. This removed the blockage. In two to three months, normal sperm flow resumed. It sounds awful, but we assume they will not be talking about zapping chimp privates in the ad campaign.

The United States product Vasalgel is currently in the animal testing phase (aka bunnies). Human testing is slated beyond that. If all goes well, the product will become available at a cost of no more than $800.

Curing Mr. Happy

Like on national TV, we can't be too salacious here. So let's talk somewhat clinically about male performance in bed. There's a whole category of drugs that help men who have trouble in this department. Blue diamond pills. Yellow oval pills. Brands that rhyme with Niagara. If you are grown up, you know what we mean here. If you are not sure, go ask your mom.

It's pretty self-evident that a man who has performance issues in the bedroom will have difficulty siring a child without some medical intervention.

Luckily, great strides have been made in the medical world to help men who have occasional or frequent trouble achieving this physiological state to make babies with a woman. It's estimated that 1 in 10 men suffer from erectile dysfunction, or ED. And, men over 40 are typically the age group that has to deal with this issue. Though, dude, if you are younger and this is a problem for you, it's cool. It happens to everyone sometimes.

ED can be caused by a series of issues. Typical causes include disease, blood circulation, injuries, surgery, or medications. Smoking, weight gain, and excessive drinking can contribute to the issue. Psychological or emotional issues can also be factors.

The latter can be treated with therapy and stress and anxiety management. Medication is typically prescribed for guys with physiological issues.

The big breakthrough in male erectile health arrived in 1991 with the discovery of the compound called sildenafil citrate. You probably know it as Viagra. The little diamond-shaped blue pill was first of a family of drugs to help men improve their sex life. Here's a summary of the family these drugs by brand:

- **Viagra:** The original erectile treatment. Lasts about 4 hours.

- **Cialis:** Called the weekend pill. A 36-hour or longer potency.

- **Levitra:** Kicks in faster, in about 12 to 30 minutes. Lasts 24 hours.

- **Staxyn:** It's Levitra, but in a form that dissolves quickly, so that it works faster.

- **Stendra:** The quickie pill. Kicks in about 15 minutes and lasts 6 hours.

The drugs are all PDE-5 inhibitors, which work by increasing blood flow to the penis during sexual activity. It works for about 80 percent of men.

Pfizer developed Viagra in the late 1990s in the UK. Their drug researchers were initially studying high blood pressure and stumbled on the surprisingly fun side-effect. We can only imagine what the human drug testers wrote on their forms.

Please report any unusual side effects:
Well, um, yes. The Missus is quite pleased.

If these drugs don't help, then there are more extreme treatments that include the following: A medicated pellet inserted into the male urethra (the tube where urine flows from); medicine injected into the penis; or the use of a vacuum device on it.

For more severe cases, a penile prosthesis is an option. It involves two inflatable cylinders that are inserted surgically. A pump is also placed inside the body and, when activated, it pushes fluid from a hidden reservoir into the cylinders and that causes an erection. This equipment is under the skin so that it's not evident in the locker room or the bedroom. Seems like a bit of a clumsy robo solution, but until there are other therapies it's a valid fix for some men. Plus c'mon, robots are cool.

 ### We're Not Making This Up

The drug Viagra is also known to reduce jetlag in hamsters. You think we're kidding? This is from research published by two Argentinian scientists from the *Proceedings of the National Academy of Sciences.*

Although there is a good selection of ED therapies available today, what's coming in the future is quite, dare we say, exciting. Here's a summary:

- **Uprima**—This cleverly named drug is a tablet that dissolves under the tongue. It works by stimulating the brain chemical dopamine, which heightens sexual interest and sensations. Sounds great. However, its major side effects are nausea and vomiting. Additionally, and perhaps unfortunately, a small number of people pass out after a dose. In late 2014, its release in the United States was (perhaps thankfully) put on hold. That said, it is currently available in Europe. Clinical trials are also currently being conducted on a nasal spray form of this drug, which might cause less nausea.

- **Gene therapy**—This technique in development by researchers will help treat and prevent a variety of diseases. It targets the genes (elements in the human biological blueprint) that cause the problems. The idea is to repair or modify the genes that might not be functioning properly in penile tissue. Replacement of these proteins would result in better erections. Gene therapy is expected to make great progress in the next decade. We're forecasting sooner than ten years thanks to the logarithmic improvement of technologies.

- **Melanocortin receptor agonists**—These are a new set of medications being developed to treat both low sex drive in women and erectile dysfunction in men. They have been shown to produce an erection in animal studies. (There is probably a banana joke here, but we'll leave it alone.) The medication impacts the nervous system instead of the vascular system. PT-141—also known as bremelanotide (or "brem", to some)—contains a synthetic hormone that acts on the hypothalamus, a part of the brain. It appears to be effective alone or in combination with PDE-5 inhibitors (Viagra and the like). Some call it the libido drug, because it acts on the brain and not specifically blood flow to help male performance. It is also known to work to boost the female libido, an issue that thus far has limited known treatment options (although note that many sufferers have turned to off-label drug use to treat it). If you click around the web, you'll find people who have tried it with great success. You can even order it online for about $30 for a small vial. Main side effects include facial flushing and nausea. Brem is not yet approved by the FDA for mass-market use, but it is being studied by Palatin Technologies, Inc., which expects to wrap up current research sometime in late 2016. It says it may then make a submission for approval to the FDA. More research is needed before the drugs can be proven to be safe.

Baby-Making Technologies

Couples, particularly single randy teenagers and young adults, spend a large chunk of their lives trying to avoid making babies, while still going through the rather fun motions.

However, when a couple is ready to make a baby, ironically, sometimes the process doesn't work. Here's good news and better news. Today we know so much more about sex, fertility, and the biological processes required to produce a healthy newborn baby. The expertise on this topic is growing at a remarkable and compounding rate.

The logarithmic improvement of technology is at play here. More than 2,000 years ago, Aristotle cracked open a few fertilized chicken eggs in his efforts to understand the nature of life.

We can thank the humble microscope for opening up the world of human fertility because human eggs are a lot smaller than chicken eggs. A microscope powerful enough for use in scientific investigation was produced in the early eighteenth century. But magnifying tools had been available long before that. Glass was invented in the first century and the Romans investigated viewing objects through it almost from day one.

In the thirteenth century, Italian Salvino D'Armate designed the first eye glass that allowed its wearer to see magnified objects through one eye. An early microscope came later as the magnifying glass was adapted to make objects bigger.

However, it wasn't until 1677 that Dutch microscope innovator and scientist Antonie van Leeuwenhoek produced a microscope powerful enough to see sperm. In 1898, fertilization was described as the union of an egg and a sperm. And only eight decades later, in 1978, the first "test-tube" baby, Louise Joy Brown, was born in England. A year later, her sister was born, also with the aid of in vitro science.

You might see the logarithmic improvement of technology in the previous short historical account. If you did, we think you're clever.

Aristotle had the curiosity but couldn't get much further than what he could observe with his eye. But then the invention of glass begat magnifying technology, which in turn provided the technology for a microscope. That was in the first 1,700 years or so. With the invention of the microscope, it took scientists only 200 to 300 years to figure out how to make babies in a test tube.

The First Test Tube Baby

If you are 40 years old (or older), you might remember a news story from the late 1970s about Louise Joy Brown. Brown was the first test tube baby, born on July 25,

1978. Her genesis was aided by the first in vitro procedure, a process where egg and sperm are introduced and combined to produce a viable human embryo in the lab.

It all happens in a petri dish, a shallow receptacle where egg and sperm are mixed. The resulting embryo is then re-implanted into a healthy woman. It in turn embeds in the wall of her uterus and she becomes pregnant.

In Vitro Fertilization

The process to create Louise Joy Brown, developed by U.K. medical pioneers Dr. Patrick Steptoe and Dr. Robert Edwards, is called IVF or in vitro fertilization. Edwards received the Nobel Prize for his pioneering work on in vitro fertilization in 2010. Steptoe, Edwards's partner, died in 1998. He was passed over by the Nobel judges because it is not awarded posthumously (somehow that seems unfair).

 Why "In Vitro?"

In vitro, from Latin, means "in glass." The first "test tube" baby was conceived in a petri dish, which is a flat-bottomed glass dish.

Here's how a typical IVF procedure is done today.

A laparoscope, a probe with a light on it, is inserted through a small incision just below a woman's belly button. It is used to reach her ovaries and extract her eggs. The eggs are mixed with sperm in a petri dish. Two days later, the fertilized egg is deposited inside the woman's uterus where it can attach itself to the uterine wall.

Not all procedures work the first time. The success rate is 28 percent to 35 percent, which, as you can imagine, is frustrating for couples who are struggling to conceive. It can also be an expensive process. Treatment ranges from $10,000 to $15,000 and is not covered by medical insurance.

 Fertility Abroad

About 20,000 to 25,000 couples travel outside of their home country each year to receive treatment elsewhere. It's a trend known as fertility tourism. According to the website IVF-abroad.org, Dubai and Israel surprisingly have a good reputation for IVF. Couples with tighter budgets go to Mexico, Panama, Argentina, or Columbia for IVF treatment, where procedures are more affordable.

Candidates seeking IVF treatment are typically dealing with infertility problems, ranging from medical issues, such as fallopian tube damage or polycystic ovarian syndrome, to older couples who have passed their best reproductive years. As of

2013, 5 million IVF babies had been born worldwide since the procedure had been developed. That's just slightly smaller than the population of Atlanta (5.5 million people in 2013).

A Swedish study, cited in *Time* magazine, found that 47 out of 100,000 infants born from IVF develop cognitive deficits, such as low IQ or problems in communicating or socializing with others, compared to 40 out of 100,000 naturally conceived children.

Artificial Insemination

Artificial insemination (AI) gives women the choice to have a baby with or without a partner. The fertility treatment is not quite as fun as the traditional baby-making process. No sex required! Sperm from a selected man—a donor or partner—is inserted into a woman's uterus by a doctor with an unsexy medical instrument. It's all very clinical and the man can be a continent away and anonymous to the mother.

This is a good option for single women wanting to have a baby, or lesbian couples. Donor sperm is usually acquired from a friend, acquaintance, or donor bank.

Male/female couples seeking this treatment are typically dealing with medical issues such as immune system rejection of sperm and dysfunctions of the cervix (such as cervical scarring) that make it difficult for the couple to get pregnant on their own.

The first AI procedure on humans dates as far back as 1770. A renowned Scottish anatomist and surgeon, Dr. John Hunter, advised a patient of his who had hypospadias (a condition where the opening of the penis is located on the underside of it, instead of the tip) to collect his ejaculate, put it in a syringe, and have his wife insert it into herself.

This was perhaps a primitive yet effective early AI operation. However, it led to more studies in fertility science. The big breakthrough arrived mid-century when Dr. Jerome K. Sherman, a then-doctoral candidate at the University of Iowa, discovered the ability to freeze sperm to preserve it for later use. In 1953 he pioneered the technique that allowed the use of frozen sperm to be used to create a viable and healthy pregnancy. The method was unveiled at the 11th International Congress of Genetics in 1963 and a decade later, the first sperm banks opened for business.

Today, there are three common AI procedures. The easiest and most common one is intracervical insemination which involves the use of raw (unwashed), fresh, or frozen semen which is deposited on the neck of the cervix to most closely mimic natural baby making.

Intrauterine insemination involves introducing washed semen (see related factoid) directly to the uterus. Women receiving this procedure should be under

30 years of age and the donor of the sperm should have a sperm count of more than 5 million per milliliter.

Lastly, there is a method called intrauterine tuboperitoneal insemination where both the fallopian tubes and the uterus are injected with 10 ml (about 2 teaspoons) of sperm. This procedure is often used for couples who are dealing with male infertility and mild forms of endometriosis, a problem where uterine cells grow outside the uterus causing pain and infertility.

WASHED VERSUS UNWASHED SPERM

You can't put a price on love but you sure can on sperm. When it comes to buying sperm, it's important to know there are two varieties. There's unwashed, or raw, sperm, and there's washed sperm.

Unwashed sperm is good old-fashioned natural sperm. It is used primarily for intracervical insemination, a procedure which often can be conducted at home.

Washed sperm involves a process called, logically, sperm washing, which separates seminal fluid from sperm cells using a medical device called a centrifuge. It helps flush away the poor swimmers, which are less viable candidates. It's also used for intrauterine insemination where it helps reduce the chances of uterine cramping.

Both types of sperm, when purchased, come in vials of either 0.5 ml or 1 ml. These amounts have been determined to be the ideal amounts to successfully impregnate a woman. Washed sperm always costs more due to the larger sperm count found in a vial. For this reason, it is the most popular of the two.

The Future of Reproduction Technology

While IVF technologies increase the chances of producing a baby after natural methods have failed, it doesn't always work or guarantee a healthy or disease-free baby. At least until now.

Next Generation DNA Sequencing

How about an IVF procedure that guarantees a woman will produce a 100 percent genetically healthy baby? Dr. Dagan Wells, a medical researcher at Oxford University in England, has pioneered a new method that not only does that, but also improves the chances that the egg will attach to the uterine wall so that a woman becomes pregnant.

With current IVF procedures it is easy for chromosomal abnormalities to be overlooked under the microscope. This can lead to problems with implantation and can cause complications in the pregnancy so that a mom could lose her baby.

In designing the new procedure, Wells and his team used existing DNA sequencing techniques, but improved them. DNA sequencing is a process whereby the blueprint inside a cell that describes a human being is checked to ensure it is genetically healthy and has no errors. It's like a spell check on a document. Or, maybe it should be a called a cell check.

DNA sequencing, where one cell is checked, is less expensive than current IVF screening procedures, but here is the catch. Scientists can only safely remove one cell from each embryo at a time without the risk of destroying it entirely. As you can imagine, this is a delicate and time-consuming process.

Wells's team solved the problem by creating a unique method that sequences cells from groups of embryos at the same time. They used a system of barcodes to keep them organized and to ensure that the correct cells are paired back with their parent embryo following the screening process.

When the cells are removed, 2 percent of each embryo's genome are sequenced. This percentage is enough for them to identify the number of chromosomes. It does not however, provide additional information with regards to specific genes.

Wells's trials were conducted at Main Line Fertility, a clinic outside of Philadelphia, where the parents of the first baby born using his technique were treated. They gave birth to a healthy baby boy named Connor Levy in 2013.

It's easy to see why Wells's method is an improvement. Prior to entering trials, the couple had produced a total of 13 embryos that developed properly, but the problem was, researchers had no idea which ones to transfer. Wells analyzed the embryos and his test selected 4 of the 13 that were actually healthy. Of that four he only needed to use one.

This is very different from current IVF procedures. They require the analysis of up to three embryos for best results. With more embryos introduced in the uterus, there is a 30 percent chance that the couple will produce multiples (twins, triplets, or high-order multiples).

Wells's technique eliminates this. There is no chance of multiple births.

Current genomic procedures cost approximately $5,000. Wells's procedure runs as little as $1,000 per treatment. And, with his accuracy rates, a genetically healthy baby can be produced on the first implantation with the first embryo.

100 Percent Guarantee

If you'd like to learn more about the work of Dr. Wells and his team, visit Reprogenetics at superyou.link/reprogenetics

Baby-Saving Technologies

If you have a sick baby, as a parent you'll do anything to make the little tyke better. Thank goodness these technologies are either here, or on the way. Of course, in some cases you have to act before the baby is born or ever gets sick.

Stem Cells from Cord Blood

Before your baby is born you should be asked if you want to save your child's cord blood—that is, blood in the umbilical cord. This procedure allows stem cell-rich blood to be collected and preserved for therapeutic medical use by the baby's family at a later date.

The process is relatively simple, it takes only ten minutes and the baby does not feel a thing. Once the baby arrives, the cord is clamped and cut as it normally would be 1 to 2 minutes following the birth. A needle is then inserted into the umbilical vein, the area of the cord that remains attached to the placenta, and 1 to 5 oz of blood is then drained from the cord into a collection bag. The bag is shipped to a blood bank where it is tested and processed using a controlled method of freezing.

The blood is stored in either a private cord blood bank for the exclusive use by family members, or can be donated to a public cord bank and used for blood transfusions. The blood from the umbilical cord is rich in stem cells, which are seed cells that can develop into many types of human cells. It can be used to help cure a myriad of diseases, such as leukemia. Without cord blood, treatment for leukemia would require a bone marrow transplant from a matching and often hard-to-find donor.

Other uses include therapies to treat these major illnesses: Aplastic anemia, thalassemia, Hodgkin's disease, non-Hodgkin's lymphoma, cerebral palsy, autism, type 1 diabetes, and host of rare and fatal metabolic disorders.

According to Babycenter.com in 2012, 33,900 cord blood units were shipped from the United States worldwide. In total there are more than 1 million cord blood units stored in United States family blood banks.

The average cost for the first year of storage ranges from $1,400 to $2,300. In the ensuing years, you can be charged up to $1,800 per year for storage.

Only a small percentage of people will ever use their cord blood. At any point in time, it can be donated to a public blood bank. Most western countries have cord blood banks, including the United States, Canada, and the UK.

We asked medical futurist Dr. Bertalan Meskó (who you will hear more from in Chapter 4, "Lifesaving Hacks: Whirring Hearts, Printed Organs, and Miraculous Medicine") about cord blood to help us decide whether authors Kay and Andy should save Carter's blood.

He told us that if he were having a baby in 2014 he would probably opt for cord blood. In another five years he thinks cord blood will not be necessary. As soon as 2019, Meskó predicts that based on the developments he has been witnessing in medical science, "We will be able to convert skin cells into any type of tissue," meaning that adults will be able to use their adult skin cells to treat their diseases.

Genome Sequencing

It's not a mainstream procedure yet, but in June 2014 the first American baby had his genome sequenced at birth. The baby was born in California, and his father, Razib Khan, was a graduate student studying genetics at the University of California.

Genome sequencing is a process by which an individual's genes are mapped. It is a detailed blueprint of what makes you you. This allows doctors to look at specific genes to determine if potential diseases will develop.

In an interview with the *MIT Technology Review*, Khan said sequencing his son in utero "was more cool than practical."

 Razib's Baby

Read more about Razib Khan and his adventures in genomics here: http://www.superyou.link/dealwithit

To get his son's genome sequenced, Khan had to first get a sample of his DNA, even though he had yet to be born. He did this by requesting a chorionic villus sampling (CVS) test, which is used on pregnant women to test for genetic abnormalities in unborn babies. Cells are taken from the placenta and analyzed to retrieve valuable information about the growing baby's DNA.

His wife's doctor sent the sample from the placenta to Signature Genomics, in Spokane, Washington, for analysis. When the results from the test came back normal Khan asked for the raw data. The request met a lot of resistance.

Britt Ravnan, one of the company's lab directors, explained why in the *MIT Technology Review*: "What was unusual in this case was that it was not the patient or the physician asking for the sample, but the patient's husband."

After Khan's wife and her doctors filled out the necessary paperwork, Signature Genomics released the information. Khan sequenced his son's genome using a free online tool called Promethease. It's a do-it-yourself genome sequencing program that is easy for anyone to use, although users have to agree to a stringent terms of use policy. They are asked to upload specific genetic information or export it from a personal genetics testing company, 23andMe.com. The software then generates a genome report. It was that easy. If Khan can do it, you can too.

However, genome sequencing is a controversial subject. If couples get their baby's genetic information before their baby is born and the information suggests a genetic disorder, the couple could choose to abort the baby.

One other wild card. The genetic information isn't guaranteed to be accurate. It's more of a predictor. Genome sequencing has limitations. It predicts possible diseases, but that doesn't mean a condition will necessarily manifest in an individual. Some outcomes, such as heart disease and stroke, don't come from a single gene. Other factors are involved including lifestyle choices and environmental exposure.

The possibility of elective abortions motivated by genetic information are why anti-abortion lobbyists and some religious organizations will fight the future approval of genome sequencing for babies, prior to birth. Khan argues that personal genomic information is a personal right. "How dare the government question [our] right to know the basic genetic building blocks of who [we] are," he wrote in one of his blog posts.

Nanotechnology for Babies

If there is one field of science that is going to have a major impact on our health and well-being, it's nanotechnology. Nanotechnology is the ability to change, adjust, or repair elements at nanoscale. A nanometer is the diameter of 5 carbon atoms, which is at the molecular level. This is why developments in nanoscience are showing great promise for the field of biology. When it comes to humans, it means we'll be able to manipulate not only tiny cells but their contents, too.

> "If there is one field of science that is going to have a major impact on our health and well-being, it's nanotechnology."

The greater application of nanotechnology is that we can build tools to manipulate the building blocks of the universe at the atomic or molecular level. Theoretically if you wanted to build a cheese sandwich from its base elements, super advanced nanotechnology tools would let you do that.

Ray Kurzweil told *Big Think*, "We'll be able to create devices that are manufactured at the molecular level by putting together molecular fragments in new combinations. I can send you an information file and a desktop nanofactory will assemble molecules according to the definition in the file and create a physical object."

Not only will we be able to modify a cheese sandwich at its elemental level, we'll be able to use nanotechnology in medicine specifically for babies. It will be useful in fertility and the actual processes that create and maintain life, ensuring that a couple can conceive and birth a healthy baby.

The University of Alberta in Canada is currently researching nanotechnology to improve the success of organ transplants for newborns.

The challenge with transplants is that the blood type of the donor often needs to be compatible with the blood type of the recipient; otherwise the recipient's immune system will reject the donor organ within the first 24 hours. This is true for adults, who require an exact blood type match. Babies (and young adults) have a developing immune system so it's possible that their body could accept an organ with a blood type that doesn't match their own. In medical circles, organs incompatible with blood type are what's called ABOi organs.

In 2014, 50 percent of babies who needed a new heart died while waiting to receive a heart that matched their blood type. This is an unfortunate statistic, especially when 60 percent of donor hearts for babies are discarded because they don't get used in the window of time before they become unusable.

The Canadian research team is currently conducting trials to see if attaching blood antibodies to nanoparticles that are sent directly into the bloodstream can increase the likelihood that a baby's body will accept an ABOi organ. In the future, vaccines could be developed that would make it possible for babies or children to receive the antibodies via an injection.

In June 2014, the University of Alberta received $1.1 million in funding from the Canadian federal government to develop this research, and studies are have started on laboratory animals. If successful, the research would transition into human testing.

If the proposed nanotechnology is effective, it could save the lives of 100,000 babies worldwide each year.

 Saving Lives, One Baby at a Time.

Each year, 20-million premature babies die worldwide. In developing countries, the mortality rates are significant due to a lack of incubation equipment.

In 2011, a team at Stanford University, led by Jane Chen (CEO) and Rahul Panicker (CTO), launched their first product—the Embrace Infant Warmer. The affordable $25 baby warmer is a cocoon-shaped pod designed to keep a baby warm. No electricity is required because the device is warmed with a wax pouch insert.

Embrace has saved babies with the product in places such as South India and Nepal, where trials first began. The company's aim is to help millions of babies in resource-constrained areas survive by developing low-cost healthcare technologies.

3D Printing To Save Babies

3D printing is an exciting prospect in the medical world. It was created in the mid-1980s by American inventor Chuck Hull who developed stereolithography, a process where an ultraviolet laser is used to heat plastic material so that it hardens into a number of horizontal layers that are piled vertically to create a specific object.

For example, in the world of engineering, machine parts can be printed on demand. Or, new inventions can be prototyped quickly without the need of a costly manufacturing process.

Because newborns are tiny, operating on them or finding donor organs or tissue to treat their diseases is not easy. Historically, organs come from donor babies—ones that haven't survived. But, all that is changing with the advent of 3D printing.

In 2012, the first 3D printed splint saved a baby boy named Kaiba. It was surgically implanted into his body to repair his trachea, the pipe that delivers air from the month to the lungs.

The baby's windpipe was too narrow, which restricted his ability to breathe. It's a rare condition known as tracheobronchomalacia.

In 2012, Dr. Scott Hollister and Dr. Glenn Green, professors at the University of Michigan, performed the groundbreaking operation. They took an X-ray of the baby's trachea and custom-designed a splint to fit his body. Then, they performed a surgery to replace the old trachea with the one they created. Immediately following the surgery, Kaiba could breathe normally.

Eventually, the splint will grow with his airways. It's biodegradable so that it will completely dissolve in his body over a period of three years.

The doctors also received a FDA approval for a second similar surgery to treat another baby boy, Garrett Peterson, a year later. His surgery was performed on January 31, 2013. Today, he is a very healthy little boy.

They are now discussing plans to start a formal study so that they can test their method on more babies. One day soon there is hope that is will become a common medical, baby-saving, technique.

In 2015, doctors at C.S. Mott Children's Hospital at the University of Michigan's Health System used a 3D printer to build a model of an unborn fetus' face.

An ultrasound revealed that the fetus had a (potentially cancerous) mass, and doctors were uncertain of the correct procedures required to deliver the baby without restricting its ability to breathe. So, they used an MRI to gather images to print a 3D composite.

The 3D model revealed that the unborn baby had a cleft lip and a cleft palette with no airway obstruction. With this knowledge, the team of doctors were able to safely deliver the baby.

Dr. Glenn Green, Associate Professor of Pediatric Otolaryngology at Mott Children's Hospital, led the team of doctors. He told CBS News "by doing the 3D modeling, we were able to tell that the mass wasn't cancerous and determined that it would be safe to not do the most advanced types of intervention."

This was the first time 3D printing was used in utero. Green told CBS News, "we think this is a real game changer". He suggests that in the future, 3D printing could help many more doctors: "OB's could potentially have a 3D printer right by the ultrasound machine."

 Print Your Fetus?

In 2016, 3D printed fetuses might be the hottest trend for expectant parents. Watch this video to learn more: http://superyou.link/3dfetus

Designer Babies

Two months into author Kay's pregnancy with Carter, we obtained an image composite from Morphthing.com to get a sense of what Andy Walker Jr. might look like. Using the program, we uploaded a picture of both our faces and clicked a button, and got a sense of what our developing son looked like. As you can see from the photo in Figure 2.4, it's hardly scientific, nor is it accurate. In fact, if you cross Kay's picture with a picture of the poet William Butler Yeats you get a similar baby with black-and-white old fashioned glasses. He looks like he could write a heck of an Irish poem, though.

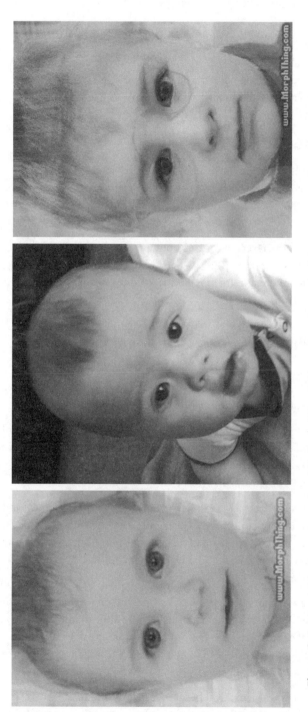

Figure 2.4 *Morphthing.com composite of Andy and Kay (left), an actual picture of Carter at 5 months (center), and a Yeats/Kay combo from Morphthing (right).*

Still, it was close as we could get to seeing what our little guy would look like before he was born.

Today, there's no definitive way to know for certain what children are going to look like before they are born, but with advances in medical science, and movements in government policy, one day parents will be able to select their traits. Yes, it's time to delve into the very controversial world of designer babies.

Gender Selection

What if you could choose to have a boy or a girl? For some, this might seem a bit unnatural. It was for Kay, who as we mentioned earlier, was pregnant while writing this book. Now that she has a boy she admits she wouldn't mind if baby number two was a girl.

Before she started the research for this book project she wouldn't have considered gender selection as an option. But, as she educated herself on the process, she became much more open to the possibility. Read the next section and see if your opinion changes, too.

Nature's Method

Let's start with the natural gender-selection process. It's important to know how things work naturally, so you can better understand how intervention works to get the gender parents want for their baby.

Chromosomes are an important part of gender selection. They are thread-like structures found in the cell's nucleus, which is the cell's command and control center. These segments of DNA contain genes and genes carry valuable information from each parent to their offspring.

Genes map out what traits a person will get. It's why baby Carter has Andy's Sasquatch feet and Kay's finger toes. Carter also has one second toe that's longer than his big toe on one foot only and he got that from his maternal grandfather. (And, we're sure, that one day he'll be mad that we told his toe secrets to the world here. Sorry, Carter!)

Otto Bütschli, a German zoologist, was the first person to identify the chromosome in the 1880s, but it took many years for us to fully understand them. It was not until 1955 that cytogeneticist Joe Hin Tjio discovered that the normal number of chromosomes in each human cell is 46. A cytogeneticist, by the way, is someone who studies chromosomes specifically.

Tijo's breakthrough was a big deal because it helped scientists understand what was normal and abnormal. They could study cells under a microscope and identify genetic abnormalities.

Now, there are two chromosomes that determine whether a human will become a boy or a girl. They are the X and Y chromosomes, and when they combine they produce a code that determines gender.

Carter is a boy, so he has XY chromosomes. Ellie, his little girlfriend down the street, has XX chromosomes, which makes her a girl.

Women carry only X chromosomes. Men carry two kinds of chromosomes (some sperm have the X chromosome, whereas some have a Y chromosome). This means it's the father that determines the gender of his offspring. This process of sex differentiation happens during week six of a baby's development. That week, the baby, now an embryo, has sex chromosomes from both parents that start to express themselves. The woman provides an X to the offspring and the man provides either his X or his Y. The result is an XX girl baby or an XY boy baby.

Based on the chromosomes, sex organs grow from an originating organ called the gonads that turn into testes for a boy or ovaries for a girl.

The whole sex differentiation process has always been left up to fate. That is, until scientists figured out the mechanics of it all and discovered new gene-screening technologies that would allow the process to be manipulated.

Before choosing the sex of a baby was even a possibility, the technology used to do it had to be established. Gene-screening technologies were originally established to screen embryos for genetic abnormalities, but once this became available it was possible to decipher the baby's gender using the same techniques.

Amniocentesis

Amniocentesis was perhaps the first procedure used to determine gender. It was originally developed to screen for disorders such as Down syndrome, cystic fibrosis, sickle cell disease, muscular dystrophy, and Tay-Sachs.

By the 1970s, it became a profitable tool when marketed for gender selection. In Asia, where some cultures value male babies more than female babies, there was high demand.

Amniocentesis involves the extraction of amniotic fluid, the liquid that surrounds the fetus. This liquid protects the baby and helps it grow. The fluid is composed of water and electrolytes at first, but by weeks 12 to 14 it also contains proteins, carbohydrates, fats, and the baby's urine.

The amniotic fluid is sampled with a syringe that penetrates the mother's belly. Once the fluid has been extracted, fetal cells are separated from the fluid. The sex can be identified by examining the cells under a microscope.

The procedure is still in use today, but not commonly used for gender selection. It's used primarily on three categories of mothers to be:

- Women who are older than 35

- Women who already have a baby with a genetic disorder

- Women who are carriers of potential genetic mutations

The Ericsson Method

In the 1970s, Dr. Ronald J. Ericsson created a new method of gender selection called appropriately, the Ericsson Method. It's a procedure that divides sperm based on gender and uses it in high concentrations of the selected gender sperm to create a gender-specific baby. Simply put, to make a girl baby, doctors take the girl sperm and put it in the mom and vice versa for a boy.

Ericsson believed that the sperm with X chromosomes, female sperm, swim slower than sperm with Y chromosomes, male sperm. When a group of sperm are released in a test tube that contains albumin, a protein in human blood plasma, or more informally, the clear white substance you'd see around a yoke if you cracked an egg (the white), some will swim slower into the albumin than others. Those sperm, the slower swimmers, are the female ones and they'll be used in artificial insemination for a female baby. The fast swimmers are used to make boys.

Overall, the method is only 70 percent to 72 percent effective for producing male babies and 69 percent to 75 percent effective for producing female babies. While it is still used in United States clinics, its lower success rates make it a choice for gender-selection treatments, but not the best one.

Sperm Spinning

Sperm spinning involves separating male and female sperm based on their genetic composition. Besides being slow swimmers, female sperm are larger, and they live longer than male sperm. Female sperm also look different when viewed under a laser. And they look great in heels.

> "Female sperm are larger, and they live longer than male sperm."

The sperm are placed inside a centrifuge where they are spun so that their contents separate. Immotile sperm (aka duds) and seminal fluid are separated from the healthy sperm and the remaining sperm are stained with a fluorescent dye so that the DNA in each cell can be viewed. The female sperm shine brighter under a laser because they are plumper. (But, warning: Never call a female sperm "fat".)

The sperm-spinning technology is called MicroSort. It was established by the U.S. Department of Agriculture. In 1984, however, the rights to the MicroSort technology were sold to a company in Virginia named MicroSort and they are still in operation.

Preimplantation Genetic Diagnosis

The preimplantation genetic diagnosis (PGD) method is currently the most effective gender-selection technology. PGD is an embryo selection technique. It does two things. First, it allows for the embryo to be sorted—male versus female. It also is a process used for genetic screening. The results are a gender-specific and genetically healthy embryo for implantation. Once the selection process is complete, it can be used in either IVF or AI.

More specifically, this is how it works:

When a sperm and an egg meet, they merge—or fertilize—which means 23 chromosomes from the man combined with 23 chromosomes from the woman generate the 46 chromosomes necessary to create the first cell of their offspring. The newly formed single-celled entity is called a zygote. The zygote cell divides into two in a natural cell-division process known as mitosis.

During mitosis, the cells duplicate themselves entirely and multiply. This has to happen for the initial cell to eventually become a human being. So, one cell becomes two, two cells becomes four, four becomes eight, and so on until you get something that resembles either Kim Kardashian or Donald Trump, or someone in between. (We shudder at the thought.)

By the third day, it consists of eight cells. At this point, one or two cells can safely be removed and used for PGD without compromising the developing embryo.

Since PGD can screen for genes, it's also been used to save potential offspring from developing a long list of genetic disorders.

In 2000, Adam Nash was the first American baby who was genetically engineered using PGD to save a sibling's life. His sister Molly had been born five years earlier with a condition called Fanconi anemia, a fatal disease that affects bone marrow. She was dying, so Dr. John Wagner, the Director of Clinical Research at the University of Minnesota's Marrow Transplant Progress, used PGD to select an embryo with the same genetic abnormalities as Molly, but one that wasn't a carrier of her disease.

The selected embryo became her brother Adam. When he was born the stem cells from his umbilical cord blood were used to save her life.

However, engineering offspring based on traits such as gender is a bit more controversial because it's not medically necessary. So, there aren't many countries that sanction it. In places such as the United States, the technique is being used because there have been no policies established to prohibit it.

Nevertheless, an average of 4,000 to 6,000 PGD procedures are performed around the world each year.

The Fertility Doctor, the Media, and the Vatican

When the media is camped out on your front lawn and the Vatican comes calling, you know you are controversial and on the cutting edge of your field.

That's the position Dr. Jeffrey Steinberg, a Los Angeles-based reproductive endocrinologist and founder of The Fertility Institutes, found himself in during early 2009.

Steinberg had been conducting research in albinism, people having little or no pigment in their eyes, skin, or hair. This is a result of inherited genes that don't make the usual amounts of a pigment called melanin (1 in 17,000 people in the United States has some form of albinism).

Steinberg's intention was to find ways to prevent eye issues suffered by people with the disorder. In the process, he isolated a gene for eye color, and that led to the discovery of additional genes related to a series of physical traits such as eye color, hair color, and hair texture. The findings opened up the possibility that parents-to-be would be able to select the physical traits of their unborn children.

Theoretically, doctors would use in vitro fertilization to create a series of embryos, and then would select one of them for genetic analysis. If this process was ever available, your fertility doctor would select the embryo that best genetically matches the traits you want. Taking things a step further, some researchers believe that as genetic science improves, specific genes could be engineered to produce the selected traits.

The media dubbed this "designer babies." *The New York Daily News* ran the headline "Custom-made babies delivered: Fertility clinic doctor's design-a-kid offer creates uproar." ABC News ran the headline "Fertility Doctor Will Let Parents Build Their Own Baby."

And there were plenty of phone calls.

"We had the usual outbursts from the radical right and radical left and even people in between. We had people that thought we were Nazi mongers (and) people who couldn't wait to get on the list to get a girl with red hair," Steinberg recalled.

> "We had people that thought we were Nazi mongers (and) people who couldn't wait to get on the list to get a girl with red hair."

Then Steinberg got a call from The Vatican. "That was the straw that broke the camel's back," he said. "They were very nice. They have scientists at the Vatican, and we discussed it for quite a while."

Steinberg decided to stand down. "I said, Yeah let's back off for now. No one is going to die if they don't get these traits. We have a list hundreds of names long, and for every characteristic you could imagine. So we backed down. I just didn't want to spend the time on it."

Controversy is not new to Steinberg. Early in his career he attracted unwanted attention while practicing in vitro fertilization.

"Anything that is new is scary," he explained." When I did the first in vitro fertilization baby way, way back I came out of my office and someone had put a hand scrawled note on the windshield of my car that said 'test tube babies have no soul.' Now you go to a party and a third of the people there are test tube babies."

The Fertility Institutes's primary business these days is gender selection. Parents can preselect a boy or a girl. Even that had its initial controversy.

"My professional society heavily criticized it, but a third of those members are now participating in it and offering it," Steinberg said.

By 2014, The Fertility Institutes had produced almost 10,000 gender-selected babies. The technique they use is PGD. Steinberg says his company does not guarantee gender, although, it has never gotten the gender wrong. However, the uncertainty couples face is in the implantation of the gender-selected embryo. Sometimes it doesn't stick to the uterine wall of the mother, and so the procedure fails to result in a viable pregnancy.

It's also a costly procedure and the majority of the research being done today in gender selection is focused on getting the cost down. In 2014, a couple who wanted the procedure would face a bill of $18,000. The cost to identify an egg as a boy or a girl is not expensive. However, the cost to ensure it is genetically healthy is.

Steinberg explained, "It doesn't do us any good to know we've got a girl if the girl is missing a chromosome because she is never going to make a viable baby. People don't want to know they have an XX embryo. They want to know they have a viable girl."

To do this, they have to select an embryo that is the desired gender and has 46 chromosomes, which produces a healthy human.

During this selection process Steinberg's team made an unexpected discovery.

In the process of gender selection, they found that some couples are "good boy-baby makers" and some are "good girl-baby makers." It's a process that is completely unique to the couple.

 Does Family Factor Into the Gender of Children?

If you are looking at your brothers and sisters or extended family to try to understand whether you'll have boys or girls, don't bother. Gender trends in your family tree won't help. Good-boy-making couples or good-girl-making couples are determined uniquely by the genetic chemistry between you and your partner.

Steinberg says that couples who are good at making boys before they come to him tend to be good at making boys during the gender selection process (and vice versa).

"When people who already have a lot of boys at home ask for a girl, we get both types of embryo in the lab but we find that we get mostly boys, and the girls we do get are never quite the same quality. This explains why they are making boys at home."

Whereas couples make both boy and girl zygotes (zygotes, remember, are fertilized eggs), the body only uses viable candidates. Women shed the nonviable zygotes through a natural selection process. Good boy baby-making couples produce more viable boy zygotes than girl zygotes, and those fertilized eggs stick more readily. This is what is happening when you encounter a couple that's had a series of same-gender children.

The other challenge that fertility specialists deal with is the incidence of abnormal human eggs. When Steinberg and his team looked at the viability of eggs from young healthy women (egg donors aged 23 to 25), they discovered 25 percent to 30 percent were genetically abnormal. Steinberg called this finding "astonishing because we never thought humans were so inefficient as far as making normal genetic embryos."

So what's the future of designer babies? Work still has to be done to get there. And there has to be a big demand for it. Steinberg predicts designer babies will be available in some form in the next decade.

Today, 20 percent of Steinberg's clients who ask for gender selection, also ask for other trait selection. He puts their name on a list and tells them "Not now, maybe later."

In 2014, Steinberg had 700 to 800 couples on a list who said they'd pay to choose their baby's eye color. The second trait most requested is hair texture. Number three? Athletic ability.

Qualities such as athletic ability and height are traits that require the isolation of multiple genes (they are called "polygenic" in scientific terms), so Steinberg said it will take a bit longer to figure those out. When athletes come in for gender selection, Steinberg requests blood samples (and permission) to study them later.

 Anyone for Tennis?

If you could genetically select athleticism as a trait in your baby, what sport would you ask for? It turns out most moms-to-be that visit Steinberg's clinic say, if it were available, they would ask for tennis players.

Genetically Engineered Babies

If selecting traits for your baby intrigues you, and you were disappointed that it's not yet available, don't worry. A lot of activity is going on in the world of genetics, especially around babies. It won't be long before it is available. There are technologies available today that will allow parents-to-be to genetically screen for diseases.

Here's the catch: The service is only (currently) available to women who want to use a sperm donor to get pregnant, because donors are genetically screened for quality before their sperm is admitted into the sperm bank.

What's in the way is politics, policy, and social acceptability.

Maybe the wait will be longer than expected. The science is there, but the societal acceptance isn't there yet.

The following option, however, is possible today.

GenePeeks is a New York-based company that has developed a technology called Matchright. This technology pairs the DNA of two hypothetical parents and virtually creates a hypothetical baby who has their combined genetic makeup. It uses scientifically designed algorithms and next-generation sequencing tools to match a woman's DNA with more than 1,000 sperm donors in their catalogue.

That's right. Women seeking to use a sperm bank can access the service.

The results from the test suggest the best genetic matches, meaning the ones that have the least chance of disease.

The company was created by Anne Morriss and Dr. Lee Silver. Silver is a biology professor from Harvard University with an interest in genetics, and Anne Morriss is a mother on a mission to save parents from ever having to deal with a chronically ill child. Her son was born with MCADD (medium-chain acyl-CoA dehydrogenase deficiency), a disease that prevents the body from converting fats to sugars. It can lead to seizures, breathing difficulties, brain damage, coma, and sudden death.

Here's how it works. Each person is a carrier for multiple gene mutations, so everyone is at risk of having a child with a rare recessive disease. When two parents carrying the same recessive gene come together, there is a 25 percent chance they will produce a sick child. Each parent themselves carries one recessive gene and one healthy gene, which is why they are perfectly healthy. The disease goes undetected until it's paired with DNA that carries the same recessive gene where it gets expressed.

Whereas most IVF clinics already screen donors for a dozen or so genetic disorders, perform regular health checks, and acquire each donor's family history, the Matchright process evaluates the possibility of 500 genetic conditions.

The service is only currently available to women using a sperm bank who want to screen out donors that are not a good match in terms of their combined genetics. Women who are interested in using GenePeeks can order a home test kit that takes a sample of their saliva and is sent back to a lab for analysis. The analysis takes about 4 weeks, and then they receive a personalized catalogue of FDA-approved donors which have been selected for them as healthy sperm match candidates.

The technology is not yet ready for couples preplanning a pregnancy, but Gene-Peeks says it has intentions to make it available in the near future.

It's currently only available in the United States, but if you live outside its borders, the company suggests on its website to call and inquire.

Testing 1, 2, 3 ... Is Your Baby Genetically Healthy?

If you are pregnant today, you might not know that your genes have been or will be screened while you carry your baby. Blood tests fall under the category of what are called "carrier tests," which are prenatal tests that screen parental genes. The most common tests, called the Triple Screen and Quad Test, determine whether the baby will develop genetic disorders or brain defects.

There are additional genetic prenatal tests available, but they are much more invasive and involve the extraction and analysis of cells from the developing fetus and surrounding structures. The procedures are for women whose pregnancies are high-risk. This is because the procedures are invasive and pose a very small risk of miscarriage (about 1 in 100 women). Women whose pregnancies are deemed high risk are 35 years or older, have already had a baby with chromosomal defects, or have genetic markers that signal potential problems.

One common test is called amniocentesis. You may recall from earlier in this chapter, that amniocentesis involves the extraction and analysis of amniotic fluid. Another common test, performed at an earlier stage of pregnancy—10 to 13 weeks—is chorionic villus sampling (CVS). Cells are taken from specific cells in the placenta that match the cells of the baby that is being formed. But, if both amniocentesis and CVS don't provide sufficient test results, a third method, cordocentesis can be used to extract cells from the umbilical cord for analysis.

These carrier tests and prenatal tests are what's available today, but soon things might look very different. Correcting gene problems before they start is the future. But, it's a very controversial topic because although the technology is available, it's social policy and public opinion that are currently blocking its availability.

There are signs of progress though.

In October 2013, the U.S. Patent and Trademark Office awarded a controversial patent to 23andMe, a Mountain View, California company that specializes in genetic testing. Though it is best known for home DNA testing kits, the patent gave it exclusive ownership of a genetic calculation technology called "gamete donor selection." If permitted, parents who use fertility clinics could select traits for their unborn children.

Traits would be selected by checking boxes on a long list of criteria and would include characteristics such as: gender, height, weight, and eye color.

When the media got all excited about the news, 23andMe.com responded—in part with a blog post that said, "We never pursued the concepts discussed in the patent and we don't have any plans to do so."

There was even more media attention when the FDA ordered 23andMe to stop selling its personal genome service in late 2013. The FDA said the $99 Personal Genome Service (PGS) that asks customers to send a saliva sample to a lab to receive their personal genome violated the Cosmetic Act.

Two years later, 23andMe resumed its operations, and this time with the FDA's approval. Today, it offers a similar test but with zero information on personal health risks. The new test provides information on carrier status, which relates to the probability of genetic mutations and disease in potential offspring. The kits also come with a higher price tag of $199.

We asked them about designer babies, and a spokesperson responded by email: "23andMe does not play a role or have any plans to pursue any technology in the field of designer babies."

Maybe not. But we think it could be offered at some point in the next few years as soon as the government gets out of the way, and public opinion shifts.

Daddy, Mommy, and Your Extra Mommy

One more pop quiz! Can three people have a baby together? Not without some scientific help. But yes, a three-parent baby is genetically possible. The procedure is intended to prevent a range of incurable mitochondrial diseases currently affecting 1 in 10,000 people.

The creation of this special baby involves three-parent in vitro fertilization (TPIVF), which uses the DNA from three people (two women, one man) to create one fertilized egg for implantation.

When a baby is made, DNA from the man and DNA from the woman come together to produce their offspring. What most people don't know is that there is a type of DNA that's given to the child exclusively by the mother. It is called mitochondrial DNA.

Mitochondrial DNA comes from the mitochondria, the "power centers" found in most of the cells in the human body. They convert chemical energy from food into energy that our bodies use to run.

Women who are carriers of the mitochondrial DNA mutation (the "MELA" A32A3G mtDNA mutation) will pass it on to their babies unknowingly because they are often perfectly healthy themselves.

Since the mitochondria perform so many critical functions in a number of tissues, there are hundreds of mitochondrial diseases. Each disorder comes with a list of different health problems. Common symptoms range from a loss of motor control, muscle weakness and pain, and gastrointestinal dysfunctions.

But that's not all. Add to that difficulties swallowing, poor growth, cardiac disease, liver disease, diabetes, respiratory issues, and seizures, to name just a few.

Scientists have found a way to intervene in the fertilization process to eliminate the potential of genetically transferred mitochondrial disorders.

In the egg with the mitochondrial mutation, the cell's nucleus is removed. Then they insert that nucleus into the second woman's egg, which has had its original nucleus removed. This hybrid egg, which now houses the nucleus from the first egg but contains the healthy mitochondrial DNA from the second egg, is then fertilized with the sperm (see Figure 2.5).

How to make a three-parent baby

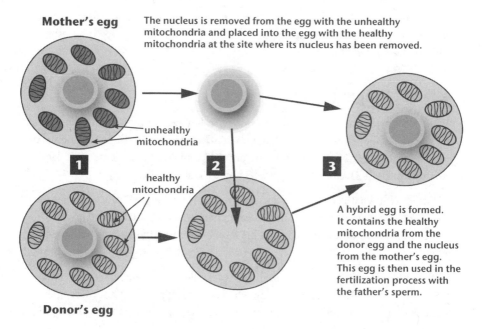

Figure 2.5 *The process of three-parent baby making is shown here.*
(Illustration provided by Cornelia Svela.)

In 2015, Britain became the first country to legalize this procedure. (So, as you read this book, the first three-parent baby might be making its way into the world.)

In February of 2014, fertility rules were changed by the UK government to make Britain the first country in the world to allow mitochondrial replacement (MR) therapy. But the rules were subject to additional approvals from the British parliament.

A CNN February 2014 report suggested that the FDA was also considering the legalization of three-parent babies in the United States.

Art Caplan, the director of medical ethics at New York University's Langone Medical Center, told CNN in an interview that the procedure to correct mitochondrial mutations is a "humane and ethical thing to do."

But, he also suggested that there is a more controversial side to be considered: "Where we get into the sticky part is, what if you start to say, 'While we're at it, why don't we make you taller, stronger, faster or smarter?'"

We are all for it. You might be also. Or maybe not.

Pregnant Men?

Though the technology exists to likely make it possible for a man to get pregnant, there would still be a lot of social, political, and legal hurdles to overcome, especially in the United States.

The First "Pregnant Man"

In 1974, Tracy Lehuanani LaGondino was born female, but by the time she reached the age of 10, she felt more male than female. Tracy started to use the pronoun "he," and on his 23rd birthday he started the physical transition to become male. He renamed himself Thomas and became Thomas Trace Beatie.

It started with what is called "top surgery," which is the surgical procedure on the breasts of transgender patients. In this case, their removal. After that, Thomas looked like a man on the outside, but his female reproductive organs remained intact.

In 2005 he married his wife Nancy and two years later the couple decided to have a baby. But, it was Thomas, not Nancy, who carried their child. Nancy was infertile and couldn't become pregnant. So they decided that Thomas would become pregnant using donated sperm.

The publicity surrounding this extraordinary parenting scenario made him a media sensation and led to appearances on *Saturday Night Live*, *The Late Show with David Letterman*, and the *Oprah Winfrey Show*.

Thomas went on to birth two more children and then he had his female reproductive organs surgically removed. (He and Nancy also divorced in 2012.)

All three birth records mark Thomas's gender as the mother with the letter M for male. And so, he remains the first man to give birth.

 Don't Be Fooled

Although the video at Malepregnancy.com seems like a true take on the first pregnant man, you can rest easy. It's not. The procedure seems plausible, especially to anyone who has no formal medical training. Give a man some female hormones, use IVF to implant an embryo and placenta in his abdominal cavity so that he can grow the baby, and then deliver it with a Caesarean section, and we were almost fooled. But with a little poking around we discovered none of it is real.

The Future of Male Pregnancy

Whereas Thomas Beatie is the first man to birth a baby, he will not likely be the last. Beatie was born a woman, so it could be argued this is not a true case of a biologically male pregnant man.

Can men really become pregnant? If you look at the animal kingdom, then the answer is yes. Male seahorses and pipefish, a closely related species, carry fertilized eggs in their pouches and birth babies after a few weeks. Nature hasn't provided the same facility for wannabe man-moms. But if we have learned anything while writing this book, it's that if something is not possible now, it will be shortly. Gene therapy lets us change the human blueprint. Stem cells allow the growth or repair of any human tissue. Bioprinting makes it possible to create new organs from scratch. And if none of that works, nanotechnologies will let us push the building blocks of life around with tiny nanoscale bulldozers.

"... if something is not possible now, it will be shortly."

The other ray of hope for Mr. Mom wannabes is this curious fact: A miracle baby born in Montreal in 2012 shows that a uterus might not be needed to bring a baby to term. Dionne Grant, a Jamaican tourist visiting Canada, gave birth to a healthy baby boy after the child grew outside of her uterus in Grant's abdomen. This is a rare condition called an ectopic pregnancy, in which a fertilized egg implants itself outside the uterus, usually in the mother's fallopian tubes. (Fallopian tubes are conduits for an egg to descend from the ovaries to the uterus.) Doctors believe the baby survived because the placenta—the meaty tube that provides nutrients to the baby during its development—glued itself to the top of the Grant's uterus, accessing blood to nourish the baby.

So could a dad carry a baby in his belly? Yes, likely. There would have to be other interventions using some medical magic, but we say give it ten years and will all be possible. Probably much less.

For you doubters out there in Super You Land, consider this: Scientists can now grow vaginas. Read on.

Lab-Grown Vaginas

It's taken two decades of work for Anthony Atala and his team at Wake Forest Institute for Regenerative Medicine in North Carolina to bioengineer and transplant a very delicate human organ—a vagina.

In 2014, a culmination of techniques used on rabbits that started in the 1990s led to the first successful vaginas grown in a lab. Four women suffering from Mayer-Rokitansky-Kuster-Hauser syndrome (MRKH) participated in the trial.

MRKH affects the development of the genitals in 1 in 5,000 women. Those women are born without a vaginal canal and are missing parts of their reproductive system. It affects their ability to have sex, menstruate, and reproduce.

All four women who participated in the study had abnormal uterus development but their vulva, clitoris, and labia were intact. They all received the operation at the ages of 13 to 18 years old, a time when the body is still maturing.

The procedure involves the removal of small part of the vulva, which is used to grow new cells. These are then layered into a scaffold made of collagen that disintegrates one month after the surgery. Each scaffold is formed using calculations made from MRI scans. This helps fit the new vagina to the shape and size of each recipient's body. The scaffold is then placed into a cavity in the abdomen. A bit of stitching holds the new vagina in place and a stent makes sure it maintains the proper form as it grows.

Atala's team waited four to eight years to publish the results of the trial to ensure its long-term success. All the women now have functional vaginas and two have uteruses. The big question is whether these women will be able to have children or whether they will experience side effects long-term.

From Birth to Forever

Reflecting on this chapter, it's clear to us that the process of procreation and raising a child in the next century is going to become a very different process to what it was for our parents. The options for a yet-to-be born child that are about to become possible are quite astounding. As parents, we will be able to set them up for clean genetics, fix any problems in utero, and birth them into a life that will be very different from previous generations. Life expectancy of your child might become limitless, as you will see in our longevity chapter. A life free of disease might also be possible. And those children might grow into humans that will have access to all the wonders of technology found in science fiction and beyond.

There is evidence that with the realization of stem cell technology, gene therapy, and nanomedicine, not only will our children be able to repair themselves through technology, but perhaps they might even elect for an extended life and choose to delay or eliminate death as an option. And if we are all of the age where we are lucky enough to enjoy this too, we will be able to live to see their children grow and the generations beyond that.

That might seem like a wild and seemingly unrealistic statement to some, but the emerging science, and prognostication among most futurists, suggests it will all be possible within this lifetime. Certainly it will be possible in the next few decades and the next decade will bring the beginnings of all this to everyday life.

And like it or not, designer babies are on their way. It will be an option for the next generation of parents, provided there is social and political will. That is the wild card. The question is do we want that option? Would we design our children's hair and their eyes and their proclivity toward certain physical skills or capabilities?

Your authors wouldn't choose these designed aesthetics, but we would choose a clean genetic palette for our children, when it becomes possible.

3

Beauty Hacks: Becoming Barbie, a Lizard, or Whatever You Want to Be

You are ugly. How do we know? Because you say so. Or at least you think thoughts like that. You might not use the word ugly in your head. It might be something such as: I'm fat, or saggy, or tired, old, wrinkly, or ... I'm young and cute, but there's that thing about my ... well, you get the idea.

Most people say harsh things about their own looks to themselves every day, even if it's only internally.

Look in a full-length mirror. Chances are your eyes fall on one or more particular flaws you dislike (or, in some cases, hate) about yourself. And your critical gaze lingers there as you think through what you can do to fix, hide, or change it somehow.

If you are female, studies show you are much more likely to be dissatisfied with what you see in the mirror than men. The consensus is the obsession with self-image is driven by the media. Open any magazine and idealized, impossible bodies (typically hyperimproved by Photoshop) stare back at you.

Studies at Stanford University and the University of Massachusetts revealed that 70 percent of college women feel worse about their looks after reading women's magazines. A 2006 study published in the journal of *Psychology of Men and Masculinity* said prime-time television and music videos make men uncomfortable with themselves.

In her summary of body image studies, Dr. Kate Fox, an anthropologist at the Social Issues Research Centre (SIRC) in the UK, said up to eight out of ten women are dissatisfied with their bodies, and more than half see a distorted image of how they really look. Although that work was published in 1997, nothing much has changed for women.

It's different for men. Fox's 1997 report said "men are more likely to be either pleased with what they see or indifferent."

Yet, according to a 2011 study of 394 British men conducted by Dr. Phillippa Diedrichs at the University of the West of England, men are now showing more concern with the way they look. More than four in five men (80.7 percent) talk in ways that promote anxiety about their body images.

Insofar as female beauty is concerned, SIRC's Fox reported that standards have become progressively more unrealistic over time.

In 1917, the physically ideal woman was about 5 feet 4 inches tall and weighed in at almost 140 pounds. In the 1970s, fashion models and beauty queens weighed 8 percent less than the average woman. In 1997, they weighed 23 percent less. Today, that number hasn't changed much, though some suggest it's higher.

This culturally rampant dissatisfaction with our looks, and the human need to look better, is fueled by evidence that beauty provides real life advantages.

In his book *Beauty Pays*, economist Daniel Hamermesh says the best looking people earn an extra $250,000, on average, during their work life, compared to their less-attractive colleagues. They are also more likely to remain employed, be promoted, and—when they apply to borrow money—successfully secure bank loans. Sounds crazy, right?

In a 2010 *Newsweek* survey, 57 percent of hiring managers reported that qualified but unattractive applicants will have a tougher time winning the job they want. More than half of the managers suggested that job seekers should spend equal time and money on improving their looks as on writing and perfecting their resumes.

Not the most practical advice.

In addition, 61 percent of surveyed managers reported that it helps for a woman to show off her figure at work. Of course most of those surveyed were men, probably the creepy kind.

> "… 61 percent of surveyed managers reported that it helps for a woman to show off her figure at work." …

Managers also placed looks above education when asked to rank nine character traits. Experience ranked first. Confidence was next. Ranking third? You guessed it, appearance. Appearance trumped the candidate's alma mater, which came in fourth.

This suggests if you are the hottie at your community college campus, you are better off job hunting than if you are a frumpy Harvard grad. Is that fair? Not at all. In an ideal world, your qualifications are always more important than how cute or sexy random people find you.

We think you're awesome, dear reader, no matter what you look like. Unfortunately, that doesn't mean that all the knuckle-dragging power mongers got the memo.

There's good news, though. If you don't like what you see in the mirror, it can be changed. And thanks to science and technology, and the accelerating rate of advancements, an ever-increasing suite of beauty-enhancing procedures are coming fast and furiously down the timeline toward the present.

It will be easier, faster, and less painful to become more beautiful in the next decade. Or at the very least, you'll have new tools to adjust those self-critical areas you frown at in the mirror.

Cosmetic Surgery Trends in America

Let's explore the weird world of beauty with an inventory of the most recent data in the world of cosmetic surgery.

The American Society for Aesthetic Plastic Surgery (ASAPS) reports that almost 13 million cosmetic surgical and nonsurgical procedures were performed by board-certified surgeons in the United States in 2015. (This was the latest data available as we wrote this book, however this data is updated each year in March on http://www.surgery.org).

Between the years 1997 and 2015, there was a 680 percent increase in surgical and nonsurgical cosmetic procedures performed on Americans. Nonsurgical procedures increased by a whopping 1369 percent. Surgical procedures also grew by 112 percent in that time.

According to ASAPS, the top surgical procedure for both men and women in 2015 was liposuction, or fat removal. The five most popular surgeries on their list for both genders are:

1. Liposuction
2. Breast augmentation
3. Tummy tuck
4. Eyelid surgery
5. Nose surgery

The five most popular surgeries on their list for women are:

1. Liposuction

2. Breast augmentation

3. Tummy tuck

4. Breast lift

5. Eyelid surgery

The procedures that trended up in 2015, and saw the biggest increases since 2012, were below the belt. That is at least insofar as women are concerned, who want cosmetic enhancements "down there." We're talking buttock enhancements, labiaplasty, and vaginal rejuvenation.

Buttock augmentation procedures increased by 20.7 percent between 2014 and 2015, but that pales with the bump of 58 percent in 2013 over 2012. In 2015, more than 20,000 buttock augmentations with fat grafting were done.

 Fat Can Be Good?

Fat grafting is the relocation of fat from one part of the body to another. Word has it that luscious-bummed celebs are driving the trend. We're looking at you, Kim Kardashian.

Labiaplasty procedures increased by 16.1 percent during 2015. Labiaplasty is the surgical alteration of the genital lips in women, which is designed to make them more aesthetically pleasing. Pornography is going pop culture with its stars walking on the mainstream red carpets, so the trend is giving every woman the social permission to have her parts enhanced for aesthetic purposes. That, as well as the trend toward complete removal of pubic hair, is exposing everything to be seen.

Vaginoplasty—the tightening of genital muscles—is also a popular surgery with women. This procedure is popular with mothers who want to reverse the stretching effects of childbirth or aging bellies. The idea is to reverse the impact of aging "down there," and to tighten-up age-related droop.

If you want improvements above the belt, more mainstream cosmetic procedures include liposuction (aka fat removal). As we pointed out earlier in this chapter, liposuction is the number one procedure for both men and women. Although women from all walks of life account for the majority of customers for cosmetic surgeons, men are less inclined to go under the knife, accounting for only 9.5 percent of all procedures performed in 2015. When men do elect for cosmetic surgeries, the top five choices are:

1. Liposuction

2. Nose surgery

3. Eyelid surgery

4. Male breast reduction

5. Facelift

SUPER PENIS?

If you read the section on rejuvenation of lady parts and wonder whether there is an equivalent surgery for men, then, here's some good news. Men who want longer penises can benefit from a surgery that can give them up to an inch to an inch and a half of additional length. A surgeon cuts the ligaments that hold the penis in its usual position, allowing the shaft of the penis to descend. Patients need special weights, or a stretching device, provided by their doctor to wear under clothing to ensure a permanent increase in length. The hazards? Possible scar tissue, an erection that points down, and a hairy base of the penis. If you want extra girth, fat injections can be given; however, the fat can clump causing a lumpy appearance.

Lizardman

Of course there are some procedures that aren't seen in Top 10 lists from ASAPS. Those are the extreme procedures that few people dare to choose, but are nonetheless very real and possible.

What if you wanted to look like a lizard? You might think we're kidding, but we promise you, we are not. We spoke to a man who chose this very look.

His name is Erik Sprague. One day in 1997, he opened the Yellow Pages, ran his finger down the listings for cosmetic surgeons and placed a quick call to one of the doctors.

A few weeks later he was in an operating room receiving a transformational surgery that would dramatically change his appearance, and how people related to him for the rest of his life.

Sprague asked his surgeon for a bifurcated tongue, that is to say a tongue split into two. To do this, a surgeon makes an incision centrally from the tip to the underside base of the tongue. This splitting technique allows for a superior control of movement where both sides of the tongue can operate as distinct entities. A drop of sugar on one side and a drop of salt on the other side can both be distinguished separately.

Through a series of additional medically unnecessary body modifications, he has taken on the aesthetic of a reptile (see Figure 3.1).

Figure 3.1 *Erik Sprague is known as "Lizardman."*
(Image credit: Allen Falkner. Used with permission.)

After a stop-and-start progression of 700 hours under a tattoo artist's needle, Sprague's skin now has a green scaly appearance. At the time of this writing, Erik's tattoo is still in the works with 100 hours left to go.

In addition to the bifurcated tongue, he has had innumerable piercings. Five Teflon horns have been surgically implanted under each eyebrow, and four of his teeth have been filed into fangs. Most recently he had his lips inked green.

And none of this happened over a weekend.

"People seem to think that one day it just happened, but there was a long development and a progression," he explained.

He said it probably all started when he was a kid. "Some kids draw on paper; some kids draw on the table or the walls; and some kids draw on themselves. I think you can guess which one I was," he said.

The body modifications started in college when he got into piercings. Then he got his first tattoo in 1993 during his senior year. It all followed from there.

The idea of a lizard sprung from a notion of challenging aesthetic norms. Sprague's aim was to redefine what we know as human. Prior to his surgeries, he was constantly working on the transformation of self through body art and the spaces in his environment.

"It was aligned with my work as an artist with costuming and body painting, by transforming myself and spaces. It was a bringing together of a lot of different elements which became a project," he explained.

Why a lizard? Sprague said, "The reptilian aspect came out of a personal aesthetic choice and a symbolic choice. It is hard to go anywhere in the world without seeing the reptile being depicted as a powerful image. The reptile plays a strong role in the use of technology, so I made use of this as an artist, as well. Also, if you are featured on a TV show and 40 million people start to call you 'Lizardman' you don't really argue with that."

Sprague makes his living doing public appearances, performing as a comedian and front man in the band Lizard Skynard.

He said his loved ones have been very supportive throughout the process. "My mom actually said that by the time I got around to tattooing my face it would have been weirder if I hadn't."

We spent almost an hour on the phone with Sprague, who holds a Bachelor of Arts degree in Philosophy, exploring everything from his reptilian look, his family, his philosophy about the world. We left the conversation feeling quite inspired. And, we can confidently say, he is quite sane.

And although his look is extreme (see a video here: http://superyou.link/lizardman), he might be the future of cosmetic surgery, at least in the sense that if you want to look a certain way, it's entirely possible. Also, your transformation likely will happen with more precision and with less effort than Sprague has had to endure.

More philosophically, he is an example of how human beings can use technologies at their disposal to transform their looks into whatever they desire.

In the 1990s, he looked at research that used luminescent proteins from jellyfish as marker cells in cancer patients. "It would be cool to get those into me and have my skin be bioluminescent," he said.

What about a robotic tail? The logarithmic acceleration of technology suggests that robotics will improve drastically in the next decade and beyond, making something like a robotic appendage within the realm of possibility. Limb regrowth technology, which we will talk about later, will also make that possible.

"If I could get a tail, I would be really tempted," Sprague said. "My biggest concern would be to get one that would not interfere with my quality of life. I wouldn't want to have to work around a tail when I wanted to sit or lie down."

He also said he'd be interested in finding a way to have octopus-like skin. "Being able to change the color and the texture. That would be cool!"

Extreme Cosmetic Hall of Fame

What Sprague has done is certainly unusual, but his extreme choices are not unique. There are a handful of other pioneers who have embarked on changing their appearance with the assistance of extreme cosmetic surgery or other body modification techniques.

Valeria Lukyanova—"Real Life Barbie"

Valeria Lukyanova is a Ukrainian model who has a haunting resemblance to a Barbie doll. She claims to come by her figure naturally, with the exception of breast augmentation surgery.

"Don't believe any stories about me having rib-removing surgeries—they aren't true," she told *The Daily Beast* website. "The only surgery I had was breast surgery. My mother's waist is as narrow as mine—I inherited that from her."

The rest of her body has been sculpted, she claims, with a simple regime of strict diet and exercise. She also uses blue contact lenses to cover her naturally green eyes. (See Valeria do her Barbie makeup here: http://superyou.link/barbiemakeup.)

As an aside, she makes a living teaching people how to connect with their spiritual selves through a process known as astral travel. She also says she communicates with aliens. Enough said.

Justin Jedlica—"The Human Ken Doll"

If a human Barbie exists, shouldn't there be a human Ken doll too? Sure enough there is. Justin Jedlica was dubbed the real life Ken doll by the media after an appearance on the TV show *20/20*.

He has undergone 149 cosmetic procedures, of which 19 were full-on cosmetic surgeries, to attain his sculpted look. They include pectoral implants, tricep and bicep implants, and a myriad of facial surgeries (see http://superyou.link/justinsbiceps). His first surgery was a nose job, 3 days after his 18th birthday.

The married, gay model didn't aspire to look like Barbie's plastic male counterpart. He says the inspiration for his look is from Japanese comics known as manga (or their animated equivalent, known as anime). Oh, and surprise, Joan Rivers and Michael Jackson are his idols.

Jedlica didn't walk into a doctor's office with a Ken doll and ask to look like the toy, he explained. It was a media label. "I don't even know if I look like a Ken Doll, but if other people want to say I do, it's flattering. As a kid, you play with Ken Dolls and kind of assume that is what a handsome guy is supposed to look like," he added in an interview with *The Daily Beast* website.

The TV show, *Inside Edition*, brought him and human Barbie Valeria Lukyanova together for a photo shoot, but it didn't go well. They reportedly "hated the sight of each other," according to the UK's *Sun* newspaper, and traded insults separately through the media.

Maria Jose Cristerna—"Vampire Mom"

Maria Jose Cristerna is not your average mother of four. The Mexican lawyer turned tattoo artist decided to take on the look of a vampire after divorcing her first husband (see her in this video: http://superyou.link/vampiremom).

She has titanium implants under the skin in her forehead that appear to look like horns. More than 90 percent of her body is tattooed and she has multiple piercings and four dental implants that look like fangs. What's next? More horns in the back of her head, she said.

In a *Huffington Post* interview, Cristerna said, "It's my dream to be immortal." Speaking of immortality, be sure to read Chapter 8, "Hyper Longevity: How to Make Death Obsolete."

Patricia Krentcil—"Tanning Mom"

The term *"tanorexia"* has been loosely dubbed the condition for an individual who participates in excessive tanning routines. It's not just a tanning addiction, the person actually sees themselves as too pale and takes drastic measures to change the pigment of their skin.

Patricia Krentcil was labeled "tanning mom" by the media (see video: http://superyou.link/tanningmom). That's thanks to her excessively tanned appearance.

She garnered media attention in 2012 when the New Jersey Division of Youth and Family Services charged her with second degree child endangerment (it was later revoked) following a visit to a tanning salon with her five-year-old daughter.

Lacey Wildd—Largest Breasts

It appears that Lacey Wildd believes "too much of a good thing" does not apply to the world of cosmetic surgery, at least as far as her breasts are concerned. Wildd has spent $250,000 on cosmetic surgeries. That includes 12 breast augmentations, three butt implants, and several liposuction fat reductions, among other procedures (see her Instagram feed: http://superyou.link/racylacey).

Her most famous assets are surely her boobs which went progressively from 32AA breasts to LLL. She even had a procedure where an internal corset was inserted to support her chest and flatten her stomach.

Each breast weighs 21 pounds, so Wildd needs to prop pillows under them when she drives and only uses the back burners of her stove to avoid roasting them when cooking. You can't make this stuff up.

She says she wants one more breast enlargement to double her existing size to QQQ which would render her breasts the size of car tires, though doctors say it will destroy her chest. She has yet to find a surgeon willing to do the job.

In a 2014 interview on the Bethenny Frankel show, Wildd admitted that her obsession is fueled by her need to support her six children. She said if someone paid her $1 million to stop, she might take the offer. So far there have been no takers.

ORLAN—Performance Artist

French performance artist, ORLAN is a woman who has been using her body as an art medium since 1964 (see video: http://superyou.link/ORLAN). Perhaps, one of her most extreme projects was "The Reincarnation of Saint Orlan" in 1990. She transformed her appearance with multiple plastic surgeries to mimic the bodies of women in sculptures and famous artistic works. Her intention, she said, is to draw attention to the man-made concept of beauty. She claimed that female beauty is constructed by men for the pleasure of men. And women live into it.

Cosmetic Enhancements, Non-Weird Edition

Becoming a human lizard or vampire mom might not be for you. Heck, going anywhere near a surgical knife might not be for you either. However, many people start their days with a healthy application of makeup. Humans have been slathering themselves with clay and natural pigments for a very long time.

Actually, makeup use goes way, way back. There's evidence that tribes from Africa, Europe, Asia, and Australia used body paint as far back as the Prehistoric era. Lawrence Barham of the University of Bristol, in England, excavated a pigment deposit from Zambia dating back 400,000 years.

And while the "paint" has arguably been refined, it's still used today by cover girls and your sister, mostly on their faces.

Cosmeceuticals

Makeup is evolving, though. Why paint the ugly when you can medicate it at the same time? Cross a cosmetic with a pharmaceutical and you get a cosmeceutical.

No really, it's no joke. It is an emerging billion-dollar industry that is evolving at a makeup counter near you.

Elle Magazine describes cosmeceuticals as "products that tread the line between cosmetics and pharmaceuticals."

A cosmeceutical cream is more than just a moisturizer. Its makers claim it can improve the skin's health and appearance because of ingredients with biologically active capabilities. These additives are not quite drugs. Think of sugar. It's an ingredient in food. It tastes good. It perks you up. But no one is putting it behind the counter at the pharmacy.

Cosmeceuticals have properties that act on the skin, presumably to enhance beauty and thwart the effects of aging.

If you use products that contain retinol or peptides, then you're using a cosmeceutical. Retinol is essentially Vitamin A, which can be used to improve skin health and vision. In high doses, it is also toxic. (Weird fact: Avoid polar bear liver stew, it contains toxic levels of retinol. Might be good on your face, though.)

Peptides applied topically to the skin can influence the formation of collagen, a protein in skin which promotes thickness and suppleness. When collagen breaks down in the skin, wrinkles form. So apply peptides, the thinking goes, and you will have more youthful skin.

Whereas cosmeceutical products can sometimes be more about marketing hype than function, there is an element of usefulness and function that holds some promise.

Companies continue to look for active compounds that might help repair skin, protect it from further aging processes, make it look or feel healthier, and as a result look younger.

The American Academy of Dermatology cautions, however, that cosmeceuticals are not subject to approval by the U.S. Food and Drug Administration (FDA). Companies promote the ingredients as inactive to avoid scrutiny. A product is either regulated as a cosmetic or as a drug. If a cosmetic company sells a product it claims to affect the structure or function of the body, the FDA considers it to be a new drug and that would require clinical research trials (studies on animals and humans) to prove its effectiveness and safety. If the claims can be shown to be true through research. and the product works as advertised, then the regulatory body will approve it—but as a pharmaceutical, not a cosmeceutical.

CONSUMERS GUIDE TO ANTI-AGING PRODUCTS

What are you putting on your skin? Here's a guide to the potions in the drugstore that may help you maintain a youthful complexion.

ANTIOXIDANTS:

What are they: Antioxidants are nutrients that stop the oxidation process of molecules, and their chain reactions, which damage cells over time.

How they work: When applied on the skin, they disarm the effects of free radicals, which are atoms, ions, or molecules that harm the skin.

What to look for: COQ10, Vitamin C, Vitamin E, and/or the word "antioxidant."

PEPTIDES:

What are they: Peptides are short chains of amino acids, or rather "small proteins," named as such because they are smaller in size relative to protein molecules.

How they work: Peptides stimulate the production of collagen, which maintains skin tightness.

What to look for: The ingredient "matrixyl," and/or the word "peptide."

RETINOIDS:

What are they: Retinoids are derivatives of Vitamin A.

How they work: Retinoids increase collagen production and inhibit collagen breakdown.

What to look for: Retin A, Renova, Retinol, or Vitamin A (over-the-counter and prescription treatments are available).

SUNSCREENS:

What are they: Protective creams that guard against harmful ultraviolet (UV) rays and sun damage.

How they work: Minimizes skin damage (of collagen and elastin) caused by UVA and UVB rays from the sun.

What to look for: Look for a product with SPF 30 or higher. Also look for ingredients such as zinc oxide; avobenzone, marketed by Neutrogena as "Helioplex," "Active Photobarrier Complex" (Aveeno); and ecamsule, marketed by L'Oreal as "Mexoryl SX".

Advice from the authors: *As you can see, when the marketers get hold of science, they bend it to their own devices. Buyer beware.*

Needle-Free Botox

If this book had been written in the early 1990s, we'd be telling you about this amazing new antiwrinkle treatment that is about to sweep the nation.

It's called Botox, a synthesized neurotoxin. It comes from a bacteria that can be sourced from nature and is found in soil and the digestive tracts of some fish and animals.

In high concentrations, the botulinum toxin can result in botulism, a severe, life-threatening illness. Left untreated, botulism can cause respiratory failure (lungs stop working) and death (everything stops working).

Here's how Botox works as a wrinkle therapy: Your doctor injects it into the muscles of your face so that they are paralyzed for up to four months. The toxed-up muscles relax and your wrinkles go away for several months.

In the early 1990s, when it was first introduced as a cosmetic procedure into the marketplace, Botox probably sounded far-fetched: a therapy only a suicidal nutjob would ask for. And yet here's the thing: More than 4 million of you out there in Super You Land had the procedure done in 2015.

It is also used as a pain control treatment for people with migraines and severe muscle pain, though word has it, it can have mixed results.

Now some of us are afraid of needles, including one of the authors who isn't Andy or Sean. You might be, too. The awesome news on the Botox front is you will soon be able to take Botox in pill form. A company called Revance Therapeutics is developing a needle-free application of Botox that can be applied directly onto the skin.

It's currently undergoing testing, and researchers say it can reduce the appearance of crow's feet, (the wrinkles on each side of your eyes) without any unwanted side effects such as, you know, death.

Guyliner: Makeup for Men?

"Hey Steve, you are looking great. Did you change your eyeshadow?"

Ok, so maybe this will never be overheard in the men's locker room anytime soon, but men's cosmetics are an industry unto themselves.

A 2012 report from *Time* magazine claimed that male cosmetics are among the fastest-growing segments in the beauty industry. During that year, men's cosmetic sales grossed an astounding $2.6 billion on its way to a projected $3.2 billion by 2016.

Now before you get your jockstrap in a knot, know that it's not lipstick and rouge being sold to dudes by the vat load. But men are increasingly interested in skincare and manly smelling balms, soaps, and toner. "Double-action face wash" gents?

> "... it's not lipstick and rouge being sold to dudes by the vat load."

It's no surprise that popular makeup brands such as Sephora, Clinique, and MAC have capitalized on what *Investor's Business Daily* reported to be an "underdeveloped" market with "great potential."

It's not just about soaps, cologne, and aftershave anymore. The time has come for classic brands such as your grandfather's Old Spice to make room for male makeup.

John Stapleton from a men's grooming video on the MAC website, sells it like this: "The world has come to a different place when it comes to men taking care of themselves … men spend more time in the mirror than women do," he says in the promo video as he applies eyeliner.

Sorry John, we don't think so. Still, there's no denying there are some men interested in exfoliating.

He says, "A tinted moisturizer reduces redness in the face. A greyish-beige powder adds masculine structure to the brow or fills in patches where hair growth falls short. A little eyeliner creates an edgy look that makes a man look tired in a hot sexy way."

We think the copywriting needs work.

Still, socially acceptable makeup for men has a solid history. Egyptian men were known to moisturize, Roman men wore coverup, and the men of Louis XVI's court had their powder and wigs.

> "In the 1980s, new wave bands gave men permission to wear eyeliner—guyliner!— at the clubs."

In the 1980s, new wave bands gave men permission to wear eyeliner—guyliner!—at the clubs. Adam Ant. Robert Smith. Billy Idol. Boy George. Andy Walker.

Then there's more recent evidence of dudes that fix their faces. Johnny Depp in *Pirates of the Caribbean* and Rolling Stones guitarist Keith Richards have both made it okay for manly men to at least consider making their eyes pop, if only when their football buddies aren't around.

Today's men dabble with concealer and skin bronzer and they use discrete anti-aging skin lotions. Will guyshadow and mickstick be next?

Electro Cosmetics

The future of cosmetics might be equally about form and function if Katia Vega has anything to do about it. She is a researcher at Pontifical Catholic University of Rio de Janeiro, Brazil and a beauty technology designer.

Her focus is a host of wearable computer products that are integrated into cosmetics. A set of nails doubles as a wireless personal identification system using magnets, radio-frequency identification (RFID) tags, and conductive polish. (See Katia here: http://www.superyou.link/ katiavega.)

Open a car without a set of keys. Go on a shopping spree without a credit card. Perform a piece of music without touching an instrument—all thanks to beauty technology.

Vega's project, AquaDJ, allows a nail-wearing disc jockey to mix tracks by gliding his hands over water. Project Abrete Sesamo (Open Sesame!) uses the same technology to open a door with a sequence of finger movements.

Conductive makeup, such as eye shadow and false eyelashes, stick to the skin and connect sensors to actuators. Actions such as blinking can be used as control mechanisms. (Anyone remember Barbara Eden's character in *I Dream of Jeannie?*) Vega's project Superhero, uses a combination of conductive eyelashes and blinking to levitate miniature flying drones.

As technology brings further miniaturization, you can imagine what Vega might invent next. Blink and the laundry gets done? We hope so.

Cosmetic Stickers

Your skin's health and youthfulness is a subjective thing. "Oh Marge, you look younger today!" But does she really?

MC10, a company founded by stretchable electronics inventor John Rogers, intends to bring some objectivity to this with what it calls "cosmetic stickers." These "epidermal electronics" don't bring color, tone, or shape to your skin, but can alert you to skin health and hydration. The wearable technology is called the Biostamp. When applied, it's like a skin sticker that can stretch and flex with the skin as the body moves.

The Biostamp can measure all kinds of physiological functions. In cosmetic applications, it measures your skin's properties while you sleep and even recommends the perfect moisturizer. Or it alerts your smartphone when you need to apply sunscreen.

The electronic sticker is made of circuits that act as a sensor. Besides skin health monitoring, it can also collect data from the brain, muscles, and the heart.

MC10 suggests that besides those concerned with their beauty, the stickers can also be used for all kinds of health and performance applications by athletes, pregnant women and new moms, as well as the elderly. Distance runners can use them to monitor hydration and other performance indicators. The rest of us can use the technology to know when to put on sunscreen.

The first product to use MC10's technology is the Reebok CHECKLIGHT (see it in this video: http:// superyou.link/checklight). It's a skull cap integrated with MC10's electronics platform and monitors impact during sports.

Tattoos

In 1771, explorer James Cook returned to New Zealand from a voyage to Thailand. He reported a skin-staining method called the "tattaw." This was the first known mention of the word "tattoo" in Europe, although the practice had been around since the Neolithic times.

Tattoos Today

Tattoos are created by an artist using a machine which resembles a dental drill. It moves a (hopefully) sterilized needle up and down between 50 and 3,000 times per minute to puncture the skin. The needle penetrates the top layer of the skin called the epidermis, and leaves a drop of ink at a depth of about a millimeter in the dermis, the second layer of skin. The dermal cells are more stable than the cells of the epidermis, so the tattoo ink stays in place and is subject to minor fading and dispersion during a person's lifetime.

By one account, 45 million Americans have at least one tattoo. According to a 2013 survey by the PEW Research Center, 17 percent of Americans say they regret their tattoo, and 11 percent eventually get them removed. There's now a burgeoning business in tattoo removal.

The cost to get a tattoo in the United States ranges from $45 for a small tattoo to $150 per hour for a larger piece of skin art. However, it costs $250 to $1,000 per laser session to remove it. In a removal session, a laser is used to break down the ink particles and the waste is washed away by the body's natural processes over time.

The tattoo removal business is brisk. In the United States, it is expected to generate more than $83 million by 2018, according to IBISWorld, a market research company.

 The Most Universally Common Tattoo Symbols of All-Time

1. Asian Characters (Chinese and Japanese)
2. Tribal Art
3. Butterfly
4. Phoenix
5. Koi fish
6. Dragon
7. Eagle
8. Lotus flower
9. Rose
10. Skull
11. Star
12. Yin and yang
13. Dove
14. Cross
15. Chains
16. Anchor

Freedom-2-Ink

Let's face it, people do dumb things such as getting the name of a lover or spouse tattooed on their skin. Perhaps you've had "I love Richard" tattooed on your arm. Later, however, you discover Richard is more of a Dick, and that he isn't the sweet, charming partner with whom you expected to get old, wrinkly, and incontinent. However, if you've used a revolutionary new removable tattoo ink, you can erase Richard forever and with a lot less fuss and pain than you'd suffer with laser removal.

In January 2009, a team of scientists from Harvard University, Brown University, and Duke University, created a tattoo ink that can be erased. Freedom-2 ink was launched in select areas of the United States. The ink is both permanent and removable. The dye which makes up the art is stored in microscopic capsules, which are inserted into the skin for life. If you change your mind, they can easily be zapped away.

The major benefits?

- The removal process is less painful and less costly than the removal of a traditional tattoo.

- There is greater health and safety with this type of ink. It's stored in capsule form, which protects you from the leakage of toxins into the skin.

Initially, it was only available in black and red ink when it was released for sale. Since then, the product and its associated technology has been acquired by a another company, Nuvilex. When we first started writing, it was unclear whether Freedom-2-Ink would be further developed. We called Nuvilex, and the person who answered sounded like a woman on a cell phone. She promised to have someone call us back. They didn't. The company has since changed their name to PharmaCyte Biotech Inc., and there doesn't seem to be any mention of the technology on their site. Too bad.

Invisible and White Tattoos

A tattoo solution for those who want a covert skin art look might be the invisible tattoo. This type of tattoo became all the rage in the rave dance scene when it first emerged. Special ultraviolet or glow-in-the-dark ink used in the tattoo is only visible under black light, and can glow in a range of colors from whites to purples. The colors depend on the ink used and the tattoo owner's skin tone. Invisible tattoos are perhaps ideal for the businessman-by-day who likes to keep his bad-boy image under wraps until the clubs open after dark.

With that said, although the ink is invisible without UV light, the scarring on the skin isn't. The other drawback is that the ink in the tattoo might lose its ability to glow in black light over time, just like those star stickers you put on your bedroom ceiling as a kid.

Another option is to go with a white tattoo, which uses white ink and forgoes the black outlines on the inked art. The effect is a ghostly image on the skin. (See the white tattoos here: http://superyou.link/whiteink.)

The problem with the white tattoos is that they can fade. They can also turn yellow and potentially look like a scar or an odd skin infection over time.

LED Tattoos

If you want something hipper than old fashioned ink, try an LED tattoo. The process, in development by Dr. Brian Litt, a neurologist and bioengineer at The University of Pennsylvania, involves the surgical implantation of a 1 millimeter silicon plate under the skin that is only 250 nanometers thick. By comparison, a sheet of paper is about 100,000 nanometers thick.

The plate is built on thin films of silk, which eventually dissolve over time, allowing the body to accept the chip.

The LED device can turn people into walking billboards that display flashing, animated skin art (see video: http:// http://superyou.link/ledtatts). However, they are not just used for aesthetic purposes. Litt sees a future use for his technology in the medical world. People with diabetes could use the device implanted in their arm to show alerts about blood sugar levels.

So far the technology has been tested on mice successfully. And no, the display did not read, "Minnie 4ever."

Gadget-Activated Tattoos

In addition to the myriad of traditional tattoo options, you can also get tattoos that interact with gadgets. In 2012, Nokia developed the concept for an electromagnetic tattoo and, in 2014, filed for a patent. Ink is placed either on the skin or in the skin. The more invasive approach helps protect the tattoo from daily wear and tear.

The tattoo looks like any other tattoo except that the metallic particles in the ink can be magnetized. It can be made to vibrate when your smartphone rings. You are then able to feel a phone call or message alert or other notification. Three short tingles means your battery is low, perhaps? Or constant buzzing can mean your mother-in-law is trying to reach you?

With the patent filing, it seems the technology will be available soon. Some experts think it won't be commercialized for a decade, however. We figure it might show up faster if Nokia finds more than just recreational uses for it.

OLD-SCHOOL BEAUTY TRICKS

Humans go to some pretty crazy extremes for beauty. A collection of some body-shaping tools that have been used throughout history to manipulate the human body without cosmetic surgery follows.

French corsets—The French get credit for the corset, a garment that holds the shape of the torso into a desired aesthetic. However, its invention has links to Britain, Greece, Italy, Egypt, and other countries. The trend was popularized by Catherine de' Medici, the wife of King Henry II of France, who placed a ban on thick waists during the 1550s. Women wore corsets for the next 350 years. Corsets are still available today but they're not popular. It's been suggested the manipulation of a waist with a corset can lead to problems with the heart, lungs, stomach, breasts, and uterus.

Chinese foot binding—Women with small feet have been the epitome of beauty in China. There are many stories to explain why this was the case, ranging from a concubine with clubfeet, to an Emperor who fell in love with a small-footed courtesan. However it happened, the practice of foot binding became a trend. Between the ages of four and nine, a girl's feet are molded, toes curled under with enough force to break the bone, and then are bound to hold their new shape. Ouch! This painful practice didn't fall out of fashion until the late 1950s, and "lotus" shoes designed to accommodate disfigured feet were available until the late 1990s.

Body spacers—Piercing the skin and stretching it with objects, known as spacers, is a beauty regime that's been performed for many years. It still happens today, all over the world.

Africa and the Amazon tribes, such as the Masai (Kenya) and the Huaorani (Amazon Basin) stretch their ears. Other tribes of the same regions use plates to stretch their bottom lips. The Kayan women from Northern Thailand, known as "long necks" or "giraffe women," use brass coils to elongate their necks. Some have up to 25 rings they can never take off. Closer to home, there's that barista down at your favorite coffee shop, who has used spacers to expand the piercing in his earlobes to roughly the size of dinner plates. (Wait until he finds out how much it costs to undo that surgically when he wants to apply for a government job.)

Push-up bra—In 1945, Frederick Mellinger returned from World War II and was inspired to start a line of lingerie his friends said their girlfriends would wear. Three years later, he invented the push-up bra, which pushes a woman's breasts together to give the appearance of a fuller bosom. He called it the "Rising Star."

Hair

Hillary Clinton once said, "If I want to knock a story off the front page, I just change my hairstyle." It's true, hair is newsworthy. If we're famous enough it makes headlines. If we're not, our friends and family notice.

And, have you ever noticed that an ordinary person with a new hair style can suddenly turn heads? It's equally true if someone you know gets a new style that is unflattering or strikingly different from what they had before.

Hair is a defining element of beauty and a sign of youth and vitality for both men and women. The hair business is a billion-dollar industry, and science is working harder than ever to give you the hair you want, or remove the hair you don't want.

> "The hair business is a billion-dollar industry, and science is working harder than ever to give you the hair you want."

The body of knowledge and the advancements dealing with hair regrowth and a cure for baldness far outweighs the science around hair removal. So we'll deal with baldness shortly. First, let's look at a few developments around hair removal.

Hair Removal

If you have hair in places you don't want, there are lots of potions, shavers, and gadgets to get rid of it in the short term. However, current solutions either smell funky, take time, or hurt like hell. There are, however, some permanent solutions. You might have heard of a few of these before, but some are very new and exciting.

Electrolysis

If you want to get rid of body hair for good, electrolysis is the only permanent, FDA-blessed solution currently available.

Surprisingly, electrolysis has been around for 135 years. It's used to remove all types of hair, even very small and fair follicles (something laser hair removal can't claim; more on that shortly). Electrolysis delivers electricity through a very tiny probe, straight to the hair follicle. This dose of electricity causes localized damage to the areas where applied, permanently stopping hair growth.

Physicians use one of three methods:

- **Galvanic**—Up to 3 amps of electricity destroy the hair-growth mechanism by causing a chemical reaction in the follicle.

- **Thermolysis**—Heat is applied to the tissue that supports hair growth and destroys it.

- **Blend**—A combination of galvanic and thermolysis are used to kill the hair-growth mechanism.

The time frame and price of treatment varies with each patient as well as the amount of hair being removed. Each session lasts from 15 minutes to an hour. The procedure is relatively painless. A tingling sensation can be felt and a topical anesthetic is usually applied to prevent discomfort.

Laser Hair Removal

Since 1997, lasers have been used to permanently reduce unwanted hair. The procedure involves a method known as selective photothermolysis, which fires a pulse of light at the hair in the follicle. Laser energy is absorbed by dark matter in the skin or hair follicles, so brown or black hair is destroyed by the light energy. However, lighter hair and skin is unaffected, which is a major downside because it works best on patients with fair skin and dark hair. The bottom line is laser hair removal is a hair-reduction technique. The discomfort experienced is similar to electrolysis, but requires fewer treatments.

Topical Gel for Longer-Term Hair Removal

Researchers might have found an easy way to keep hair off for longer periods of time by halting growth, at least for a few weeks. A research team from the University of Pennsylvania have discovered that Cidofovir—a widely used drug to treat viral infections of the eye—is also useful for temporary hair removal, and in high concentrations, useful in the treatment of AIDS.

To test it as a hair reduction therapy, the researchers added the drug to a rub-on gel. Sixteen men, with the ability to grow dense beards, were recruited for the study. The group tested two concentrations of the gel—some men used a 1 percent solution and some used a 3 percent solution—on just one side of their faces. The men used a dummy gel on the other side, then shaved both areas as normal. In order to measure the amount of hair that had grown back, they were asked to stop shaving 48 hours before the assessment.

The results of the study were published in the *Archives of Dermatology*. The findings showed that the men who used the 3 percent topical agent had a significant reduction in hair growth. They were able to stop shaving for six weeks before their facial hair grew back.

According to a 2012 report by *Glamour* magazine, a product such as this can save us all a lot of time and money. On average, a woman will spend 7,718 hours shaving and $10,000 on shaving products in her lifetime.

A survey on Askmen.com reports that men spend six months of their life shaving. This is based on a daily routine that starts at the age of 15 and goes until the age of 75.

Hair Loss

People spend a lot of time getting rid of hair in places where they don't want it, but they also worry about losing it in places where they do want it.

(That's true of fat, too, don't you think?)

Let's face it, hair loss is not a pretty thing to endure. One day you're teasing a set of luscious locks to create your Duran Duran hairdo and, then, in a seeming blink of an eye, you are middle-aged, and have a George Costanza cut.

The medical term for hair loss is alopecia. It's happening right now as you middle-aged men read this book. And it's a problem not limited to men. A large proportion of women suffer from alopecia, too. Women often start to notice hair loss in their 50s and 60s (though some suffer hair loss much earlier).

A 2013 study from the International Society of Hair Restoration Surgery, reports that 35 million men and 21 million women in the United States deal with alopecia each year.

Currently there is no cure for hair loss, however, science aims to change that.

A Short History of Baldness

Perhaps the earliest baldness "cure" was chronicled in the *Edwin Smith Papyrus*, an ancient Egyptian medical text. The remedy involved a pomade of animal fats extracted from a lion, a hippopotamus, a crocodile, an ibex, and a serpent. And, because you can't walk into a pharmacy today and purchase a bottle of Dr. Nefertiti's Croc Tonic for Lustrous Hair, it's easy to see this (nor any of the other myriad of natural remedies) didn't actually work.

So, people such as King Louis XIII of France, who lost his hair at the age of 23, had to take matters into their own hands. He used a white powdered wig to conceal his problem, which consequently became a major French fashion statement.

It was not until 1868 that the link between baldness and genetics was published in *The New York Times* and it still took many years for a medical procedure to be developed to solve the problem.

In 1959, New York dermatologist Norman Orentreich discovered that he could surgically implant hair grafts from the back of the head to the front.

Since then, there have been a variety of efforts to magically stop hair loss or regrow hair. Laser hair therapy—also known as light therapy—is one technique that is supposed to stimulate hair growth. Some doctors argue it can be effective for some, others are more skeptical. Either way, bald men are not lining up for the solution.

Topical creams have thus far been the cheapest and some of the most effective ways to remedy hair loss. Here's a summary, plus several more techniques that folliclely challenged men are turning to these days.

THE TOUPEE THAT CAUSED A SUPER ACCIDENT

How far might you go to cover up a bald patch? Would you risk the success of your career to protect your physical appearance? These were questions that tennis player Andre Agassi considered before he removed his hairpiece. In his book *Open*, the world-renowned tennis player shared the truth with the world when he said: "I wore a wig…and it cost me the French Open."

In case you don't remember the incident, Agassi lost the final in the 1990 French Open to Andres Gomez of Ecuador because of a wig fiasco.

As his story goes, the wig fell apart the night before and, although he repaired it with a bunch of clips, he had to scale down his performance to keep it from falling off.

Good thing he married (his now former wife) supermodel Brooke Shields in 1997. Her fashion forward advice to Agassi was to shave his head, an action he likened to "going without teeth."

In the end, the new hairdo, or rather the lack of one, was a clever move for the sports star. When he retired in 2006, it was suggested by the BBC that he was, "Perhaps the biggest worldwide star in tennis history."

Rogaine

The active ingredient in Rogaine is minoxidil, which was originally an oral treatment for high blood pressure. However, researchers discovered it had a useful side effect—hair growth. The reason for this is not fully understood, however, what scientists do know is the drug acts as a vasodilator (a fancy word to say that it opens blood vessels, allowing more oxygen, blood, and nutrients to the hair follicle).

Rogaine works best for men with early-stage hair loss in small amounts. It's an over-the-counter medication and costs about $30 at the local drugstore. Approximately one third of men who take Rogaine see results.

Propecia

Finasteride, sold as the brand Propecia, offers a better success rate. Studies show up to 9 out of 10 male Propecia users see good results. At $60, it costs double the price of Rogaine and a prescription is required. It's a bit different than Rogaine because it blocks the creation of the hormone, DHT, which causes hair loss. Even though Rogaine is unisex, Propecia is only for men. In fact, women shouldn't even handle Propecia as it is known to cause birth defects, if she is exposed to it during pregnancy. Other possible side effects for men include impotence and a particularly nasty version of prostate cancer.

Hair Transplant Surgery

Medications only work for small amounts of hair loss and at early stages. For a more permanent solution, hair transplantation is the only procedure currently available.

Hair follicles are taken from a "donor" site (on the patient's own head) and grafted to a "recipient" site. Each hair graft is composed of strands and skin taken from parts of the body genetically resistant to balding.

There are two popular methods for the removal technique of the grafts, which is the crucial step in the procedure. Hair follicles grow on a slight angle to the skin's surface, so the transplant tissue must be removed in one of the following ways:

- **Strip harvesting:** A strip of skin of about 0.5 inches by 6 inches (1 centimeter to 1.5. centimeters by 15 centimeters to 20 centimeters), is removed from an area of good hair growth on the scalp. Hair follicles are removed in bunches, and are inserted in puncture points at the recipient site. This method leaves a small scar, usually at the base of the head (at the donor site).

- **Follicular unit extraction:** Individual units of one to four hairs are removed and inserted at the recipient site. It is a more time-consuming procedure and not everyone is a good candidate. Hair transplantation is an outpatient procedure requiring mild sedation and local anesthetic. The cost ranges from $4,000 to $15,000. The price varies with the amount of hair transplanted.

The most immediate potential risks with these hair transplant procedures are bleeding and infection. In some cases, patients suffer a disease called folliculitis which involves inflammation on the head at the site of new hair growth. The good news is that folliculitis is easily relieved with antibiotics. Another potential risk is shock loss, where the new hair falls out.

Robotic Hair Restoration

Once in a while, a patient who undergoes hair transplantation has to deal with hair growth that looks unnatural. For these patients, robotic hair restoration is a good choice.

The anti-retroviral treatment and access to services system (ARTAS), which uses high-resolution digital imaging, is a robotic hair transplant device used by a doctor to identify and harvest hair follicles from the back and sides of the head. The follicles are then relocated by the doctor to where they are needed. The procedure produces no scarring.

Hair Cloning

A team consisting of researchers from Durham University in the UK and Columbia University in the United States, believe they might have solved the problem of hair loss using a method they have dubbed "hair cloning." And no, there are no sheep involved.

The process involves three steps:

1. A small strip of skin and hair is removed from the back of the head.

2. Cells from the sample are extracted and grown upside down in a petri dish.

3. The patient receives hundreds of small injections of these cells into the bald area.

The method has proven successful in mice studies. Human hair cells were removed and grafted on the backs of mice. The hair produced was white, although it is suggested that with more testing it could become the same color as the hair extracted from the donor's head. There's no word on whether the mice were happy with the results.

In a 2013 journal article published by the *Proceedings of the National Academy of Science*, Durham researcher Dr. Colin Jahoda, said, "There are still a lot of technical hurdles to cross before using this as a cosmetic treatment, but this is a very important step forward."

The step forward Jahoda refers to is the process of growing the cells upside down, which helps the cells clump together and multiply.

Human trials are next with hopes that both men and women will benefit from the future therapy. It's expected to range in price from $10,000 to $13,500.

Hair Loss Gene

Dr. George Xu, professor of dermatology from the Perelman School of Medicine at the University of Pennsylvania, made headlines for a major advancement in the study of hair loss. (His research was published in the 2014 edition of *Nature Communications*.)

Xu's team found a way to artificially grow a group of epithelial stem cells that are necessary for hair growth. These cells are found near the root of the hair follicle, in an area medically termed the "bulge." When tissue at the root of the hair follicle is damaged, the epithelial cells help it heal. When they are absent, hair growth is impossible.

The scientific magic used in this breakthrough was to employ a specific type of stem cell known as induced pluripotent stem cells (iPSCs). In basic terms, iPSCs have the ability to morph into any type of cell in the body.

Using gene therapy, Xu's team turned human skin cells into the iPSC variant of stem cells. Then they used the iPSCs to create epithelial stem cells, which were implanted onto a mouse. The result was a patch of human hair growth on the mouse. What you just read is a simplified explanation of the process. As you might imagine, the actual science is much more complex.

The bottom line (Xu said in an interview with the *Penn Current*, the university newspaper) is that while the achievement is a step forward in the cure for hair loss, we may be still ten years away from a solution that can be used to restore hair loss in humans.

Weight Loss

It's true what they say that beauty isn't skin deep. You can slap yourself silly with creams and anti-aging elixirs, dress your face with rosy cheeks and red lips, and put foam on your head to coiffe a set of luscious locks, but none of that will ever make you thinner. When it comes to the shape of your body, its fate is in your hands. And, to be honest, most of us are looking for a quick fix to get skinny.

Two-thirds of all Americans are classified as either fat or obese. We spend $40 billion on dieting on average each year, which suggests that most of us have been chubby at one time or another. Perhaps you tried The Atkins Diet in which you ate a hefty plate of carb-free eggs and bacon, and a daily steak. Maybe you tried the famed ultra-low calorie Cabbage Soup Diet and slurped up gallons of soup. Or perhaps you've been to Weight Watchers, tried the South Beach Diet, or replaced meals with chocolate shakes using Slimfast products.

Unfortunately, it's likely that none of those diets have produced any permanent results for you. However, the scientists and technologists are furiously working away to find a permanent slimming solution. Here are a few pioneers we discovered along the way.

Soylent

In 1973, Charlton Heston starred in the science fiction movie, *Soylent Green.* The movie tells the tale of Detective Robert Thom, played by Heston, who uncovers a horrific secret. Food wafers manufactured to feed a starving (and futuristic) New York are made of human remains. "Soylent Green is people!" cries Heston in the movie's last moments.

> "... there is a product on the market today called Soylent, and it might be able to make you thin."

Of course, the future of thin is not cannibalism. However, there is a product on the market today called Soylent, and it might be able to make you thin.

Its inventor is Rob Rhinehart, an entrepreneur who developed the food-replacement product while he was living as a starving college student (see Figure 3.2).

Figure 3.2 *Rob Rhinehart, creator of Soylent.*
(Photo copyright Rob Rhinehart. Used with permission.)

He calls it a "pretty unremarkable" and an "off white, brownish" shake that tastes "kind of bland." Not to worry though, there are no human remains in this concoction.

The FDA-approved meal replacement is a combination of powder and oil that when mixed with water contains all the essential nutrients the human body needs to survive. The oil component is made from canola and fish oil, whereas the powder component contains a list of 33 ingredients in total. But, it's not as scary as it sounds. It's a simple list of recognizable vitamins and minerals—ingredients you can pronounce.

Rhinehart created Soylent thanks to a personal dilemma. He was studying engineering at the time, and as a bachelor, he was broke, busy, and eating very poorly. His habits were affecting his health, but, he admits, the truth is that he didn't want to take the time to prepare nutritious meals.

To create Soylent, he used the same process he used on hardware in engineering, a process of optimizing and breaking down machines to see how efficiently and simple he could make them run. "I took the same approach to food. What is the most basic essential thing you could live on?" he explained.

To find out what humans need to survive, he looked at what humans are made of and assessed all the components of the human body. Then he did a bit of self-education in nutritional biochemistry. The result was a nutritionally balanced simple, elemental shake.

There was a lot of testing involved to make sure he engineered the correct balance of nutrients. Then he used himself as a test subject and lived on his invention for 30 days.

Rhinehart says his health improved. He felt a dramatic increase in energy and he woke up feeling more refreshed and with more stamina. He also says that he felt sharper, his mental performance improved by 20 percent—results he obtained from measurable tests. His muscle to fat ratio improved as did his blood levels.

It took some work to get the formula right. The testing phase, he admits, was a fascinating process: "It was kind of cool to see. As I would change something in the formula, it would affect my body. With different levels of potassium I would get heart arrhythmia." (His heart beat irregularly.)

"Adjusting the ratios of fatty acids would affect my skin, Rhinehart said. " If I dropped sodium levels I became mentally foggy."

It took 15 tweaks for him to arrive at the final product, a refinement process that helped him find the most effective combination (see Figure 3.3).

Figure 3.3 *Soylent. Is this the future of food?*
(Image copyright: Rob Rhinehart. Used with permission.)

Today, Rhinehart's diet is 80 percent to 90 percent Soylent. Now regular food, he said, is more for fun and enjoyment. Food tastes better because he eats less of it and he craves fresh, healthy ingredients. He chooses sushi instead of pizza, for example, when he goes out with friends.

Soylent also saves him time and money. The effort required to make Soylent is minimal—simply stir powder and oil with water. Each meal costs $4.

Though he admits living without the pleasure of food is hardly romantic, it provides a practical solution to real world problems. It can help with famine in underdeveloped countries. For those who lead hectic lifestyles, it provides a healthy meal in a few minutes.

Soylent's initial audience has been young male urbanites. "People who can't be bothered with cooking all the time or can't afford take-out all the time," explained Rhinehart. "Then again everybody eats. And everyone has some difficulty with food. I would really like it to be a commodity, a utility."

Truck drivers have also shown curiosity about the product. It's a market that caught Rhinehart by surprise. Truckers sit for long hours with no exercise, have poor food options on the road, and sometimes can work more efficiently if they don't stop.

Rhinehart has also had an inquiry call from NASA, which sees Soylent as a way to feed astronauts on missions. He's also been called by non-government organizations

(NGOs) who want to use Soylent to fight famine. Soylent could also help fight global waste. Food and its associated packaging accounts for up to 50 percent of household garbage.

In the future, Rhinehart suggests that food could be designed for the experience of eating. He suggests that we won't have to worry about our health because we will be able to modify food for its nutrition. He also thinks food will be unrecognizable, although it will have similar traits: "If you love really fresh fruits because they're crunchy, colorful, and sweet we will have things like that, but without the issues of having to grow it, having it spoil and having it be expensive. Food will be designed for pleasure and it will be as practical and efficient as possible."

A Starbucks white chocolate mocha with all the health benefits of a piece of wild salmon? That sounds good to us.

Weight-Loss Surgery

If you are not yet willing to give up food (we don't blame you), and you are looking for a major weight-loss solution, you should consider weight-loss surgery. There are several options today available to people who struggle with their weight.

Gastric Band

The laparoscopic gastric band, lap-band, A-band, or LAGB, as it is variously known, is an adjustable device used to constrict the top portion of the stomach. It restricts the amount of food that can enter the digestive system, so you don't over eat.

The lap band was originally approved for people suffering from extreme weight gain—those who doctors consider morbidly obese. In 2011, it was further approved for generically fat people. Those people who tip the scales at a 30 to 40 rating on the body mass index (BMI).

The band is opened and closed using a syringe. A port under the skin, located under the ribcage or lower on the torso closer to the hip, is accessed with a needle. A doctor or nurse adds saline solution and the band gets tighter. Or removes saline, and the band gets looser. This impacts how much food you can eat.

The band does not guarantee you'll lose and keep off the weight, but it's a tool to stop you from overeating. After all, you can melt chocolate into a liquid, drink it all day, and gain weight regardless of the band's design to restrict food.

We know all this because Andy, one of the authors, had one installed in 2010 and lost 60 pounds with it. He still has the band today. Insider tip: If you go on a cruise, avoid the all-you-can-drink milkshake bar. The band is not going to stop you from gaining 10 pounds on a cruise.

Gastric band surgery costs on average about $15,000.

THE FUTURE OF THE GASTRIC BAND

Banders (as recipients of the gastric band are sometimes called) require ongoing adjustment to the tightness of their gastric bands. This requires a visit to their doctor or trained health professional. Complications with the device can occur when the band is overtightened, so a better way to adjust the band would be ideal. Enter: self-adjusting bands.

These bands would include a port and a computer-driven pump, which could be controlled wirelessly and alter band settings as needed. This "smart" band could monitor pressure and deflate automatically during times of severe obstruction (when food gets stuck). The band could also tighten to stop a patient from eating after a certain amount of time. This auto-adjustment would prevent excessive progressive pressure on the stomach and esophagus, which can cause complications. A patent filing in 2012 shows that Allergan, a gastric band manufacturer, has certainly considered such an improvement. However, as of this writing, it has yet to arrive on the market.

Gastric Bypass

Gastric bypass surgery is a surgical procedure prescribed to treat morbid obesity, which, in clinical terms, is a BMI of 40 or more. The goal of surgery is to dramatically reduce the size of the stomach and thereby decrease the amount of food that can be consumed in a sitting. An extra-large, all-dressed pizza and a postsurgery stomach is incompatible.

There are several surgical techniques that can be used, but all of them divide the stomach into two pouches, a small thumb-sized top pouch and large lower pouch. The two sections are either partitioned or divided completely into two separate compartments. Then, the small intestine gets rerouted so that it connects to each of the newly created stomachs.

Food enters the small pouch when it is eaten, and then bypasses the main stomach into the small intestine. The larger stomach pouch never receives any food but continues to flow digestive juices into the small intestine. (See a picture at http://superyou.link/gb.)

The small pouch is now a stomach that is 90 percent less than its original size. Patients can only eat one quarter to one half cup of food at a time, at first. By the end of the first year, they can eat one cup of food. When the patient eats, the wall of the small pouch gets stretched, sending a message to the brain that triggers a feeling of being full.

If the patient tries to consume more than they are physically capable of, they immediately throw up. Over time, the stomach stretches slightly, but only after subsequent weight loss is achieved.

After gastric bypass surgery, patients no longer have the ability to eat large quantities of food. They lose 65 percent to 80 percent of their body weight. Conditions such as type II diabetes, hypertension, and sleep apnea improve greatly, or in some cases, are eliminated.

It's a pricey procedure, costing $15,000 to $35,000. But, then again, it's a surgery that can save your life.

Sleeve Gastrectomy

Like the gastric bypass, sleeve gastrectomy is a surgical weight-loss procedure in which the stomach is downsized to about 25 percent of its original size. The result is a stomach that has a sleeve or banana-like shape. The surgery permanently reduces the size of the stomach, although it can stretch over time.

GUT SLEEVES

Speaking of sleeves, one experimental procedure that might eventually become available is the "gut sleeve." A study conducted on rats and published in late 2013 describes a nonsurgical approach to weight loss which places a thin flexible silicone sleeve near the end of the stomach. In humans, the sleeve would be inserted through the mouth and down the esophagus into the stomach, just above the small intestine.

The researchers found that the sleeve significantly reduced body weight. Placing the sleeve directly below the stomach, in the upper part of the small intestine, changes the way that the body senses food and how full a person feels. Research using this technique on humans was apparently in the works in late 2013, but no progress has been publically reported since.

The Future of Thin

Dr. Martin Fussenegger, a professor of biotechnology and bioengineering at the Swiss Federal Institute of Technology in Zurich, has created the first weight-management device that regulates metabolism. By mid-2014, it had been successful in studies with mice. Human trials have not yet been conducted. However, the technology used in the device was also used to treat psoriasis in mice in 2015.

While it's been described by some media as a microchip, Fussenegger said the device is completely bio-based. He describes it as a group of "tiny, caviar-like capsules, 400 micrometers in diameter, which have been fused together as a synthetic network." This network is small enough to be injected into the body with a syringe.

The balls, what he refers to as "designer cells," are sensitive to a wide variety of fatty acids and fatty acid mixtures. This allows the balls to constantly monitor the body's blood fat levels. When blood fat levels rise—a process which happens naturally after we eat a meal—the designer cells release a hormone that sends a signal to the brain to stop eating.

To fully understand the real magic of these little miracles, it's important to know how blood sugar works in the body.

When blood sugar levels rise, the body produces a hormone called insulin, which converts blood sugar into the energy the body requires to run. A surge in insulin tells the body there is enough energy, which makes the body start storing fat instead of burning it.

For example, if you are starving and eat a sugary food (such as a chocolate bar) your blood sugar levels spike and your body pumps insulin into the blood stream. This sends a message to your metabolism to store energy as fat, instead of burning it. Because a chocolate bar won't be enough to satiate you, chances are you will keep eating and everything you eat at that point turns into fat.

Now, imagine you have Fussenegger's implant in your arm. If you eat a chocolate bar your blood sugar levels will still spike, but you will stop eating because you won't be hungry, which means less fat is stored. Note to investors: Sell Hershey stock now.

Mice who received the treatment lost a healthy amount of body weight. They did not starve or die, which means that Fussenegger's capsules are far more effective than a time-release drug, which releases a drug in intervals but doesn't actually react to changes in the body.

One of the next steps is human testing. Candidates will receive a port-like implant just under the skin, which will need refueling every three to four months to prevent the blood from clotting. However, as Fussenegger suggests, it will take up to ten years before we see this technology come to market. Of course, if you talk to futurist Ray Kurzweil, he'd likely shorten that prediction. The algorithmic improvement of technology always suggests that any technological prediction is shorter than we think it is.

When this technology does come to market, it's likely to be a huge help for anyone struggling with their weight. He says "the metabolism will be treated at the molecular level right where the problem occurs."

He also sees this system being used to treat illnesses affecting blood levels, such as diabetes and immune disorders. Treating different illnesses will require different sensors, which is something Fussenegger's team continues to explore.

Zerona Laser

If you can't wait ten years for Fussenegger's implant, then you might want to consider a Zerona laser treatment. It was featured on *Dr. Oz*, where Dr. Jamé Heskett from the Wellspa in New York City introduced it as a weight-loss solution that "guarantees at least three inches of weight loss (all over the body) and up to 11 inches in two weeks."

Zerona is a cold laser that perforates fat cells and makes their contents leak, but leaves blood vessels and other cells in the same area intact. The contents of the damaged fat cells are eliminated by the body's normal detoxification process. What's left of the fat cell shrinks. And you do, too.

The procedure is noninvasive and completely painless. Most importantly, the process doesn't destroy the entire fat cell, which can otherwise affect endocrine function.

The time requirement for Zerona treatments is no more than your average workout regime with a major bonus—less effort! For 40 minutes, three times a week, all you have to do is lie in a chamber. Then about three weeks later you could see a reduction of up to 3.5 inches (a little more than 9 centimeters) on your waist, hips, or thighs.

Zerona treatments cost between $2,000 to $4,000, a price that varies with the number of treatments needed.

A very small group of studies suggest that most patients lose inches off their frames. A 2012 article published on Shape.com, explains that results vary based on your current size, diet, and exercise regime. It is best used to enhance an already healthy program of diet and fitness. It typically works especially well once your body has reached a plateau, where weight loss is much harder than it was when you started.

Shape.com writer, Charlotte Hilton Andersen, underwent the Zerona treatments so she could accurately write about them for the magazine. She says one of the most surprising results for her was a major increase in energy. And, she says, it wasn't as simple as lying in a chamber: "I was cautioned to eat very clean, work out moderately, wear compression undergarments, and drink so much water that if peeing were an Olympic sport I'd win gold." However, the cleansing effect led her to an unexpected side effect—crazy energy.

Although there are accounts of people losing up to 3.5 inches, Andersen only lost one-quarter inch on each thigh and 1 inch on her waist.

Andersen might have proven that Zerona does work, but perhaps not as well as advertised.

Liposonix

Liposonix is a quick, noninvasive treatment using ultrasound technology to eliminate that last inch of fat that won't go away with diet and exercise. Liposonix.com claims the procedure is a "one treatment, one hour, one size" weight-loss solution.

It uses a technique called high-intensity focused ultrasound. Ultrasound technology is used to fire high frequency sound waves at unwanted fat cells, which are destroyed leaving surrounding cells and tissue unscathed.

Liposonix targets and eliminates fat cells from the treatment area and are released from the body by its natural elimination process. This process takes 8-to-12 weeks.

Candidates must have a BMI of 30 or less, and be able to "pinch an inch" of fat in the area receiving treatment.

The company claims a 91.3 percent success rate, saying that "subjects responded that the flatness of their abdomen had improved after just one treatment."

We took a look through their before and after photo gallery and while there are some results, they seem very minor. Once again the promise seems intriguing and actual results may vary.

Cosmetic Surgery: A History

In ancient times, if you were a cave person who didn't like their looks, there was not much you could do outside of basic skin adornments such as tattoos and piercings and slightly more uncomfortable modifications such as foot binding and teeth filing.

Makeup, as we discussed earlier, has been around for a long time, and has perhaps been the most accessible adornment for those that want to improve their looks or enhance their beauty.

The concept of rearranging human flesh surgically into new shapes to enhance or repair one's looks is something that has been available for centuries from self-appointed surgeons (and sometimes questionable barbers). However, it's not something that has been safely or readily available to the mass market until fairly recently.

The ancient Egyptians performed plastic surgery on their dead royalty to ensure they looked good in the afterlife. Cheeks were stuffed with cotton and noses were enhanced, but these procedures were reserved for the dead. They figured you'd need to be recognizable and kinda sexy in the afterlife.

One of the first references to reconstructive plastic surgery on the living is found in ancient India. Sanskrit texts from 600 B.C.E. describe procedures that repaired noses or ears lost in battle or as punishment for crimes committed.

The Romans in the first century are said to have practiced plastic surgery extensively on the face, and particularly on the nose, eyes, and lips. There's even evidence of circumcision reversal, male breast reduction, and blepharoplasty (correcting droopy eyelids). But one of the most common procedures was scar removal from the back. Scars on the back suggested a soldier turned away from battle, so evidence of back injuries was shameful for the soldier.

Of course, all this was rather painful without the availability of general anesthesia. Antipain potions that used ingredients such as cannabis, opium, and alcohol were variously used through the centuries for pain control, but by most accounts early surgeries were horrendously painful. General anesthesia wasn't discovered and adopted until the eighteenth and nineteenth centuries.

In the late fifteenth century, Italian Gasparo Tagliacozzi wrote the first plastic surgery textbook: *De Curtorum Chirugiau* (1597). Later dubbed the father of plastic surgery, Tagliacozzi saw the need for plastic surgery to correct injuries from fights and skirmishes in the street, but perhaps more notably to repair noses, which were destroyed by syphilis.

It took the wars of the 1900s to bring new technologies and techniques into play to allow surgeons to repair looks with surgery. It started with plastic surgery in the early 1900s, where surgeons developed techniques to repair the devastating injuries suffered by soldiers of World War I. Trench warfare made heads and necks particularly vulnerable to injury. Popping up to have a quick look at the enemy could result in a face full of disfiguring lead.

Airplane crash victims also produced candidates for the repair of serious face injuries. And the arrival of the automobile and its glass windshields also required surgeries to repair cuts to the face when glass windshields would shatter in accidents.

 Plastic versus Cosmetic

Although often used interchangeably, the terms plastic surgery and cosmetic surgery are not the same. Plastic surgeons correct birth defects, repair damage from accidents, tumors, and disease. Cosmetic surgeons perform elective surgery, which is medically unnecessary and for aesthetic purposes.

The knowledge of aesthetic restructuring came as a byproduct of World War I. As is often the cause, the heartbreak of war drives technology innovation.

During that time, Sir Harold Gillies, later dubbed the "father of modern plastic surgery," developed

"As is often the cause, the heartbreak of war drives technology innovation."

techniques to treat soldiers with disfiguring impairments. It included skin grafting, the process of transferring skin flaps from one part of the body to another for the reconstruction of injuries such as burns and disfiguring facial injuries.

The need for procedures to repair injuries sustained by soldiers was a driver for new technologies including the invention of the dermatome instrument in the 1930s. It's a knife-like device that gave surgeons a tool to harvest long thin slices of skin of a consistent thickness for use in skin grafting, which helped produce better results when treating burns or repairing traumatic damage.

Cosmetic Surgery Today

Today there is a menu of safe and successful options when it comes to cosmetic surgery. Following are some of the most common surgeries that will take the ugly away.

Nose Job

A nose job, or more formally rhinoplasty, is a procedure to enhance, reshape, or repair a nose. Nasal skin and soft tissues are detached from the nose cartilage and bone (doctors call it the "osseo-cartilaginous nasal framework") and molded into a new position.

Patients might need a rhinoplasty because they have a working issue with their nose, such as nasal trauma or injury, congenital defects, or respiratory impediments. Others just want to look and feel better.

Caution: It's among the most difficult cosmetic surgeries, so pick your surgeon wisely.

Liposuction

Liposuction is a surgical procedure that removes fat from under the skin. It's used to take out pockets of fat that don't disappear naturally with diet or exercise.

Candidates for the procedure are not overweight and lead active and healthy lifestyles. Having good skin elasticity also helps so that the skin shrinks to fit the new, trimmer, you.

Eyelid Surgery

This surgery, more formally called blepharoplasty, modifies the eye region of the face, and more specifically, the area from the eyebrow to the upper portion of the cheek. The process involves eliminating excess skin from the eyelids, smoothing underlying eye muscles, tightening support structures, and/or resecting (partially removing) and redraping the excess fat around the eye.

This surgery is for people with bags under their eyes or drooping eyelids.

Tummy Tuck

Abdominoplasty is its medical name but most people know it as the "tummy tuck." It's the removal of excess fat and skin from the midsection or lower abdomen. The muscle and fascia, the connective tissue surrounding the muscle, are then tightened, resulting in a firm stomach and waist. Incisions are made under the pubic bone from hip to hip and at the belly button.

Good candidates for this surgery have loose or sagging skin around the belly as a result of dramatic weight loss or pregnancy.

Facelift

A facelift, or rhytidectomy, helps remove visible signs of aging in the face and neck. It can include a number of changes: removal of sagging in the midface, reduction of deep creases, tightening to areas where fat has fallen or is displaced, or the tightening of loose skin. It doesn't change the fundamental appearance of the face, nor does it stop the aging process (too bad!).

Dermabrasion

Dermabrasion and dermaplaning involve the surgical scraping of the skin to remove irregularities leading to a softer appearance. A surgeon uses tools such as a wire brush, a piece of sterilized sandpaper, or salt crystals to perform abrasion (think sandpaper on wood) to the upper- and midlayers of the skin.

It's used to treat people with severe scars from accidents or surgery. It's also used to remove acne scars and to refine wrinkles.

It should not be confused with the term "microdermabrasion," which is a nonsurgical procedure that removes the top layer of mostly dead skin, and is typically performed by beauty salon personnel or via self application of creams or lotions.

Cosmetic Dentistry

The term "cosmetic dentistry" includes a variety of aesthetic enhancements to the teeth:

- Adding materials to your teeth, such as porcelain veneers and crowns.
- Removal of materials, such as enameloplasty, which is the reshaping of teeth by removing a small amount of enamel.
- Orthodontic procedures that involve straightening the teeth.
- Teeth whitening.

Anyone looking for a set of perfect pearly whites can have these procedures done.

Vaginoplasty

Sold often under its commercial name "vagina rejuvenation," vaginoplasty aims to tighten a vagina that has loosened due to childbirth or aging. It can be performed alone or with labiaplasty, which is the altering of the lips surrounding the vagina.

Claims have been made that a tightened vagina leads to increased sexual pleasure, but the American College of Obstetricians and Gynecologists (ACOG) say it is purely an aesthetic treatment.

Both the ACOG and ASPS do not endorse this procedure and say it requires more scientific exploration.

Breast Surgery

A surgical procedure to enhance or reduce breast size. Enhancement done with implants or fat is called a transfer breast augmentation.

In clinical terms, it's called mammaplasty, and it's been rated the second-most common cosmetic surgery procedure in the United States in 2015. Because of this reason, we include a deeper exploration next.

It's a great option for women who have lost breast volume due to weight loss or pregnancy, or breast tissue due to cancer.

Breast reduction surgery is a procedure to make large breasts smaller. It involves the removal of excess fat, skin, and tissue. The areola may also be reduced if it has been stretched. The surgery results in a smaller and perkier bust. Pain caused by large breasts in the back, shoulder, and neck is also eliminated.

Breast Augmentation

Since the late 1800s, women have been enlarging their breasts. Back then, the materials used were, perhaps, somewhat creative. Until the 1960s, surgeons used things such as ivory, glass balls, ox cartilage, plastic chips, paraffin wax, and ground rubber to enhance the female bust.

In the 1940s, Japanese prostitutes injected their breasts with sponges and silicone stolen from the docks of Yokohama to attract the attention of American servicemen who were happily distracted by buxom pinup girls, such as Betty Grable and Jane Russell.

Unfortunately, the ladies often suffered a nasty side-effect known as "silicone rot" where gangrene set in around the injection site.

Post-war America was obsessed with the female breast. Pop culture teamed with images of tight sweaters, low necklines, and bullet bras. It was the decade of Marilyn Monroe and the first *Playboy* centerfold. And in 1959, the first buxom-molded Barbie was created.

When the first silicone breast prosthetic hit the market in 1962, it made perfect sense. Previously, silicone had been used as an injectable by cosmetic surgeons, but it caused problems. Women suffered inflammatory reactions and hard breasts.

Thomas Cronin and Frank Gerow, two American plastic surgeons, solved the problem by putting the silicone in a bag. They called it the Cronin-Gerow implant, a product which they had manufactured by the Dow Corning Corporation.

Their first patient was Texas housewife Timmie Jean Lindsey, who was visiting Jefferson Davidson Hospital in Houston to have a large tattoo removed from her chest. At the hospital she met Dr. Gerow, who was seeking a candidate to perform a breast implant trial. He asked her to participate, and she said yes. Today, at 82, she still has the implants.

In 2012, Lindsey was featured in a *New York Daily News* article on the fiftieth anniversary of the silicone implant. The paper labeled her a "perky grandma." In the article she reports that her implants never leaked or ruptured although they have calcified somewhat.

"I'm so proud that it's available to so many women," she told the paper. "It's not vanity getting reconstruction. I think it's necessary. It puts them back whole again."

Although Perky Grandma kept her pioneering chest a secret for 20 years, she says she has no regrets when it comes to, what she calls, "her minor role" in the history of breast surgery.

Bigger Boobs...Today

Breast implants today are made from either silicone or saline. Both have an outer shell made of silicone but differ in a few ways. Table 3.1 provides a basic comparison between the two types of implants.

Table 3.1 Saline Versus Silicone Implants

	Saline	Silicone
Contains	Sterile saltwater solution.	Viscous silicone gel that is sticky and slippery to closely resembles the feel of human fat.
Insertion	A saline implant requires a smaller incision than silicone. They are inserted empty and filled once in place. This means that saline implants can be placed through smaller entry points, such as the armpit or belly button, which reduces visible scars.	They are inserted full, which means they require larger surgical incisions than saline implants.
Appearance	They are said to appear more stiff and unnatural than silicone implants. There is also an increased chance for wrinkles and ripples on the skin.	Silicone looks and feels more like a natural breast.
Ruptures	The contents of a ruptured saline implant are absorbed by the body without posing health risks. Surgery is still required to remove the silicone shell.	A ruptured implant will not pose long-term health risks but will cause short-term pain in the breast and problems such as the hardening of breast tissue. Although silicone implants were taken off the market from 1992 to 2006, due to health concerns, studies have since shown there are no long-term risks associated with silicone in the body. Like saline, surgery is required to remove the silicone shell.
Age restrictions	Minimum age is 18 years old for breast augmentation and any age for breast reconstruction. There is no age restriction because saline implants can be adjusted as the body matures.	A woman must be at least 22 years old to receive silicone implants for breast augmentation. This leaves enough time for the breasts to finish their maturation process. Silicone implants cannot be altered, like saline, so they can't be changed if the body changes.

The ASPS cautions that breast implants are not guaranteed to last a lifetime. Developmental changes in the course of a woman's life, such as pregnancy, weight loss, and menopause, can lead to changes in surrounding tissue, which can augment the appearance of the augmented breasts.

As with any cosmetic procedure, there are risks to consider, such as infection, bleeding, reduced sensation in the nipple, scarring, or implant rupture. Capsular contracture—the formation of scar tissue which hardens the breast—is another risk to consider. An experienced surgeon can minimize all the risks. They will evaluate which implant suits the patient best and they know how to choose the best insertion points from a selection of five possible areas (see Figure 3.4).

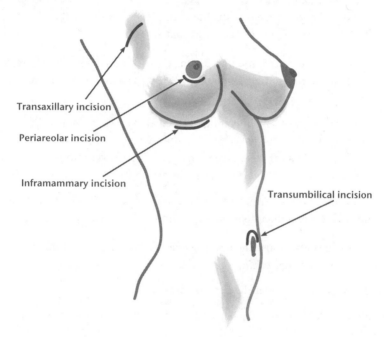

Figure 3.4 *The various breast implant insertion points are shown here.* *(Illustration by Cornelia Svela.)*

Here are the three most common insertion points:

- **Inframammary (IMF)**—An incision is made in the inframammary fold located directly under the breast. It is the most popular option for silicone implants. Entry through the IMF allows for the most precise positioning of the implant, although it tends to leave the most visible scars.

- **Periareolar**—A 5-centimeter incision is made and the implant gets fed through the nipple. Scars are left on the inferior half of the areola's circumference where they are not easily visible. It can affect the milk ducts and the nerves to the nipple resulting in problems later with breastfeeding.

- **Transaxillary**—This approach involves entry through the armpit (the axilla) where the implant is pushed across and positioned in the chest, leaving no visible scars. Unlike the other incision points, the muscle does not get cut and remains strong, pushing the implants upward. This means that following surgery the patient usually wears a stabilizer band—a "bando"—to push the breasts in place. One of the authors, Kay Walker, has saline breast implants and thinks that this is the best option if you are worried about scars.

The preceding three insertion points are the most common. However, here are two more that can be used, although they are less popular and require precision placement for success.

- **Transumbilical**—The implant gets channeled upward from an insertion point in the navel. This procedure is for saline users only. Silicone implants are too large to insert and maneuver into place without rupture. This technique is complex and not as popular.

- **Transabdominal**—A two-in-one operation. The implants are inserted through the abdomen following a tummy tuck procedure.

A physician will also choose the best placement for the implants into one of two pocket options (see Figure 3.5).

- **Submuscular**—Under the pectoral muscle.

- **Subglandular**—Over the pectoral muscle and behind the breast tissue.

The best placement is discretionary and based on a combination of implant type, desired size, and body type.

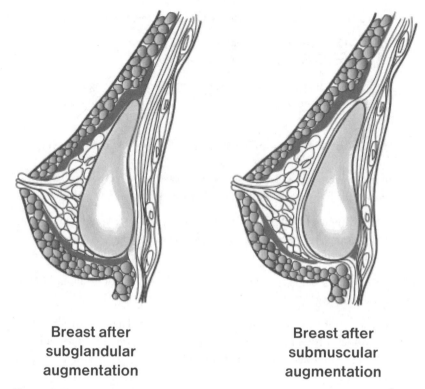

Breast after
subglandular
augmentation

Breast after
submuscular
augmentation

Figure 3.5 *The two breast implant placement locations are shown here.*
(Illustration by Cornelia Svela.)

Fat Transfer Breast Augmentation

It is possible to alter breast size without implants with a procedure called fat
transfer breast augmentation. It involves a combination of liposuction and breast
enlargement, both performed on the same day. Liposuction is used to extract fat
from areas where there is excess, such as from the abdomen, inner thighs, buttocks,
or hips (aka the "saddlebags"). Then the fat is redistributed to the breasts.

Recycled fat? Oh, yes. Except it's yours. And it's moved from a place on your body
you don't want it to a place you do. Yes, ladies, hips get skinnier and boobs get
bigger. Why couldn't nature do that for us? Life's not fair.

The fat redistribution procedure involves a three-step process:

Step 1—Fat removal: Fat is removed carefully by hand with a cannula (a special collection tube) connected to a syringe.

Step 2—Fat cleaning: The harvested fat is processed using a machine called a centrifuge, which separates it from unwanted biological components.

Step 3—Fat distribution: The fat is injected into tiny incision points.

A cosmetic surgeon might also use a tissue expansion system before the operation, which makes the body grow additional skin to support the new growth.

Because the method uses fat from your own body, it is a popular option for women looking for a completely natural breast augmentation alternative. There is no need for implants and the rippling that can occur is also not an issue. Scarring is also minimal.

The cost? About the same as implants. In some cases, it might be about $1,000 more if your surgery is done in the United States.

But the majority of women are still opting for implants. The fat transfer procedure is not for women who want big Pamela Anderson breasts. The best candidates are women who want moderate adjustments such as contouring or a slight size increase of one cup size.

There is another major downside. The injected fat can get reabsorbed by the body resulting in a significant loss of the size increase, up to 50 percent to 60 percent. This usually happens six months after the procedure. Our question is, does the fat migrate back to your butt? If so, boy is nature a bitch.

Ideal Implants

The Ideal implant is America's newest innovation in breast technology, that was released to the market in September of 2015. (You can locate a surgeon authorized by Ideal Implant in the U.S. and Canada by zip or postal code here: http://www.idealimplant.com.)

The implant is a marriage of the silicone look using the saline fill. Externally, it resembles your average saline implant, but inside it is very different. The saltwater solution is contained by shells that are nested together providing structure to the liquid. The result is an increased control of movement, less bouncing and wrinkling, and less chance of possible collapse. The edges of the implant have also been lowered to better contour to the chest wall.

A group of 500 women were chosen for clinical trials in 2010. They have signed a 10-year follow-up agreement so that information can be gathered to monitor the long-term safety of the Ideal implant.

DO BREAST IMPLANTS AFFECT A WOMAN'S ABILITY TO BREASTFEED?

Dr. Stephen Greenberg, a board certified plastic surgeon and the author of the book *A Little Nip, a Little Tuck*, told *ABC News* that most women with breast implants can breastfeed.

"Women with breast implants don't understand and think that they won't be able to breastfeed and are surprised when they realize they can. The overwhelming majority of women who have breast implants can breastfeed whether they're using a silicone or saline implant."

Incisions made under the fold of the breast (inframammary incision) or through the armpit (transaxillary incision) rarely cause trouble for a mother-to-be. The same goes for transumbilical incision procedures, which are done via the belly button. (This surgical approach is less common and more prone to error when the implants are placed.)

Surgeries in which an incision around the lower part of the nipple (peri-areolar incision) are performed should be avoided if you want to breastfeed, say the experts, as damage to the milk ducts might result.

While we are on the subject of boobs, we will talk for a moment about sex reassignment surgery.

Sex Reassignment

Some would agree that cosmetic surgery is a life-altering procedure for many adults, especially those with gender dysphoria (GD). People with this psychological disorder experience discontent with their birth gender. They are commonly referred to as transgendered individuals.

There are many theories around the cause of GD. They range from behavioral causes to biological causes, as well as hormone abnormalities that occur during prenatal development.

The treatment for GD is twofold: A psychological component and a biological component. They are most effective when paired with one another, although some individuals choose to work with only one method.

The first record of sex reassignment surgery dates as far back as 1921.
Dora R. castrated himself as a child and was finally given a procedure in 1930 to reconstruct his genitals, changing them from penis to vagina.

Christine Jorgensen was the first American to get a series of sex transitioning surgeries. She received treatment in Copenhagen shortly after World War II, which made headlines in *The New York Times*. When she returned home, she became the first advocate for transgendered people in America.

As defined by the World Professional Association for Transgender Health, sex reassignment covers a wide spectrum of surgical procedures, such as:

- **Hysterectomy**—Removal of female sex organs

- **Mastectomy**—Removal of breasts

- **Phalloplasty**—Construction or reconstruction of the penis

- **Chest reconstruction**—Breast implants

Hormone treatments are also administered because they are necessary for small physical changes, such as growth of hair and tone of voice, which can't be altered surgically.

There are some exciting advances in the future for sex reassignment, such as lab-grown vaginas. Also see the future of breast enhancement later in this chapter.

One promising development around sex reassignment might be the discovery of what's being called the "sex change gene."

A research team, led by David Zarkower and Vivian Bardwell of the University of Minnesota's Department of Genetics, Cell Biology, and Development, have discovered that an important male development gene called Dmrt1, causes male cells in mouse testes (on the street, we'd call them mouse balls) to become female cells.

The study determined that in mammals, sex chromosomes—XX in females and XY in males—determine the future gender of an unborn animal while it is still an embryo. However the researchers found that when Dmrt1 is lost in mouse testes— even in adults—many male cells become female cells, and the testes show signs of becoming more like ovaries, the organ in females that produce eggs.

Earlier research revealed that taking away the Foxl2 gene from the ovaries of a mouse caused female cells to become male cells. The removal also resulted in the ovaries becoming more like testes.

The implication for transgender and intersex humans is that sex determination could be changed in mammals. Understanding these processes might lead to new therapies. Scientists might be able to learn how to reprogram cells to take on different identities, which might also mean that one day soon, a therapy could be developed to allow an adult female body to be reprogrammed to become male (or vice versa).

Weird and Wonderful Cosmetic Surgery

If you think surgery to change your gender is pretty wild, just wait until you see what other procedures are being performed today. Welcome to the weird and wonderful world of cosmetic surgery.

Palm Alterations

In Japan, unwanted fortunes discerned from palm reading can now be corrected with cosmetic surgery. A surgeon uses an electric scalpel to redraw lines in the hand, ostensibly to forcibly redraw the client's future.

According to the website *The Daily Beast*, 37 surgeries were performed between January 2011 and May 2013 at the Shonan Beauty Clinic in Tokyo.

Men and women in their 30s are the most common buyers of this cosmetic surgery. Men are in search of emperor line modification to increase business success and finance, while women, on the other hand, seek to better their odds of marriage by modifying their romance lines.

The procedure takes about 10 to 15 minutes to perform and about a month for the wounds to heal. Clients can have up to five lines modified in a session.

Double Eyelid Surgery

One of the more popular cosmetic surgeries requested by some east and central Asians is double eyelid surgery. This procedure modifies the epicanthic crease, a skin fold of the upper eyelid, that covers the inner corner of the eye.

The surgery modifies the crease that mimics the Caucasian crease of the eye that most Asian women do not naturally have. Dr. Seo, a surgeon from Seo Jae Don Plastic Clinic, sees this trend predominantly in Korean women. His clinic is one of the most successful in a city dubbed the "plastic surgery capital of Asia."

Iris Surgery

Iris surgery was originally pioneered by Dr. Kenneth Rosenthal to address medical eye issues such as heterochromia and ocular albinism. Now the technology is used cosmetically to electively change the patient's eye color.

The artificial iris is a thin, nontoxic prosthesis made of the same silicone used in artificial lenses for cataract surgery.

The surgery has become so popular that BrightOcular has taken advantage of this business opportunity. The company describes the procedure as rapid and minimally invasive. It takes 15 minutes per eye, so if you feel like changing your eye color on your next lunch break it is certainly doable! And don't worry, if you are not happy with the result it can be as easily reversed.

The largest market for plastic surgery in the world is Korea. Korean women, long to fit the Western ideal of fair skin and blue eyes. In fact, one in five women in Seoul have gone under the knife to change their looks. In 2010, 5.8 million cosmetic surgery procedures were performed compared to 4.5 million in the United States.

Foot Surgery

Are you one of those people cursed with ugly feet? Perhaps you have a second toe longer than the first. Or, maybe you are having issues fitting into those stiletto heels because your feet are too wide. Don't fret, Dr. Oliver Zong, the Director of Surgery at NYC Footcare, can help. He specializes in elective foot surgery and "foot facelifts." Whether you want fat removed from your toes, or have your toes shortened, it's all possible.

NYC Footcare is part of a growing trend, an area of cosmetics which is considered a $45 million industry.

Limb Lengthening

Perhaps the most remarkable cosmetic surgery procedure on the market is limb lengthening. This isn't the reinvention of the rack from the dungeons of the Middle Ages. It's about becoming taller, voluntarily. With the procedure, shorter people can become taller. That said, the radical procedure can be costly. At $85,000, it's a tall price to be paid. (Sorry, we couldn't resist. That line was begging to be written.)

Perhaps the reason for its costly nature is the highly invasive surgery in which the leg bone is broken. A telescopic rod is used to slowly pull the bone apart at a rate of roughly 1 millimeter per day. Arteries, tissue, and skin regenerate as new bone grows around the rod and repairs the break.

Of course, with the advent of stem cell therapies, as well as gene therapies, discussed later in this book, this barbaric procedure should become obsolete. These therapies, as you will soon see, will allow you to look any way you want. Just hang on for a decade or two.

The Future of Beauty

While there are a lot of new and exciting procedures to help you enhance your beauty, the future of cosmetic and plastic surgery can be described in two words, "stem cells." As we investigate the future of cosmetic surgery, all roads lead there, as do many of the conversations that relate to healing, curing, and rejuvenating the human body.

We talk more about what stem cells are, the different types, and the controversy around them, in Chapter 7 "In Hacks We Trust? The Political and Religious Backlash Against the Future." For the purposes of this discussion, we will take a quick look at stem cells here.

So what is a stem cell? Simply put, it is an undefined cell that can form into any kind of cell. Stem cells are the foundation for every organ and tissue in your body. They are found in all parts of the human body.

They are special because they have the ability to replicate themselves into additional stem cells, or transform into any specialized cell, such as nerve or blood cells. This makes them incredibly powerful because they can be involved in most medical treatments, including cosmetic and plastic surgery.

Dr. Robert Murphy, former president of the American Society of Plastic Surgeons, says stem cells are critical to the future of cosmetic surgery. In an interview for this book, he explained, "What makes this exceptionally interesting technology is that if you can isolate stem cells and put them in the appropriate environment, you can essentially grow tissue."

These cells, when isolated in the lab, can morph into whatever a doctor needs, so they can use them in any therapy.

For a cardiologist, that could be "cardiac tissue for people with heart damage," said Murphy. "In the cosmetic realm, they can morph into tissue that can potentially either enhance damaged tissue—for people who have had radiation injury when treated for breast cancer—or cosmetically, to augment breasts."

Murphy went on to explain, "There are some studies and some trials (being done) to see if we can enhance or actually grow breasts using stem cell therapy."

That means no more implants. Stem cell science, when it matures, will allow doctors to inject stem cells into their patients to help them grow a new set of breasts sourced from their very own body.

The developing technology is so profoundly useful that it can be used to make our bodies malleable. For example, it could be used to rejuvenate your face. "We could use it to give you a much younger more youthful appearance by filling both volume and improving the quality of your skin," said Murphy.

These tools will be available to cosmetic surgeons soon, Murphy explained. "In the next 3 to 5 years we are talking about being able to extract stem cells, process them in a laboratory and re-implant them."

Thereafter, perhaps closer to five years out (perhaps later in this decade), things get exciting. "Once we get the science down, it's almost gene therapy, in that you can recruit stem cells in the body and get them to replace tissue or regrow tissue like a starfish that can regenerate a bud," Murphy said.

REGENERATE LIKE A STARFISH

If you doubt the possibility that humans will be able to regenerate tissue and even body parts, look to the animal kingdom for evidence that it might be possible.

Take the starfish, or sea star, for example. It is among a small population of animals with a magical capacity for body regeneration.

This is how it works: When a starfish loses a limb, that portion of the body closes up and the animal's stem cells kick-start a process of regeneration and rejuvenation. Slowly, the starfish will begin to form a bud where the old leg was lost. This bud grows into a new leg, or sometimes, will become bifurcated, that is, more than one leg, possibly two, will grow into the empty space.

The starfish can lose up to three legs and still have the capacity to survive and grow itself back into its original form.

In the 1940s, Canadian east coast oyster fishermen made the mistake of trying to eliminate the starfish, which is a predator of their precious commodity. Starfish that were found tangled in the oyster nets were maimed by cutting off a limb. Thanks to their regenerative capabilities this produced an overpopulation of starfish and an unprecedented decline in the oyster population during that time period.

Murphy cautioned that a lot of research still needs to be done to make this possible in the next decade and beyond (see Chapter 4 "Lifesaving Hacks: Whirring Hearts, Printed Organs, and Miraculous Medicine"), but it will very likely be a reality very soon.

As evidence of the progress, he pointed to the number of research papers published about adipose stems cells. Those are stem cells isolated and derived from human fat. While still only theoretically possible, adipose stem cell therapies have been demonstrated in the lab, but have yet to be brought into mainstream therapies.

In 2002, there were 12 research papers published on adipose stem cells. In 2012 there were 699. Checking again for 2015 there were 1,400. You can bet that a closer look at this knowledge would show a logarithmic growth in knowledge and understanding. This is why people such as Dr. Murphy can confidently predict therapeutic cosmetic uses for stem cells derived from human fat by the end of this decade.

At this point, you might be asking, can I get the nose that I want using stem cell technologies? Or the bum? Or the belly? Can you grow me nice pectorals and biceps? Can you remold someone's face to look like Scarlett Johanssen? Or her body to look like Megan Fox? With stem cells, all these things could be entirely possible one day soon.

The science behind extracting stem cells from fat and the ability to refine them for use in therapies is being refined now. Murphy said certain clinical trials in places such as the University of Pittsburgh are showing promise. Outside the United States, such as in Germany and France, adipose stem cells are being aggressively studied. He said what we need now is "scientific rigor."

"We need to develop techniques that have defined applications and accepted risks, benefits and outcomes, Murphy said. "Expect that in the next 5 years."

Remember, stem cells can turn into any kind of cell—skin, blood, bone, and everything else that gives us our appearance. Scientists just need to figure out how to source them safely, coax them into place, and have them grow into the right shapes.

"If you Google "stem cell facelift" you will get 100,000 hits right now," Murphy said, "but in fact, there is no true stem cell facelift now. In the future, you could put small amounts of fat in areas and have a withered sagging face grow into a youthful appearance, with a more rounded face, with a much better skin texture."

What about fixing a deformity? Could you take someone born with an ear deformity and inject it with stem cell and mold an ear? "Very possible," said Murphy. "Stem cells have tremendous implications because they are such a malleable type of cell. If we can learn how to mold to the appropriate environment, it has unlimited applicability."

There you have it. This technology is not a one day, some day possibility. It's right there, almost ready for you and I, to fend off the ugly we so desperately want to conquer.

4

Lifesaving Hacks: Whirring Hearts, Printed Organs, and Miraculous Medicine

At some point in your life, it's a simple fact you're going to get sick. It might be as simple as the common cold, in which case it'll probably be best if you curl up with a bowl of delicious, genetically engineered chicken soup and a good personally customized nasal spray, loaded with a cold-killing payload of nanobots. That will be the trick in a few years.

Of course the number one killer of Americans today is heart disease. So you can be sure that people you know and love will likely deal with heart issues in their lives. You might be one of them. Let's face it, there's a good chance you might end up on an operating table with your chest splayed open getting your heart replaced by a jet engine.

Wait, what?

Yeah, that's for real. We will get back to that in a bit.

As with many other areas of life, technology has inserted itself firmly into the world of medicine, a process going back thousands of years, back to the time when the first caveman dabbed some mud onto a bug bite and invented the health spa. Okay, that last fact might have been completely fabricated, but it is true that humans have made many breakthroughs in health technology along the way, including surgery, the invention of pharmaceuticals, and eventually designing new organs and body parts out of metal, plastic, and a meat-spraying inkjet printer.

I know we are probably freaking you out here, but it's all about to come true. Go take bathroom break because this chapter will not work for you on a full bladder. And we don't want you to tinkle on yourself.

When we considered what to include in this chapter, there was an almanac full of diseases and conditions that are being remedied by incredible new medical technologies that we could have included. Breakthroughs in cancer treatments and associated tumor fighting therapies alone could fill an entire book.

Instead, we looked at the number-one killer of Americans: Heart disease. And how an amazing team of researchers came together to develop an artificial heart that could save a half-million American lives every year. And millions worldwide.

We will tell how the "jet engine heart," developed by a team that includes Australian researcher Daniel Timms and Americans Bud Frazier and Billy Cohn of the Texas Heart Institute, came to be.

It is just one miraculous story driven by accelerating technology improvement. But we also examine the macrotrends that are driving cancer therapies, organ replacement, and diseases that would otherwise kill the humans they plague. Those breakthroughs, within nanotechnology, genetics, and robotics, are revolutionizing medicine today.

First, let's take a peek at the medical past.

A Brief History of Medicine

The basics with your health are generally pretty well understood today. If you are feeling sick, you pop into the pharmacy for some medicine, and if your illness is somewhat worse, you see your doctor for stronger measures (usually drugs). If you are really, really sick, a specialist is likely in your future. They will do their best to fix up anything going wrong inside you, usually using a surgical procedure.

There's nothing new here. There is evidence people knew how to suture wounds all the way back into prehistoric times, and that humans were aware of surgical intervention as early as 12,000 B.C.E.

That ancient intervention, by the way, was called trepanation, which was the opening of a hole in your skull to relieve pressures inside the head caused by other injuries or illness. As you can imagine, surgical instruments in 12,000 B.C.E. weren't exactly state-of-the-art, so opening a hole into the skull sometimes meant actually scraping a hole into the skull with a rock. We are no doctors, but common sense says you'd probably want to avoid a procedure like this.

Ancient procedures also included tooth drilling more than 10,000 years ago. There's further evidence of dental surgery in Egypt 5,000 years ago. There's even ancient evidence of splinting broken bones.

You are probably familiar with the Hippocratic Oath—especially the section that apparently says "above all, do no harm." (As an aside, this actual quote doesn't appear in the Ionic Greek version, and is attributed to a 19th century surgeon). Thematically, however, the oath and modern updates of it support its intention.

Still, Hippocrates, for whom the oath was named, is considered the father of western medicine. Born in Greece in 460 B.C.E., he formalized some of the rules and ethics of medicine—no longer was it acceptable to gouge someone's head with a rock to relieve pressure in the skull.

Hippocrates espoused a theory that medical ailments were caused by your lifestyle and not because the gods had cursed you. This allowed doctors to treat patients using a diagnosis based on observation, rather than superstition.

> "Hippocrates espoused a theory that medical ailments were caused by your lifestyle and not because the gods had cursed you."

One of the major side effects of cutting holes into people is that it really, really hurts, so it's no surprise that for the last few thousand years, people have been trying to find ways to dull the pain when undergoing such procedures. This meant herbal concoctions of varying effectiveness, and good old alcohol, opium, and related narcotics. There's also the case of early medicine men in Peru spitting into wounds after chewing a mouthful of coca leaves. In the 1800s, this technique was refined somewhat by distilling the coca leaves into cocaine.

For major procedures, most people prefer to be rendered unconscious. Apparently, people agreed with this principle as far back as the 1100s, when Arab surgeons would knock their patients out with a sponge soaked in an opium cocktail (that'd be opium, hemlock, and mandrake with a twist of lime).

It was during the eighteenth and nineteenth centuries when general anesthesia started to gain traction. This happened with the refinement of nitrous oxide (aka laughing gas, the funniest general anesthetic), and the introduction of morphine and chloroform. Eventually, that led to the modern complement of gases and painkillers that knock us out before surgery today, and then keep us numb throughout the procedure.

Dubious Moments in Medicine

In the late 1800s, morphine was one of the key pharmaceutical agents used to help people control pain during and after surgery. One problem: it was wickedly addictive. So scientists got to work on a new non-addictive painkiller to use in place of morphine. The result was a drug called diamorphine, which has since become better-known by the trademark name Bayer gave it: Heroin. Oops!

The funny thing, though, is that it's not so much the surgery that'll kill you as it is the germs. No matter how skilled your surgeon is, bacteria can get into your incisions and cause terrible infections that can kill you in an equally nasty way.

There have been some attempts to counteract infections through the ages, but it wasn't until Alexander Fleming discovered penicillin in 1928 that there was a truly recognized pharmaceutical option. This was the real start of the antibiotic revolution, and shortly thereafter, there were specific formulations that would tackle post-surgery infections, tuberculosis, and other bugs that kept people from recovering from surgeries. Antiviral medicines followed, and rather than killing the viruses outright (as an antibiotic would kill bacteria), antivirals rather prevented viruses from reproducing themselves, halting their spread.

Now, of course, today there's a wide variety of consumer products featuring antimicrobial properties, from over-the-counter medicines to hand soaps to footwear. In the past, if you got a tiny little cut on your finger, you could get a massive infection and lose your arm … you could even die. Now, you head down the block to the pharmacy and buy a tube of Polysporin for a few dollars, and soon enough you're back to chopping onions as if nothing happened.

Ultimately, keeping people from dying from complications due to small nicks, cuts, and large-scale incursions like modern surgery is what brought us into the medical modern age, allowing us to do even bigger and better things.

There's a Future in Plastics

It's hard to overstate the effect of the introduction of plastic on the world of medicine. For the longest time, most medical instruments were made out of metal or rubber. While that made them generally quite durable, their reusability meant they had to be constantly cleaned to avoid spreading infection from patient to patient. With the introduction of plastic in the operating room, it became more cost-effective for surgical tools to be switched over to single-use sterile equipment. Even for some of the devices that still came in metal, like lancets or needles, plastic packaging enabled these items to remain sterile right up until the moment of use, again helping prevent the spread of germs.

Beyond the operating room, plastic is becoming essential for some of the things left behind after surgery. In the past, hip replacement surgery might have involved joints where there was metal-on-metal contact. Now, with some of these parts cast in plastic, they're lighter than before and there's often less need for repeat surgery to replace worn parts. In one Canadian study, those with metal-on-metal hip joints were 1.6 times more likely to require a repeat surgery than those with metal-on-plastic joints.

Plastics have also taken a feature role in many prosthetics. Replacement limbs like legs and arms, previously made of wood or metal, are now custom-molded from plastic to better fit the recipient. For limbs with robotic components, plastic is also used to minimize weight, for better agility and lower cost. In fact, some enterprising do-it-yourselfers are even using 3D printers to make replacement hands out of Acrylonitrile Butadiene Styrene (ABS), which is a low-cost engineering plastic easy to machine, fabricate, paint, and glue.

Plastic will almost certainly play more of a role as people do more self-modification projects down the road, too. (We'll take a closer look at how DIY biohackers are modding themselves in Chapter 6, "Franken-You: A Better Life Through Cyborg Technology".)

Taking a Peek Inside

While the birth of diagnosis-based modern medicine dates back to the time of Hippocrates, it's worth noting that a lot of diagnoses through the ages have been sheer guesswork when it came to anything that was malfunctioning inside the body. In some cases, there were reasonably obvious signs that made their way to the outside of the body, such as swelling or discoloration, but some maladies without outward indications were a bit trickier to diagnose without surgery.

That changed at the end of the nineteenth century when Wilhelm Röntgen discovered the X-ray. As part of his experimentation, he directed a stream of X-rays through his wife's hand toward a photographic plate, and noted that they were able to pass through flesh and other human tissues, but not through bone or metal. And so, radiography was born, and medical practitioners quickly adopted the technology. With this tool in their arsenals, doctors could confirm that bones were fractured, or locate a metal fragment caught under the skin. It was a pretty nifty invention.

While X-ray imaging continues to be used today, its use has tailed off dramatically, thanks to the discovery that radiation produced by X-ray equipment could actually harm human tissue, including the increased risk of cancer. Old-school single-shot X-ray exposures tend to be very short and very focused, and are only undertaken after determining that the benefit outweighs the risks of radiation exposure.

On the other hand, CAT scan usage has grown through the years, despite using X-rays as their method of looking inside the body. CAT is short for computerized axial tomography (sometimes referenced as a "CT scan"). It uses computers to create cross-sections of the body by stitching together information taken from different locations and angles, as the patient slides through a diagnostic ring on a moving platform. CAT scans allow doctors to look at the body in 3D, providing a much better view than the previously available flat version, where a lot more was left to interpretation.

Radiation exposure from a CAT scan is many times the exposure level of a single X-ray and there are concerns that the cancer risks might outweigh the medical and diagnosis benefit.

Magnetic resonance imaging scans, also known as MRIs, have also become popular. Because they use a combination of radio waves and magnetism to create a three-dimensional scan of the patient's internal structure, they're often preferred over CAT scans. However, they have limitations compared to CAT, including higher cost, and the fact that they can't be used with patients that have cochlear implants, pacemakers, or some neurostimulator devices because of the metal in those implants. (The magnet in the MRI can tug on the metal pretty fiercely.) There's also some concern that extensive MRI usage could cause a breakdown in the patient's DNA itself, which is another reason that MRI hasn't taken over completely from X-ray-based scanning.

All these advancements have gone a long way to increasing our lifespan in ways that would have seemed unnatural hundreds of years ago, when people were dying of old age at lifespans that would now be just edging us into midlife crisis territory. Average human life expectancy has actually increased dramatically over the last 60 years. In the mid-1950s, people in the developed world could expect to live until sometime into their 60s, but now expected lifespans are north of age 80. That's because we are now able to catch a lot of the things that would have otherwise put us into an early grave, like random infections or tumors. (You'll learn more about longevity in Chapter 8, "Hyper Longevity: How to Make Death Obsolete.")

Have a Heart—Pacemakers, Transplants, and Artificial Hearts

Physicians have known since 1889 that the heart reacts to electrical stimulation, when John Alexander MacWilliam discovered that application of electricity could cause a heart's ventricle to contract. Repeated application of electrical impulses through electrodes placed into the heart muscle could, in fact, cause a heart to "beat" at a more-or-less regular rate. Research continued on and off through the first half of the twentieth century, but it wasn't until 1958 that Arne Larsson received the first implanted pacemaker designed to regulate his heartbeat, kicking in when his own heart stopped beating at a normal rate.

Larsson, a Swedish citizen, went through 26 more pacemakers during his life, and lasted to the ripe old age of 86. He succumbed to skin cancer in 2001.

The technology continued to change and improve: battery life improved; construction materials changed, facilitating longer life with less danger of corrosion or moisture damage. Also, newer pacemakers could adjust their rate based on the level of physical activity, oxygen and CO_2 levels, temperature, and the respiration rate of the user.

One of the issues with pacemakers through the years was the fact they required wired leads between the pacemaker and the heart, and damage to those leads could cause the pacemaker to malfunction, or result in other medical complications such as infection.

In April 2016, the FDA approved Medtronic's "vitamin-sized" Micra pacemaker, which "sits inside the ventricular cavity," attaches to the muscles inside the heart using tiny little metal claws, and delivers pacing impulses by an electrode located at the end of the device. With a battery meant to last around ten years, younger patients can expect to replace it during their lifetime, or at least until we can replace the faulty heart with a jet-engine version or one grown in the lab.

The Micra can be installed or retrieved via the femoral artery, meaning the procedure is "minimally invasive."

Sometimes, unfortunately, a pacemaker isn't enough to kick-start the heart, it may need a heavy-duty device to help it pump, or it may need a full replacement.

The history of organ transplantation goes back a while, but just how long is a matter of some debate. Although there have been reports of leg and heart transplants dating back thousands of years, these accounts seem dubious at best; reports of skin grafting seem to be more credible, though it's hard to say just how successful those operations were. In 1883, Swiss surgeon Theodor Kocher performed a thyroid transplant, widely regarded as the first human organ transplantation; since that point, we've learned to transplant corneas, kidneys, bone marrow, intestines, livers, lungs, spleens, and hearts, among other things. In fact, we've started to transplant hands and even faces with varying levels of success. One unfortunate man even had a penis transplant, but his wife rejected it.

And then there's the experimental work that's been done where it was shown to be somewhat feasible to transplant heads using animals. In 1970, Neurosurgeon Robert White, for example, successfully transplanted the head of a monkey onto a second monkey's body. The procedure was successful to an extent, with the animal being able to smell, taste, hear, and see the world around it. It was also able to bite.

> "We've started to transplant hands and even faces with varying levels of success."

Italian neurosurgeon Sergio Canavero, has also said he has a 30-something Russian candidate selected for a possible human head transplant in 2017. The man has a progressive and incurable wasting disease called Werdnig-Hoffmann disease. Canavero says he needs $100 million for the operation and is calling on billionaires to help fund the project.

"I need your help and I need your assistance," Canavero is quoted as saying. "Be Americans."

Critics aren't enthusiastic about the idea. The most difficult part of the transplant is restoring the spinal cord. White failed to do that resulting in the reheaded monkey's eventual death. That should be resolved using stem cell therapies, which are in development, but not likely by 2017.

Let's not get ahead of ourselves here, though, and get back to issues of the heart. They are a different story. While we've been able to successfully transplant hearts since 1965, the donor unfortunately has to be dead in order for the transplant to take place, and people with healthy hearts just don't die in sufficient quantities to satisfy the needs of people waiting for a replacement heart.

Although a mechanical heart machine was first used as a temporary replacement for the patient's own heart during surgery in the early 1950s, the bulky automobile-sized machine could barely be considered a suitable alternative for the heart. The patient was tethered to an operating room for the rest of his or her life.

Through the 1950s and 1960s, further research led to testing of more compact artificial hearts in calves, and even temporary artificial hearts in human patients waiting for a permanent human heart transplant.

The first truly promising replacement to gain international attention was the Jarvik 7 artificial heart implanted into Barney Clark in 1982. The Jarvik 7 had a component that was implanted right into Clark's chest, but it had to be attached to a large, mobility-limiting and noisy external compressor unit.

Clark survived 112 days with the Jarvik 7 implanted, but died following a series of complications. Another Jarvik 7 recipient, William Schroeder, lasted 620 days, and was even able to leave the hospital with a portable compressor unit, but his remaining days were also plagued with complications, such as a series of strokes.

Since Clark and Schroeder, there have been a number of patients implanted with either artificial hearts, or what are called ventricular assist devices. The latter devices don't replace the heart outright, but provide assistance to the patient's own heart.

Until fairly recently, the problem with both these types of devices was that they were mechanical, so they had durability issues. They were designed to simulate the organic human heart. Their mechanical components allowed them to "beat" roughly 85,000 to 140,000 times a day—or tens of millions of times per year. The problem was that the components inside the pumps eventually wore out and failed.

If the device was assisting a weakened heart and it failed, there was a chance the patient could undergo surgery for urgent replacement of the device or for an emergency heart transplant. If the patient received an artificial heart and the device failed, death was inevitable.

Consequently, these artificial units were often regarded as what Cohn calls "bridges." They give a patient time, until a human heart from a donor becomes available for transplant. Patients who can be supported with an assist pump alone generally have disease primarily involving the left ventricle, the main pumping chamber of the heart. The right ventricle, the one responsible for pumping the dark venous blood to the lungs to pick up oxygen, needs to continue to function; otherwise, the assist pump doesn't function well.

The older assist pumps (like the artificial hearts) would fail after a year or two as described above. But all that changed 15 years ago. In the 1980s, physician inventor Rich Wampler visited Egypt and witnessed local workers using a hand-turned Archimedes screw to pump water up a river bank for irrigation. (See video of the screw: http://www.superyou.link/archscrew.)

He realized that if a screw could move water against gravity, perhaps it could move blood against pressure. So he designed a prototype that used an external motor outside the body to spin a small screw the size of a pencil eraser 25,000 times per minute inside the body.

Perhaps, he reasoned, it could pull a meaningful amount of blood out of a weakened ventricle to get a patient through a short period of heart failure.

Wampler showed the device to O. H. "Bud" Frazier, chief of transplantation and assist devices at the Texas Heart Institute in Houston.

Frazier was initially concerned that the rapidly spinning screw would destroy fragile blood cells, but multiple animal experiments showed it wasn't an issue. So, in 1988, the device was used successfully to save a dying patient.

Thanks to this innovation, and ongoing work by artificial heart researcher Rob Jarvik, a number of durable, permanent, screw-like assist pumps were developed over the next few years. Unlike the assist pumps that beat like the human heart, these rapidly spinning pumps were much smaller and much more durable.

"These continuous flow pumps have no flexible membranes, no cams, no cam followers, no high torque rotors," said Frazier's partner Billy Cohn. "The new pumps eliminate all that mechanical complexity … they're small, they're quiet, and they're very durable."

In these earliest heart-assisting devices, the spinning screw, sometimes also called an "impeller," was supported on each end with a mechanical bearing that was impervious to wear and was kept clean by the rapid flow of blood.

"There was one moving part, so nothing to fail," explained Cohn.

With the newest generation of assist pump, the spinning impeller is suspended magnetically. Now, even the mechanical blood-washed bearings have been eliminated.

Enter the jet-engine heart.

One of the first of these spinning assist pumps, or continuous flow devices as they are commonly called, was the HeartMate II. This was designed to assist a patient's weakened heart by pulling blood directly from the left ventricle through a metal tube inserted through the heart wall.

"HeartMate II assist pumps are now being implanted daily, just like the one Vice President (Dick) Cheney had," Cohn told us in an interview.

There is also a version called a HeartWare HVAD pump. It is another continuous flow assist pump. Both are generally used in patients who have failure involving only the left ventricle. But what about patients with failure affecting both ventricles?

Many heart centers have tried to implant a pair of assist pumps; one in the left ventricle, and a second in the right ventricle.

"Putting two pumps in has proven to be challenging," said Cohn, explaining that two controllers are needed, plus the enlarged diseased heart stays in place and interacts with the pumps. And, there is limited space inside the chest.

"We have tried to use this strategy in carefully selected patients, but results have been mixed," said Cohn. "Other institutions have tried to use this two-pump strategy as well, but it is often not ideal."

To address these challenges, Cohn and Frazier began using animal studies to investigate whether it would be possible to remove the heart completely and replace it with a pair of rapidly spinning pumps.

"Removing the enlarged heart certainly would free up a lot of space, and decreasing the different flow paths the blood could follow maybe would be beneficial," said Cohn.

The idea of completely eliminating the heartbeat was not without precedent in Cohn and Frazier's experience.

Although a significant percentage of patients implanted with continuous flow assist pumps lose their pulse, the pressure in the arteries still rises and falls with each heartbeat—just not enough that a doctor or nurse can feel it with their fingertips.

In contrast, if the heart is removed and replaced with spinning pumps, the arterial pressure waveform is a flat line—though this has been seen only on rare occasions.

In one case, Frazier was surprised to discover that a patient who had received the HeartMate II no longer had a pulse.

The patient's heart had stopped beating altogether, yet blood was passively flowing through the lungs without help from the pumping action of the heart's right ventricle.

That was a surprise to the doctors. Yet the unexpected result was duplicated in another patient from New York State, whose HeartMate II was providing the only blood pumping mechanism in her body. In each of these patients, the flow of blood was absolutely continuous, like water in a garden hose, without a hint of a pulse.

Who Needs a Pulse, Anyway?

Cohn and Frazier's idea was validated through extensive research in the large animal facility at the Texas Heart institute. It showed conclusively that calves—there were more than 50 involved in the research—could be kept alive with continuous flow devices.

The animals would eat, sleep, and grow like normal calves, and would interact with caregivers and exercise on a treadmill, and yet they had no heartbeat, no EKG, and—that's right—no pulse.

In 2011 Craig Lewis, a gravely ill man, arrived at the Texas Heart Institute. He was in profound shock. His heart, liver, and kidneys had been infiltrated with a substance called amyloid, which was produced by a rogue population of bone marrow cells. He was too sick for heart transplantation and his heart was too small and contracted to allow safe placement of assist pumps. The other challenge: He wasn't a candidate for a temporary mechanical artificial heart, because he wasn't a transplant candidate.

After extensive discussions with his family, Lewis's diseased heart was replaced with a pair of HeartMate II pumps. The research at this point was solid. The team had been exploring the option for six years.

The device kept Lewis alive for more than five weeks, and it even allowed him to sit up and interact with his family after a prolonged pre-operative period, where he was in a coma.

Lewis died, however, because of progressive failure of some of his other organs. Still, the heart replacement showed it was possible to live without the original organ for longer periods.

Through all this work, Cohn and Frazier spent considerable time talking about their research at numerous professional society meetings and at many speaking engagements attended by the public, including a talk at TEDMED in 2012 (see video: http://superyou. link/tedmed2012). As a result of that exposure, they met Daniel Timms, a scientist living in Australia. Timms' father had died of heart failure while he was working to earn his PhD.

"...the heart replacement showed it was possible to live without the original organ."

Timms was motivated to try to find a technological solution for patients like his father, who would otherwise die without some sort of intervention.

He independently came upon the idea that a continuous-flow device was the way to go. His unique design addressed many of the challenges caused by a pair of spinning pumps. And it wasn't unlike the technology being used by Frazier and Cohn.

The device Timms conceived contained a single moving part suspended in an electromagnetic field. The single spinning disc had impeller vanes on each of the two faces. One side took the bright red blood returning from the lungs and pumped it to the rest of the body. On the other side of the same spinning disk, an assembly of vanes drove the blood returning from the body to the lungs.

"This one spinning disk performed the function of both pumps," explained Cohn.

Even better, Timms had designed the spinning disc to shift slightly from one side to the other, based on changes in pressure. This resulted in a change in the relative strength and efficiency of the left and right pumps.

"This device can respond to physiological changes faster than a natural human heart," said Cohn. "Although the strength of the electromagnetic field is adjusted 20,000 times each second to keep the disc spinning in space, 20 times a second it says, 'wow, the right side pressures are getting a little higher than the left, I need to move to the right.' Or, move to the left to accommodate exercise or whatever. So at 20 times a second it's re-evaluating the two pumps and adjusting for it."

The only problem: Timms didn't have the resources, institutional experience, and support to get it to work—and he was out of money. So, after seeing Frazier present at numerous conferences, and watching Cohn and Frazier's TEDMED talk, he got on a plane with the intention of showing them the technology.

Cohn said they weren't expecting much when he showed up in Houston. (He and Frazier had already seen a lot of unsolicited crackpot ideas.)

"He was unshaven, and had on blue jeans, but he seemed like a nice young man, so I was going to let him down softly when I figured out why his device was tragically flawed, as soon as I figured it out," said Cohn, recounting the meeting.

But Timms' pitch was brilliant. "He starts describing this device. I sort of cross examine him. I say 'What about this? What about that?' He had great responses to everything," said Cohn. "We had scheduled a 30-minute meeting. And after talking to him for three hours, I realize this guy is one of the most brilliant men I've ever met, and his device addresses many of the challenges and shortcomings of our twin turbine concept."

Specifically, it had the potential to be much smaller. It automatically balanced the left and right sides of the circulation. It would only require a single controller. And, most importantly, it wouldn't wear out after a year.

"So I beg him to stay at Texas Heart and let us implant one," recalled Cohn.

In the end, Timms' team moved to Texas to work on the project. The Texas Heart team was able to leverage their network to help Daniel and his team raise the necessary money, and leveraged their experience with heart replacement and their large animal lab to help keep the project moving forward.

Between 2013 and 2016, a number of calves had been implanted with the devices, and after the surgeries, some were able to exercise on a treadmill.

The results so far have been extremely encouraging, so much so that Frazier and Cohn are fairly convinced that the device will be the first practical permanent mechanical replacement for the failing human heart.

The first few generations of this new heart module will be powered by a large driveline exiting through the skin to a vest that holds a computer controller and several large batteries.

Ultimately, the device will be powered by technology called Transcutaneous Energy Transfer System, or TETS.

It is essentially a short-range wireless power transmitter. TETS uses external batteries that generate a high-intensity oscillating magnetic field, which is beamed into the body by an external coil-shaped antenna, attached to the patient's skin. Inside the patient's body a second antenna receives the transmitted energy. And this setup powers the heart-replacement device in the patient's chest.

The implant has minimal moving parts, for increased reliability, and it is powered by a vest that can be swapped out as easily as you might change jackets.

Each year 500,000 people die of heart failure in the United States alone. And yet, surgeons only have access to 2200 donor hearts.

For authors Kay and Andy, this hit pretty close to home. As we finalized this chapter, Kay's dad Kris, who did substantial research for us in several of this book's chapters, suffered a heart condition that almost took his life. It required a valve replacement and triple bypass to save his life.

> "Each year 500,000 people die of heart failure in the United States alone. And yet, surgeons only have access to 200 donor hearts."

So a permanent mechanical heart replacement that has only one moving part is a miracle of science and technology. It has no flexible membranes or complex mechanisms that could fail, and it is small enough to implant in most patients, including, perhaps one day, Kay's dad. It will truly change the way end stage heart failure is managed.

As you can see, the accelerating improvement of technology saves lives. Just think about the technologies at play in this one story. Scientists building on each other's breakthroughs over several decades using evolving tools, multi-disciplinary expertise, and the power and reach of the Internet, to produce a device that can replace a human heart—for a lifetime.

And the technology goes from helping extend a few people's lives to dozens, then hundreds of lives. And today, as it nears production, it has the potential to save a half million lives a year. It's quite incredible. Especially, when one of those people is someone you love.

It's Now, and It's New: The Future Frontiers of Medicine

If you think medical advances in the last 100 years have been amazing, wait until you see what's coming next. Let's peer into the future of medicine.

Nano, Nano

Nanotechnology is something people have been talking about for a long time, but what does it actually mean, especially when it comes to medicine? Essentially, nanomedicine involves targeting illness at extremely small scales. This can be anything from extremely small medicine-delivery methods to tiny little robots swimming through the bloodstream.

Biomedical engineer Mark Kendall has been working on a new technology called the Nanopatch, which uses nanotechnology to get around the fear many people have of the needle (see it here: http://superyou.link/nanopatch).

While many people have no particular love of injections from a needle and syringe—a technology that's been around for more than 160 years—there have been precious few alternatives. That is until now. Enter the Nanopatch, a spring-loaded applicator that shoots a small vaccine-loaded patch measuring roughly 0.4 square inches (1 square centimeter) at the skin.

When the patch hits the skin, roughly 4,000 tiny projections breach the skin, and vaccines located on each of the projections make their way into the bloodstream. After a short period, the Nanopatch can then be peeled off and discarded.

Even better, argues Kendall, the Nanopatch offers a better level of protection than the standard needle-and-syringe combo, allowing effective levels of protection at a fraction of the cost.

The Nanopatch can also potentially boost the efficacy of vaccines that don't currently work all that well when delivered by needle. It also allows the vaccines to stay viable without the same level of refrigeration required by standard methods. That makes them useful in areas where refrigeration isn't available or widespread. Consequently, the Nanopatch can be useful in parts of Africa where traditional vaccination methods currently don't work.

Nanotechnology has already made its way into the world of pharmaceuticals, with a number of drugs in the marketplace making use of nanoparticles, most notably for targeting cancer cells. Nanotech can also repair tissue, do genetic detection, and purify blood.

Some see an even larger role for nanotechnology in the world of medicine. Take medical futurist Dr. Bertalan Meskó, who refers to himself in his LinkedIn profile as "a geek physician with a PhD in genomics."

Meskó looks forward to a day when tiny robots course through your bloodstream waiting to spring into action at the first sign of trouble. For example, they could "release oxygen during a heart attack," or collect information and then provide alerts when we are at risk of developing a disease.

"I've had a bunch of genomic tests just to see how they work in action," Meskó told us. "They say that, 'I have this kind or increased risk for this kind of medical condition,' which, based on my genetic profile tells me I might develop a disease within the next five to ten years, but with nanorobots we might be able to get the notification even before the first symptom arrives."

In his book *The Guide to the Future of Medicine*, Meskó notes, "These microscopic robots would send alerts to our smartphones or digital contact lenses before disease could develop in our body." He even foresees a future where blood-borne microbots could identify and treat cancer cells.

But Meskó is cautious about the time frame for this type of technology entering the mainstream. When asked how soon he believes it will be commercially viable, he said "The nanorobot technology? Not ten years. At least 15 to 20. And I'm considered an optimistic futurist."

There's nothing wrong with that prediction. It's pretty much on par with Ray Kurzweil's forecast. Kurzweil told the *Wall Street Journal* in early 2014, "We'll have millions—billions of blood cell-sized computers in our bloodstream in the 2030s keeping us healthy, augmenting our immune system, also going into our brain, putting our neocortex on the Cloud."

SMART PILLS: NO NANOBOTS YET, BUT CLOSE

Proteus Digital Health, a company in Redwood City, California, is building smart drugs that can regularly chart their progress. The pills have hidden millimeter-sized silicon sensors that are harmless to ingest. They contain copper and magnesium, two elements that interact with stomach acid to produce electrical signals. The signals get sent to a patch worn on the shoulder, alerting the user to the presence of medication in their body. Additionally, they track skin temperature, activity levels, heart rate, and sleep patterns. All the information is relayed wirelessly from the patch to a computer or smartphone app.

The sensors were approved for testing in the United States in 2012. Since then, one trial tested 1,000 types of drugs, and showed them as 95% effective.

In 2015, Proteus Digital partnered with pharmaceutical Otsuka, the manufacturers of Abilify (aripiprazole), a drug currently approved by the FDA to treat serious mental illnesses such as schizophrenia, bipolar I, and major depression. The team submitted the first application to the FDA seeking approval to release a new version of Abilify using Proteus's technology.

Digital smart pills will have many benefits. Doctors could use data obtained from them to help find a drug that works for each patient, and in a shorter timeframe. Proteus.com reports that these drugs could save the United States an estimated $100 billion to $300 billion in healthcare costs each year. And, most importantly, the drugs could save many lives.

These tiny robots will also attack cancer cells (see Figure 4.1), cleanse the bloodstream, and scrub the plaque from our arteries.

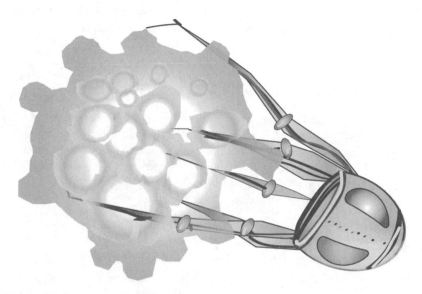

Figure 4.1 *Our artist's rendition of a microscopic nano-robot attacking a cancer cell. (Illustration by Cornelia Svela.)*

That said, If you think nanobots in your bloodstream is a way out concept, watch this short interview with Metin Sitti, a roboticist and Professor of Electrical Engineering at Carnegie Mellon University: http://superyou.link/nanobots.

I Sing the Body Electronic!

Electronics have been part of medicine for a long time now. For decades patients have had monitoring equipment of varying shapes and sizes sitting beside their hospital beds. They automatically measure heart rate, blood pressure, and other vital statistics. Electronically controlled medication drips automatically delivering doses at precise intervals. These devices, as well as more specialized equipment, are used in operating theaters around the world to help surgeons and their assistants keep an eye on the progress of an operation.

These days they include tiny cameras that can be inserted directly into the body through the stomach, through arteries, or other passageways that would have been impossible in years gone by. Surgeons are now able to see what's actually going on inside the body so they can provide precise treatment. Procedures that once required opening a patient's sternum can now potentially be performed using small cameras inserted alongside tiny robotically operated surgical tools, through a small incision. As you'd imagine, these procedures are less traumatic to the patient and they also speed up recovery times greatly.

Other technologies have started to make their way into operating theater, including tablet computers, augmented reality headsets, and heads-up displays.

SHOOT CANCER CELLS DEAD!

Young patients with chronic, non-curable diseases like cancer might feel a bit less helpless in their fight to survive thanks to HopeLab, a non-profit organization established by Pam Omidyar, an immunologist and closet gamer. The effort brings together software developers and doctors who create interactive games that educate and empower young patients to help them understand their illness, so they feel more in control.

In 2006, HopeLab produced a video game for young cancer patients, cleverly named Re-Mission. The game allows the player to go inside the human body where they can destroy cancer molecules at the cellular level.

A randomized trial showed that kids and young adults who played the game were more likely to take an active role in chemotherapy treatment than a control group that played non-health related games.

In 2012, the American Cancer Society (ACS), similarly used a video game designed to empower smokers to quit smoking. In a game called Zombie Smokeout, a player goes head-to-head against smoker zombies. The player wins the game by extinguishing zombie cigarettes.

For now, more research is necessary to assess the effectiveness of medical games; but in the near future, it's quite possible that your doctors will be prescribing mobile apps along with your regular treatment.

In one application from Germany's Fraunhofer Institute for Medical Image Computing MEVIS, a patient's liver was scanned using a CT scan. The information was imported into an iPad app. In the operating theater, doctors were then able to point the iPad's camera toward the patient's exposed liver and the app then overlaid the image of the liver with a virtual image of the patient's blood vessels, using information collected during the CT scan. The idea: Give the medical team quick access to as much info as possible, reducing the possibility of inadvertently cutting into vital blood vessels, while removing a tumor.

Straight Outta Fraunhofer

If you think you haven't had any experience with technology from the Fraunhofer Institute, you're probably wrong: It's also where the MP3 was invented. Pump up those beats!

Of course, slipping an iPad into the middle of surgery may not be ideal; the tablet requires two hands to use, and generally has to be packed up in a protective coating to make it suitable for use in a sterile medical environment. Instead, heads-up displays have a more promising future.

TURNING UP THE HEAT WITH HIFU ULTRASOUND

A revolutionary non-invasive surgical method that uses high-intensity focused ultrasound (HIFU) might soon be available for patients with chronic diseases—from cancer to various neurological disorders—who require complex surgeries. Patients who receive a HIFU operation could be in and out of the hospital on the same day they have their surgery.

Sound waves from a focused ultrasound device are used to heat and burn tissue. The energy produced by high-intensity ultrasound is harmless until it's concentrated on a specific area of the body. So using the technology, no incisions need to be made and no anesthetic is necessary. Also, any tissue surrounding the organ being treated would remain unharmed.

HIFU is not a new idea. The first trials were conducted on cats at the University of Columbia in New York in the 1940s. They were marginally successful and subsequently, the idea was abandoned. However, the invention of MRI technology—which targets areas of tissue just fractions of a millimeter across—makes it now possible to operate with greater precision.

In 2015, clinical trials using HIFU on Parkinson patients began at the University of Maryland and the University of Virginia. The first patient to undergo surgery was Kimberly Spletter. She went from being unable to dance at a wedding, to running down hospital hallways soon after receiving her surgery. See Kimberly's story in this TedX talk: http://superyou.link/hifu.

It's probably not surprising that one of the technologies already making its way into the operating theater is Google Glass. Not only does it have an eyepiece to allow surgeons to view charts, medical scans, or other relevant information, the device's voice control can enable a surgeon to get information immediately without the need to put down tools in their hands. The built-in camera can enable surgeons to livestream video of their surgery to off-site experts—or simply to provide a record for later analysis.

Dr. Rafael Grossmann, based in Maine, was the first to use Glass inside the operating room, and the experience has made him a fan of the technology. During an initial trial, he used it to livestream his point-of-view to an iPad located elsewhere in the surgery room, which then sent it to a Google Hangout. In a blog post following the announcement from Google that there would be a new generation of Glass, Grossmann wrote that his wish list envisioned a version of Glass that could be used for remote consultations, which could stand up to "field-conditions no matter the type of field! (think of doctors from Doctors Without Borders using them in disaster zones or in the middle of a conflict."

> "... the [Google Glass] built-in camera can enable surgeons to livestream video of their surgery to off-site experts. ..."

Grossmann noted during one of his *TED Talks* that heads-up technology like Glass can play a big part in increasing safety in the operating room, by reducing the possibilities of sometimes-fatal medical mistakes, such as wrong-site surgery (i.e. amputating the wrong leg). Coupled with modern apps that replace the old-school pen-and-paper checklist, surgeons can minimize the risk of entirely preventable errors which can cost patients their lives.

Meskó thinks this type of technology is going to continue to play a big part in surgeries of the future. He was one of the remote viewers of Grossmann's livestreamed operations.

"I told [Grossmann] that it's great and I believe all these operations will be recorded and that all the recorded videos will not be checked for errors by humans, but actually by the IBM Watson supercomputer," said Meskó.

But he also sees a future where the device can be controlled through the power of thought, using technology similar to that found in the Muse brainwave-sensing headband, and then have results delivered to next-generation heads-up displays: "I've worn Google Glass a few times but I had to communicate with the device by using my hand or by voice: 'OK Glass, take a picture,' and it does. But with EEG, we actually think about doing things and we'll get the same information through the digital contact lenses."

The future of electronics in medicine is going to go well beyond what surgeons strap onto their heads during an operation.

Some researchers are already starting to use electronics as a stand-in for actual human organs in prototyping both medications and health outcomes.

In a 2013 *TED Talk*, Geraldine Hamilton of Harvard University's Wyss Institute showed an "organ-on-a-chip" device that that simulated the function of a "breathing, living human lung."

The chip contains human lung cells arranged in a layer that can then be manipulated in a way that simulates the way an actual lung would behave.

The device also features a channel that behaves like blood. That is combined with channels that can simulate some of the forces real organs would experience in the body. Foreign material, such as bacteria, can also be introduced to see how the lung will react.

The organ-on-a-chip could also simulate the function of a human intestine, which could help model a disease such as irritable bowel syndrome (IBS). Other models have simulated liver, heart, and bone marrow tissues.

Each organ-on-a-chip structure is important in and of itself, but it's when you link them to each other that is when the magic starts to happen. It can show what happens when each system interacts with other systems.

This becomes important when prototyping new drugs. It's all well and good to know how a drug might affect the specific organ it was intended to treat, but if it has an adverse reaction on one of the other important body systems, it can be problematic. By linking these chips designed to simulate body parts, you can tell that a drug designed to treat asthma might actually have an adverse effect on the heart or liver (or vice versa), all without having to test the drugs on living, breathing humans that may have malpractice attorneys on speed dial.

Or, for example, it can be used to test reactions to chemical exposure (such as that found in a household cleanser, for example) on human skin.

Ithaca University's Michael Shuler has also worked with similarly linked biological systems; in the article "Honey, I Shrunk the Lungs" in the March 2015 issue of *Nature* magazine, Shuler details a test his group conducted on the effects of naphthalene, a chemical in mothballs.

In the group's testing, introduction of naphthalene into the linked system caused the death of lung cells, but that cell-death disappeared when the liver module was removed from the linked system, indicating that the lungs were affected not by the chemical itself but by a byproduct of the liver's attempts to neutralize the chemical in the linked system. How cool is that?

Hamilton noted in her *TED Talk* that the most important application that organ-on-a-chip might provide is individualized response. You provide your own cells for inclusion in one of these chips and doctors can start personalizing treatments that avoid some of the pitfalls or current one-size-fits-all treatments.

"Individual differences mean that we could react very differently and sometimes in unpredictable ways to drugs. I myself, a couple of years back, had a really bad headache, just couldn't shake it, thought, 'Well, I'll try something different.' I took some Advil. Fifteen minutes later, I was on my way to the emergency room with a full-blown asthma attack."

With the organ-on-a-chip setup, you can tailor new individual treatments without risking sending anyone to the hospital.

"We can simulate how organs work," said Meskó. "And, if we can do that now with a lung, that means in five or ten years time we can do that with other organs," he said.

Drugs will not be administered to patients without knowing the outcomes. Instead, drugs can be administered to the simulators, and the results will show what would happen if they were given to real patients.

> "With the organ-on-a-chip setup, you can tailor new individual treatments without risking sending anyone to the hospital."

This will have a profound effect on the speed of clinical trials. "Now you have tens of thousands of people and half of them will be on placebos. But you will be able to do the same thing with simulations. You will get the same quality. It frightens the pharma industry because they're not ready for these kind of changes. But they will have to change their business models for sure," Meskó explained.

Up Close and Personal with Your Genome

The thing that determines a huge chunk of who we are, what we will become, and the maladies that will ultimately afflict us is our DNA, which is a fairly unique sequence of nucleic acids that determines our genetic makeup.

When a daddy and a mommy make a baby, their DNA mixes together to produce a new person, which means that one offspring might be very different from his or her sibling. Consequently, each person has a different genetic signature that might make them prone to different diseases or conditions, and might make them react to chemicals or treatment in differing ways.

Researchers believe that understanding an individual's DNA makeup will provide valuable information for keeping them healthy, including preventative treatment, individually tailored medicines, and custom therapies. For the longest time, that was easier said than done, as the ability to decode DNA remained elusive, containing as it did some 3.3 billion chemical "base pairs." A project to decode the human genome started in 1990, but didn't yield an effectively complete sequence until 2003.

Following the completion of the project, the big factor preventing people from blasting ahead at full speed in decoding the genome was cost. In 2007, the cost to sequence a single genome was somewhere around $10 million. Now, the cost for sequencing an individual's genome is down to four digits, a drop that a chart from Genome.gov notes is well ahead of the curve Moore's Law might have predicted. In 2017, it will likely drop into the three digits, below $1,000. (See the chart at http://superyou.link/1kgenome.)

Extrapolating out from this existing price curve, no doubt some people are looking forward to a time in the very near future where you can simply send a DNA sample away in the mail with $20 and receive your complete genome by return mail a few short weeks later. Or someone might make a kit akin to the pee-on-a-stick pregnancy test but for DNA profiling. Remember, once upon a time, a rabbit's ovaries were injected with a potential mother-to-be's urine to test for pregnancy. So the idea of a drug store DNA test isn't that far-fetched.

With this type of information on hand, people can start to understand what types of conditions they might be genetically predisposed to develop.

During a 2012-released Stanford University study detailed in the book *The Cure in the Code*, geneticist Michael Snyder discovered that his genetic profile indicated that he might develop type II diabetes, despite the lack of any family history of the disease. Consequently, he was able to track the progress of the disease in his body, and how it responded to treatment. Following the study (where he was a project leader), he speculated on what this could mean for people as the price of testing drops. "I think people who are at risk for certain diseases could do a simple home test. You could probably monitor yourself every month so you can catch diseases early," said Snyder.

Meskó believes this scenario could come sooner than you'd think. "The Oxford Nanapore genomic company came up with a device, a USB-based device which I can connect to my laptop, and with a blood drop I can actually sequence parts of my DNA. With a laptop! So it's actually quite cheap based on the whole genome sequencing. So maybe in a few years' time we will be able to sequence our whole genome just with a USB stick and a laptop. It's not so futurist anymore."

See, we told you! Though we won't bet on the "laptop and USB stick." By then your "computing" device will likely be something completely different.

The device Meskó references is known as the MinION. It can identify DNA, RNA, micro RNA, and proteins, and at the time of this writing (mid-2016) is available from Oxford Nanopore for an "access fee" of $1,000 (see: http://superyou.link/minion). It's about the size of a compact remote control, and sits beside your computer to process material you put onto its sensor.

Here's how it works, for those of you that want to show off to your mom at brunch next weekend: It measures how molecules disrupt a current flowing through a protein micropore inserted into an electrically resistant membrane. As different types of materials flow through the pore, the signal that's generated can help analyze the different types of material—for example, different types of RNA strands—and in what concentration they appear in the sample.

It's not exactly a complete genomic sequencing, but for those who want to keep an eye out for certain genetic markers, it's certainly less expensive than building a complete genetics lab.

As the technology for peering into your own genetic makeup continues to become more available and less expensive, Meskó sees it becoming a part of your life right from the start.

"In a few years—three, four, or five years' time—every newborn will get their genome sequenced," he told us. "So, they will start their life with this information. We might not be able to do much about this information from a medical perspective. But at least they will have it. Their genome will not change throughout their life."

That can prove extremely valuable down the road when it comes time to develop treatments for any diseases or conditions that might transition from simple predisposition into hard, cruel reality.

We already know that drugs that might be effective for one person might do nothing at all for another person with the same condition, and might even have life-threatening consequences. That's largely because everyone has a slightly different genome, and as a result, different types of drugs have the potential to interact slightly different for different people.

That's one key reason why there are so many different drugs (and drug types) for a fairly common condition like hypertension. Some patients respond perfectly fine to the old-school diuretics, while others might see no effect. Still others might benefit from beta blockers while others might find they cause lung problems. In the end, treating even relatively well-understood conditions can involve a lot of trial-and-error when trying to pin down an effective pharmaceutical treatment.

Patients should be able to bring the power of their own genetic information to this process. After getting your genome sequenced, doctors can use modeling to have a better understanding of how your specific system works, and would then potentially have the ability to reverse-engineer a medicine that will be fantastically efficient for your specific condition, rather than throwing a series of pharma options into your system just to see what sticks … or what sticks long enough to be considered "good enough."

We bet the role of your pharmacist at your local Walgreens or Duane Reade will become increasingly important in the near term and then super tech-savvy, and then suddenly redundant. Sort of like the guy that used to develop the film from your camera at your local Photomat 15 years ago.

With your genomic information in hand, it's easier to see how to hit the bullseye, especially if you start to pair the design of the custom pharmaceutical with the organ-on-a-chip modules detailed earlier in this chapter. Or, as author Peter W. Huber puts it in *The Cure in the Code*, "We are moving swiftly toward systems that can design an anti-molecule to match any molecule found in the vast, complex, diverse, mutable library of biochemical code that defines humanity and all the rest of life on earth."

DIY Body Parts

If you've walked through a mall in the last decade, you might have seen one of those storefronts where you can build your own teddy bear using various custom parts.

If you've ever lamented the fact that you can't get replacement parts for your own body that easily, you might be surprised to know that this is coming. (Whether it makes its way to a mall storefront is another question entirely.)

One way technology and biology have come together during the last decade is in the hot field of 3D printing (see Figure 4.2).

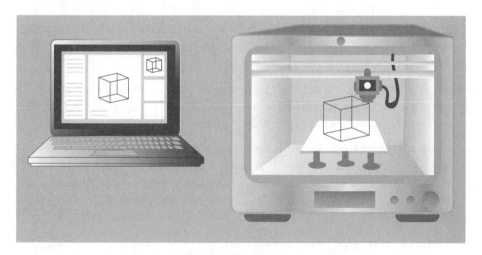

3D PRINTING

Figure 4.2 *A 3D object is designed on a computer (left) and then is printed out by a machine (right) that layers materials slice by slice to form an actual three-dimensional object.*

(Illustration by Cornelia Svela.)

In addition to 3D-printing custom pharmaceuticals, this technology can also be used to create replacement biological body parts.

Using information gleaned from CAT and MRI scans, labs are now able to use industrial 3D printers to create replacement parts, such as titanium hip joints custom-designed to fit seamlessly into the patient's own existing bone structure. Other teams have been using 3D printing to replace lost bone, enabling both facial and skull reconstruction customized specifically to the patient.

The technology can also prove extremely useful for diagnostic purposes. In one case at Brigham and Women's Hospital in Boston, a scan of a patient's heart was turned into a full-sized three-dimensional model. After examining it, a cardiologist preparing for a surgery on the patient realized the procedure would not work. In another case, following an ultrasound scan, a blind woman was able to "see" her unborn child with the aid of a 3D print made from the scan information.

Anthony Atala, Director of Wake Forest Baptist Medical Center's Institute for Regenerative Medicine, is looking for ways to regrow organs that might have already become nonviable, or as he refers to them in his 2011 *TED Talk*, "deceased."

After procuring a tissue sample from the expired organ, Atala's team was able to grow tissue that would function in the same way as the original tissue. He noted that these newly created heart cells would even "beat" the way original cells would have beaten. Atala also noted that his team was able to grow a new bladder from a section of the patient's own cells taken from the original organ that was "less than half the size of a postage stamp."

But Atala isn't content with simply regrowing the patient's own organs in a lab environment. He wants to 3D-print new organs using cells instead of ink. (See 3D bioprinting technology here: http://superyou.link/3dprinting.)

While some of this can already be done using a standard desktop printer— equipped with special cartridges, of course—Atala envisions a future where specialized equipment can scan a patient's wound and print new cells directly onto the patient in the proper layers of tissue that would have occurred in the original, uninjured form.

"Ninety percent of the patients on the transplant list are actually waiting for a kidney," Atala said in his 2011 *TED Talk*. "Patients are dying every day because we don't have enough of those organs to go around."

The solution? Scanning the patient's body to get the detailed dimensions of their kidneys, layer-by-layer, and then print a new one, layer-by-layer, using actual cell tissue. At the time of his talk, the equipment they had could print a prototype kidney—not a fully functional kidney replacement, mind you—in about 7 hours.

In early 2016, Atala and his team announced they had successfully printed an ear, along with bone and muscle structures. When implanted under the skin of rats and mice, the living printed tissue maintained its shape, grew, and developed blood vessels.

One researcher has even found a way to print tiny blood vessels using a $40 cotton candy machine (see http://www.superyou.link/cottoncandy).

As the technology exponentially improves, we can expect printing speed to improve (as well as the accuracy), leading to a viable kidney replacement within reasonable time frames. Within ten years? We are betting on it.

It's not just organs that we're creating from scratch, though. Researchers in South Korea have also recently developed what they're calling "smart skin." The skin, which features sensors embedded into a soft material known as polydimethylsiloxane (PDMS), can measure pressure when stretched or when pressure is applied, providing some of the same feedback that would have been provided by actual human skin. That includes feedback you don't really think much about, such as measuring heat and cold, or whether an object is wet.

You'd know when your dog is licking you, or when you are resting your prosthetic arm on a hot stove so you would be able avoid this scenario.

"Funny smell. Hmm. What's burning in the kitchen?"

"Oh crap, it's you, honey!"

This artificial skin can theoretically be applied over artificial limbs such as hands and legs, including artificial limbs with articulating robotic parts such as wrists and ankles, allowing an amputee to regain some of the sense information lost with the removal of the original limb. The big hurdle, as you might expect, is somehow interfacing the information being gathered by this synthetic skin and transmitting it to the brain in a way that makes sense.

The synthetic skin might not be immediately viable because there is still the problem of wiring the recipients' brains with the sensors. However, the British National Health Service has also been working on synthetic blood that might become useful faster. The artificial blood gets its start from either the umbilical stem cells of newborns or from bone marrow from adults. These cells are then cultured in the lab until they become full-fledged red blood cells. The process so far has resulted in modest success, but not quite enough to be a practical replacement just yet. However, the National Health Service (NHS) hopes to do full clinical trials in 2017 using the lab-grown synthetic blood. Again, provided this technology proves viable, it's possible that someone could have a few bags of their own blood grown before surgery, or dip into a pool of synthetic blood when donations are running low.

Then again, you know that donor blood as we know it won't really be needed for much longer. In (and probably within) ten years, we'll grow our own supply through a mass production synthesizer of some sort. It's not hard to imagine with the NHS breakthrough.

While we're becoming good at building replacement parts out of metal and plastic (hips, artificial limbs), and transplanting others (hearts, lungs), we might be on the verge of being able to plug someone's genome into a vending machine in the operating room and printing out a new kidney, knitting some sensors into a roll of replacement skin, and dispensing a jug of blood (lids and straws are located to the right of the machine). Compare this to the agony of sitting around on a waiting list (weirdly) hoping the right organs become available from some unfortunate motorcycle driver who miscalculates an eighteen-wheeler's trajectory.

Stop the Bleeding

Speaking of blood, it can be a huge problem when you suffer a deep wound that results in a rapid bleed-out that kills you.

If you've ever watched one of TV's many medical dramas, you've no doubt seen one of the show's heroes barking something along the lines of, "We've got to stop the bleeding, dammit!" or "If he loses any more blood, he'll die!" It doesn't matter now much artificial blood you have at the ready, if the patient is making like a fire hose, they are pretty much a goner.

Naturally, blood contains agents that will start to coagulate from a liquid into a gel when you receive a cut. This coagulation, or clotting, allows a scab to form, creating a protective layer that keeps the rest of your blood inside your body, allowing the tissue underneath to heal.

Generally speaking, this process works pretty well, with most cuts needing little more than a bit of pressure or a light bandage to aid the process. For patients with more extreme wounds, two recent technologies will likely make rapid bleeding a thing of the past.

An Oregon company, called RevMedx, has a solution designed for fast treatment of "gunshot and shrapnel wounds on the battlefield." Called the XStat, it's a syringe-style applicator with a 30-millimeter diameter (about 1.2 inches) that gets inserted directly into the wound. When the plunger is depressed, the XStat releases a number of tiny sponges which quickly expand into all available space, stopping the blood from flowing out of the wound until it can be properly attended to. Each sponge is tagged with an X-ray marker, enabling all the inserted material to be properly scanned and removed when the foreign object that caused the original wound is removed. The XStat was shipped to the United States military early in 2015 for use in the field after receiving FDA approval.

Joe Landolina, who is 17 years old, discovered another effective alternative as a freshman at New York University, which he now offers via his company Suneris. Also delivered to a wound using a syringe-style applicator, Veti-gel is comprised of algae-derived polymers which reconfigure their structure when they are applied to tissue, essentially super-gluing a wound shut by creating a structure that's similar to flesh, an organ, or skin.

According to the Suneris website, "Immediately after application, our gel stimulates the clotting process by physically holding pressure in the damaged blood vessel. The gel then rapidly activates the accumulation of platelets, which bind to the site of the injury to create a platelet mesh. Our gel completes hemostasis by accelerating the binding of the clotting protein, fibrin, to the platelet mesh, resulting in blood coagulation and a stable clot." Veti-gel is now in trials with veterinarians and could be applied to human wounds in the near future.

A Sensor of Wonder

Before you can begin medical intervention, you need to know that something is wrong. A bleeding wound is a pretty good indication, but a lot of the conditions that plague us have less obvious indicators. We already have a lot of technology that can help diagnose problems without a lot of inconvenience. A basic oral thermometer can show if our internal temperature is out of whack. An automated cuff can quickly show that someone's blood pressure is dangerously high. But these tools can't diagnose all ills.

As flexible electronics become more commonplace, we're starting to see new types of devices that can wrap around other objects—objects such as body parts. That opens up a new generation of health-related devices taking forms that were previously inconceivable, such as the heat-sensing synthetic skin that we covered earlier in this chapter.

Flexible electronics also enabled the creation of one of Alphabet's new wonder gizmos, the glucose-sensing contact lens prototype.

 The ABCs of Google

Alphabet is the new parent of Google, as of the summer of 2015, and holds all the non-search engine technology parts of the company.

A condition such as diabetes requires constant monitoring, including regularly pricking the skin to take a small blood sample. Unfortunately, a lot of people prefer to avoid the ouchy pinprick and consequently a lot of people self-monitor less regularly than they should.

Also, blood sugar can vary wildly through the day, and even those who regularly test their own blood might find themselves with dangerous blood sugar levels as they eat or exercise, and they might not always know that they're about to pass out from low blood sugar until it's too late. That's where the glucose-sensing lens comes in.

The contact lens prototype is striking to look at. Even though it otherwise looks like any other contact lens on the market, tiny antennae run rings around the outside of the lens (See video: http://superyou.link/superlens). They are connected to a tiny sensor that can measure glucose levels found in the wearer's tears once every second. That information can then be transmitted to a receiver, such as the wearer's smartphone. It enables monitoring in more-or-less real time. According to the blog post on the lens, the company is even exploring such options as tiny lights that would illuminate if the wearer's blood glucose level trended too high or too low. That gives them more warning that something is wrong. It would also make them look like a cool cyborg in the process.

We can see it now: "Hey dad, you've got glowy monster eyes. Do you want a sugar cookie?"

While the project was simply a prototype at the beginning of 2014, Verily (the new name for Google Life Sciences) continues to reach out to partners to bring a final product to the market. Stay tuned.

CLOTHING WITH A "SICK" SENSE?

Being healthy has never looked so good. Thanks to a new trend toward smart fabrics. Soon the clothes you wear could issue a report on how healthy your body is. Similar to wearable wristbands, clothing lines that are both fashionable and functional are being developed in places like New Zealand, Canada, and the United States.

Hidden sensors are woven into lightweight fabrics that are comfortable to wear, easy to wash, and inexpensive when produced in large quantities. Companies like Footfalls and Heartbeats in New Zealand use conductive yarn to measure compressive force, tensile force, and temperature.

The garments could provide information on:

- How hard muscles are working
- How many times your heart beats per minute
- Your breathing rate
- And even how airborne toxins could be affecting your body

The company is working with the University of Nottingham in the United Kingdom, to prototype socks that will tell diabetics how blood is circulating in their feet and alert them to diabetic ulcers. In 2010, U.S. soldiers were already using the technology. The University of California created a line of undergarments to track their vital signs. Battlefield briefs? You betcha!

At the engineering labs of UCLA, Aydogan Ozcan is also working on a sensor of the future, or rather a series of attachments that can turn your smartphone into a series of powerful sensing devices. Using a 3D printer to create housings that can snap to the back of a smartphone, Ozcan's team pops in various electronics designed to measure biometric parameters, often leveraging the phone's own built-in camera. These sensor modules can scan for viruses in a blood sample, measure toxins in water, or detect trace amounts of a known allergen inside a bit of food, and then send information to an app installed on the phone.

While a *Smithsonian* magazine article recently compared Ozcan's creations to *Star Trek's* mythic tricorder device, the comparison isn't yet 100 percent. For one thing, doctors on *Star Trek* would simply wave the tricorder in the air in front of the patient and instantly diagnose them, but Ozcan's creations still require the user to place a physical sample onto a sensor or in front of a camera lens. On the other hand, a tool that can identify a virus in the field is still pretty damned powerful, even if you have to insert a blood sample into the device. Combined with the smartphone's connectivity to the Internet, it's more like a portable bio-lab than a tricorder, a fact that only becomes more impressive when you consider that running a full lab can cost millions of dollars, while one of Ozcan's prototypes cost something like $10.

That low cost of entry theoretically puts the capabilities of a full lab into the hands of anyone with a smartphone. Locals could test the quality of their water supply with their own phones instead of having to call in the government, or doctors in the field could immediately detect the presence of a pathogen without having to send a sample to the lab.

Such tools could have huge democratizing power, even in locations where local infrastructure is weak to nonexistent, allowing people across the world to take better control of their own destinies.

Just think how this could have helped the residents of Flint, Michigan with their fight against a government who insisted that tapwater that looked like root beer was okay to drink.

Hi, I'm Your Robot Doctor ... Wait! Come back!

Robots have been in the operating room for quite some time now, as assistive devices for actual surgeons. These specialized pieces of equipment have several advantages:

- They are more accurate, reducing the need for extensive cutting and minimizing blood loss.

- They allow doctors to sit down while working, reducing fatigue.

- They contain corrective algorithms that would remove tremors found in the surgeon's hands, preventing it from being transferred to work done on the patient.

- And, the robotic tools and cameras can be positioned in ways that were just not possible for a human surgeon to accomplish using old methods.

Robotic devices in surgery can also perform automatic tasks quickly and precisely, such as certain types of incisions or in-surgery bone manipulation, using a prestored algorithm and feedback provided by the tools as they do their work. They've also been much-used in certain types of procedures where a miniaturized touch would be more appropriate, such as heart surgery or certain types of catheterization.

FIRST ROBOT SURGEON

A company in Lincoln, Nebraska called Virtual Incision has created the first robot surgeon. It is intended for emergency medical procedures in space, where medical professionals, facilities, and life-saving equipment are not immediately available.

The current model is 0.4 kilograms and has two arms equipped with medical tools. It can enter through the belly button into an astronaut's body that's been inflated with inert gas. Once inside, it is guided by a video camera, watched by a human that is monitoring the machine remotely from Earth.

In a New Scientist report, Dmitry Oleynikov from the University of Nebraska Medical Center explained: "Even something as simple as putting a Band-aid down on a table is difficult in space". Without gravity, substances like blood float freely. The risk of contamination in confined spaces is also a top concern.

In 2016, the robot surgeon was used on its first human patients. It performed successful colon resection operations at a hospital in Asunción, Paraguay. In an interview with MedGadget.com Oleynikov explained: "Virtual Incision's robotically assisted surgical device achieved proof-of-concept in highly complex abdominal procedures."

Virtual Incision predicts that in the future astronauts will need to be trained to operate this type of equipment. As space exploration moves towards planets like Mars and further away from Earth the lag time in communication increases dramatically so they will need to be able to operate on each other without consulting a doctor back on Earth.

The one thing robots have not been able to accomplish, though, is completely autonomous diagnosis followed by delivery of the appropriate corrective procedure. Human input is still always required. It's also a pretty good bet that if you asked most people out there if they'd be comfortable with a robot performing surgery without any human input whatsoever, most would be reluctant.

Maybe that's an odd position to take, considering the trust we are now putting in autonomous cars to make their way through traffic without hurting or killing anyone. At the time of writing, the majority of accidents that have occurred involving self-driving automobiles have been the fault of human drivers in other vehicles. Taken to its logical conclusion, handing the wheel over to your robot driver seems like the safest course, leaving you free to catch up on your reading, knitting, or yoga.

On the other hand, the stakes seem low if a self-driving car were to malfunction. You might be hurt if a car drives into a guardrail or even into the back of another car, but hopefully the car's other safety functions (such as airbags and seatbelts) will prevent any truly serious injuries. And, theoretically, if you notice something going wrong, you can take manual control of the vehicle.

That's not the case when you're going in for surgery, of course. While it's true that a malfunctioning robot doctor can result in little or no actual damage before a trained surgeon steps in, it's easy enough to imagine a nightmare scenario where a scalpel started cutting a number of things that weren't really meant to be cut. But what's the likelihood of something like that happening? Given the record of self-driving cars, it seems that autonomous robot surgeons could well be the safest option—after all, it's more likely that you'll get sliced and diced by a bleary-eyed doctor with a hangover and an unsteady hand.

And if you are still unsure, most of you let robots handle your money. ATMs are robots. And tellers can have hangovers, a bad day, or sociopathic tendencies.

For the record, your authors discussed this very topic on one of our official Weekly Author Conferences Online (yes, WACOs) as we developed this chapter.

> " ... most of you let robots handle your money. ATMs are robots."

Andy says he would feel perfectly happy trusting his vasectomy operation to a robot, and has no worries whatsoever that his manly bits could be snipped off and turned into a coin purse by a renegade robo-surgeon with a wonky algorithm. (Programmers have bad days too.)

For the record Sean is not so confident. While he's pretty sure that the latter scenario wouldn't happen, he's still not convinced that he wants a robot digging around "down there" without any type of human supervision.

Kay, when she heard the other two talking about such things, rolled her eyes.

Perhaps there's a middle ground.

While Meskó doesn't believe that robots will be performing autonomous surgeries any time soon, he believes that computers and robots could well work in concert with human beings, with human doctors having the final say. And the reason? Computer technology can provide a much greater wealth of information about the surgical procedure immediately at hand than any human surgeon ever could.

"If you take the best professor in the world in a specific medical specialty, he or she might have 40 to 50 medical papers in mind at once. IBM's Watson can check 200 million of these in 1 second. So we cannot compete with that. But that's all right," he said, noting that doctors could always use Watson in the decision-making process. That's something that has already been going on at Memorial Sloan Kettering Cancer Center. "Based on textbooks and the 200 million papers in the world and the patient's medical history, IBM Watson makes a few suggestions to a doctor and based on this, says 'you might want to think of this test or this treatment or this diagnostic'. And doctors, they love this method because they felt more sure. For humans it's impossible to keep in mind all these millions of papers."

"This collaboration could also be used in concert with information about the patient's genomic information to provide a more informed direction for treatment," says Meskó. "Imagine, as a doctor, I can do a search because I have a 40-year-old male patient (with these types of symptoms) and data points and biomarkers in their blood, and what can I expect? What if I can look at 10,000 same-gender, same-age, same-population people with the same health parameters and lifestyle and I could see what happened to them after using this treatment? So, I could make the best-informed decision in the world because I would have all the information I would possibly need."

In the short term, robot assistants are starting to perform some tasks around the hospital that are less productive for some human staff members. Vecna has created a mobile robot called the QC Bot. The staff at France's Centre Hospitalier de Beauvais, where the robot was deployed, refer to it by the much friendlier name "Diane."

"Diane" and her QC Bot relatives can serve multiple functions. Like the robotic mail carts that roll around many large office buildings, QC Bots can carry supplies from point-to-point around the hospital (see it in action here: http://superyou.link/qcbot). It also comes with a large screen and an onboard camera, enabling it to serve as a self-serve station for patients—people can check in for their appointment and follow the QC Bot right to the proper room where they'll be seen by the doctor. The robots also come

> "Robot assistants are starting to perform some tasks around the hospital that are less productive for some human staff members."

with the capability to kick into telepresence mode with a remote health provider when required, allowing health providers to consult with experts in other parts of the hospital, or for patients to talk to doctors.

All these tasks can help free up doctors and nurses, according to Daniel Theobold, Chief Technology Officer of Vecna, who explained this in an interview on the *Healthcare Tech Talk* podcast.

"In the pharmacy, when they prepare a cart full of medications that needs to get to a particular patient ward, it is literally a 15-minute walk from the pharmacy to that ward to just deliver that medication. That's 15 minutes that a nurse is spending not providing care to patients. So when they can load those medications up in the robot with confidence, know that the robot is not going to lose them along the way, know that nobody is going to get into the robot and take them, and know that that medication is going to get to where it needs to be, when it needs to be, and now I can turn my attention to actually doing what I was trained to do as a nurse, and that is provide care to patients. That's a win-win for everybody."

This type of interaction requires a lot of research, he notes, as robots don't see the world the same way we do—it's all just a bunch of pixels to them. They also need to be trained in the kind of social cues humans use with each other when they intend to interact with one another, such as eye contact or gestures, because there are many variables to track. While the QC Bot has mastered some of that type of interaction, there's still a long way to go before it would be even remotely capable of rolling into an operating theater and performing Andy's vasectomy.

> "... there's still a long way to go before [the QC Bot] would be even remotely capable of rolling into an operating theater and performing Andy's vasectomy."

DIY Health

We know what you're thinking. With all these medical technologies rolling out, we might be entering an era where you can finally take full control of your health right from the comfort of your own home. Almost certainly that last sentence causes some panicky moments for anyone who's had to deal with a loved one who's freaked out after taking a look at *WebMD*, and then become convinced that they had the rarest disease ever known to mankind.

"Honey, I think I've got ataxia telangiectasia. I read up on it on *WebMD*. I've got the shakes and little bumps on my skin."

"Umm, I don't think so. You have a hangover and you drank pomegranate-tinis when we ran out of cranberry last night. You're allergic, remember?"

The good news—the future is a bit rosier than that. That's thanks to inexpensive new options such as the $1,000 genome sequencing module we looked at a earlier in this chapter. With the ability to scan your own DNA at a reasonable cost—the immutable law of diminishing technology prices are in play here—you're entering a world where you can get actual readings of what's going on in your genetic makeup, rather than simply misreading a list and running around in circles thinking you're going to succumb to a rare childhood disease.

Now imagine, with your genomic information in hand, having the ability to calculate what pharmaceuticals you should be taking for conditions such as hypertension or hypothyroidism, and then printing the drugs out on your home 3D printer as you need them. There's no need to haul your carcass down to the clinic just to convince your doctor to renew your only semi-effective prescription yet again. All that information would be gathered by your home scanner and, potentially, verified by a giant remote computer in the Cloud—such as a descendant of IBM's Watson—to be sure there aren't any errors in the formula of your new home-brew drugs.

Or, imagine that you have a robot at home that can put you under with a dose of sedatives, open you up, and 3D print a healthy new kidney right into you. No more waiting around for some poor sap (who was otherwise healthy) to die in an unfortunate car accident.

We've already heard of people who have started to 3D print their own prosthetic limbs at home, complete with robotic servos inside to move fingers with the proper muscle input where the prosthetic attaches to the body, but this will almost certainly become more commonplace.

It's relatively inexpensive now to buy 3D scanning equipment to create a three-dimensional model of your body, but it's easy enough to envision a future where an amputee can scan the end of their arm or leg to get exact specifications, and then 3D print a prosthetic that fits perfectly onto the end of a limb that was partially amputated. All of this could happen at a fraction of the cost of having a new prosthetic limb cast at a medical lab, especially a limb containing robotics.

The whole DIY ethic is nothing new to a whole new generation of tech-savvy people who are trying to take control of their health. A growing number of people are starting to track their daily step-count and their heart rate through wearables such as the Fitbit or the Apple Watch. With the use of these easily available devices, combined with manual tracking of diet, sleep, weight, and other specialized exercise, people can build up a pretty robust profile of their general state of fitness.

And then there are people who are willing to go a step further and implant themselves with diagnostic devices such as temperature sensors or radio-frequency identification (RFID) chips that could contain medical information for emergency use. In the future, these implants could well expand to devices that measure your heart rate, breathing, or neural activity. At that point, people could do self-diagnostics with a wand they wave over various parts of their body, and if they receive errant readings, book an appointment with the doctor for a tune-up. We'll take a further look at these self-implanting maniacs in Chapter 6, "Franken-You: A Better Life Through Cyborg Technology".

Whether you end up going to futuristic new hospitals, or end up taking care of all your health needs from the comfort of your own Barcalounger, one thing is exceptionally clear. The future of medicine will be extremely personalized.

5

The Human Computer: How to Rewire and Turbo-Boost Your Ape Brain

According to a 2013 Gallup poll, 70 percent of Americans hate their jobs. A 2015 survey showed 70 percent of Britons dread the week ahead.

Across the pond in the UK, Kevin Warwick isn't one of those people. He is a professor of cybernetics at the University of Reading (England) and when he goes to work he gets to do things such as grow miniature brains from mice neurons, and put them into robots to see what happens.

Yes, really.

How cool is that?

But wait, there's more.

Besides that, he's done incredible things such as connect his nervous system to his wife's, allowing the couple to remotely communicate using their thoughts.

Warwick is at the forefront of brain science. He's also working with a team of researchers to develop a new brain stimulation device that will treat neurological issues, such as the very hard to solve Parkinson's disease.

Then there's this cool thing. In 2015, he witnessed the first computer beating the Turing test, impressively showing its capability to fool human judges into thinking it was human.

His industry colleagues across the world are doing equally impressive research in brain science. Dr. Adam Gazzaley is a leading neuroscientist developing technologies that work on the principles of brain plasticity. He is also Professor of Cognitive Neuroscience at the University of California, San Francisco and makes therapeutic video games. Sounds oxymoronic, but his amazing suite of brain-training video games are designed to be used to improve or reverse neurological disorders and build upon human cognitive abilities.

Yes, video games that make you smarter. Go figure.

Then there is our friend Ray Kurzweil, who we introduced in Chapter 1 "The Emergence of (You) the Human Machine." Kurzweil is building a synthetic neocortex, the part of the brain that makes humans the smartest species on the planet.

If that does not amaze or excite you, the rest of this chapter will because it's all about what happens when the brain—the most complex human organ—gets wired up with technology to make it better than ever.

Here's the bad news though.

As intriguing as all this is, brain science is pretty new and has really only made inroads since the late 1900s, and the big developments have really only happened in the last two or maybe three decades.

Until brain imaging came about in the 1970s, it was the Neuro-Dark Ages. And that kind of kicked off any real understanding of how the gray goo between our ears worked.

But let's get real here. Quite frankly the top experts still don't know much about how the human brain works. When we asked Gazzaley how close scientists are to understanding how the brain works in its entirety, he said, "Without knowing exactly what the end point is, it's hard to predict what percentage we're at along the way."

 ### The Unmapped Continent

Scientific American eloquently spelled out the progress (or lack thereof) in neuroscience in an article in 2012:

"There is one largely unmapped continent, perhaps the most intriguing of them all, because it is the instrument of discovery itself: the human brain. It is the presumptive seat of our thoughts, and feelings, and consciousness. Even the clinical criteria for death feature the brain prominently, so it arbitrates human life as well."

So what do scientists know? Well a lot about the brain of the *Caenorhabditis elegans*, a dumb little worm. In 2011, the connectome, the map of the worm's neural pathways, were published. The roundworm is 1 millimeter in length (think: the width of a strand of spaghetti). Its behavior is basic. It travels from place to place, inching forward and back. It lives in soil or rotting vegetation and it feeds on microbes such as bacteria.

  Worms Our Heart

See a useful video on on the C. Elegans worm: http://superyou.link/dumblittleworm

This puts it into perspective, however: The *C. elegans* worm possesses 0.3 percent of the intelligence of humans, though subjectively it might seem like more when compared to some of the candidates in the lead-up to the most recent presidential race. The *C. elegans* has 302 neurons and 7,000 synapses. Humans have approximately 100 billion neurons and 100 trillion synapses. Humbling, yes.

> "The *C. elegans* worm possesses 0.3 percent of the intelligence of humans."

This is not to say mapping this tiny worm's brain was not a major breakthrough in the grand scheme of neuroscience. We don't want to diminish the achievement here.

But scientists still have a long way to go to understanding (and mapping) the human brain. It's still really early in the history of neuroscience. If this was the history of flight, we'd still be perhaps at the equivalent of crossing the Atlantic for the first time. Let's rewind even further back and inspect the beginnings of neuroscience, because looking back we can see the accelerating technologies develop on a timeline and it informs us how fast new developments will come as the time between each breakthrough shortens.

A Brief Early History of the Brain

Approximately 200 million years ago, a pivotal evolutionary shift gave man the ability to learn about himself and his behavior. Before that time, it simply wasn't possible. Man was not intelligent enough to study … well, anything, including himself. This required a change in the structure of his brain's anatomy. It grew, along with his forehead, to house an essential new structure called the neocortex (you might recall this from Chapter 1 "The Emergence of (You) the Human Machine").

It's the most recently evolved structure of the mammalian brain. And it's an important one. It serves the major functions of sensory perception, spatial reasoning, thinking, and language. This allows for social behavior, tool making, and high-level consciousness.

The neocortex prompts humans to learn and study. To create science. To create art. And, most importantly to create language.

Language was a crucial technological development: "The first thing we invented was a communication technology, called spoken language," said Kurzweil. "This gave us one way of expanding our neocortex beyond the 300 million pattern recognition system modules. It allowed us to solve problems with people or a community of people by communicating with each other."

Early records of man's notions about how his biology and behavior were connected is speculative. Ancient cultures had no real-time technologies to help them understand the mechanisms of the human "machine."

In fact, most of what was known about the brain came out of a need to treat brain injuries. The oldest medical treatise in history, the Edwin Smith Papyrus from 1700 B.C.E., contains the first written record about brain injury.

The true early pioneers of neuroscience were the Ancient Greeks. Between 460 and 379 B.C.E., Hippocrates's book, *On Injuries of the Head*, was the first to suggest that the brain controlled the body. They discovered that an injury to the left hemisphere of the brain would cause an injury to the right side of the body and vice versa.

These early discoveries were based on an ill-conceived theory of medicine known as humorism, which suggested that the body was made up of four bodily fluids—blood, phlegm, yellow bile, and black bile. The theory suggested when each one is in balance, the body functions normally. An overproduction of black bile, for example, was linked to the brain disease known as epilepsy. Today, of course, we know that is ridiculous. Unfortunately, humorism governed Western medical thinking for more than 2,000 years.

Greek philosopher Aristotle, a student of Hippocrates, took a different approach. In 337 B.C.E., he proposed that the heart controlled all mental processes. The heart moves constantly, is centrally located, and is the first organ to develop in the embryo. So he thought, for these reasons, it controlled man, and was responsible for epilepsy.

Luckily this hypothesis was abandoned. In 177 A.D., a Greek surgeon named Galen of Pergamon began cutting heads open to understand the brain's anatomy. (There's really no pretty way to phrase that)

Galen is known as the "Father of Anatomy." He performed studies on the brains of animals and human cadavers. He discovered that the brain has four fluid-filled cavities that contain major communication networks. They're known as ventricles. Each one contains a protective fluid—cerebrospinal fluid—that protects the brain by bathing it in nutrients, while also eliminating waste.

Galen figured the fluid contained "animal spirits," weightless invisible miniature ghosts with their own wills and intentions.

He said the spirits were also the culprits responsible for psychiatric disorders. This was long before *Scooby Doo*, so in theory these animal spirits got away with it because there were no meddling kids to stop them.

Finally, in the late seventeenth century, physician Thomas Willis decided to map out the brain's physical anatomy. He's the "Father of Neuroscience" and the man who coined the term "neurology." He also published six books on brain anatomy. His research was revolutionary for its time.

Based on the structural differences of the human brain relative to other animal brains, Willis believed man had an immortal soul because he had a higher level of cognitive function than any other species. Said another way, Willis figured we were all pretty smart and this was his way of explaining consciousness, a phenomenon that scientists today still do not understand. (We will get to that later.)

Other scientists of that time took a similar approach. They believed in something called "dualism," which is a view that the mind is a spiritual entity. People who bought into dualism believed the mind was something ethereal and godly whereas the body is purely physical…and they are distinctly separate.

French philosopher René Descartes believed in dualism, but was the first to view the human body as a machine. He believed the mind—whatever the "mind" was—controlled the human body. And the human body fed the mind information it collected from the environment.

 Cogito Ergo Sum

French philosopher René Descartes is the man who coined the famous phrase, Cogito ergo sum, or "I think therefore, I am."

This was progress. He based his theory on his observations of mechanical statues in the royal gardens at Saint-Germain in France. Powered by water, these statues would move in ways reminiscent of humans, even though the internal parts were springs rather than muscles watching the statues made him realize that human beings were in essence complicated machines and he realized there was something else far more complex animating us.

Nevertheless, Descartes was the first to use technological principles to try to explain how the nervous system functioned. His use of technology to understand the functions of the brain was carried into the research of scientists John Walsh and Luigi Galvani. Their pioneering work contributed to the discovery that the brain is an electrical system.

One of the greatest (and most notoriously overlooked) breakthroughs in neuroscience occurred in 1773 when Walsh used strips of tin foil to generate a spark from an electric eel. The discovery was never documented, but Walsh was awarded a medal for his work. It's also worth noting that with a bit more tin foil, he could also have discovered barbecued eels. But that's another story.

Walsh's work encouraged scientists such as Galvani to further understand the electrical nature of the brain. In 1781, Galvani was an Italian scientist who made a frog's severed legs move using a current from a static electric generator.

From that point on, brain function started to become better understood. German physiologist Johannes Peter Muller discovered that although all brain cells communicate by electrical impulse, the systems of communication for each sense and bodily function were different.

And so, French physiologist Pierre Flourens invented experimental ablation, a research method where a specific region of the brain is interfered with or removed to learn how it's connected to human behavior.

This new research technique allowed new discoveries on the function of specific brain regions. Neuroscience was gaining clout as a true science backed by empirical data, and not merely assumption.

However, for some strange reason brain science took a U-turn with the study of phrenology as it gained popularity between 1810 and 1840. The now-pseudoscience suggested that measurements of the head could determine characteristics of an individual. The theory was a fad and there was no real evidence to support it. That didn't stop people from using it to make baseless and sometimes extremely unfortunate judgments about a person's character. In fact, as late as 1928, a phrenologist's testimony was used to help convict a woman accused of murdering her husband. (Never trust bumpy headed people!)

By the late nineteenth century, developments in brain imaging technology improved neuroscience dramatically. New devices gave scientists the ability to view the brain in real time. Scientist Angelo Mosso invented the first neuroimaging technique in the late 1880s. His device, called the "human circulation balance" machine, measured the redistribution of blood to the brain in real time. This was the genesis of more advanced noninvasive brain imaging devices.

Notice here that it took a long time to get to this point from the time when the Ancient Greeks started their work. This technology, however, further accelerated the understanding of the human brain.

Fast forward to the 1970s through the 1990s where there was an influx of brain imaging technologies. Many of them are still in use today 20 to 25 years later, and they are instrumental in helping scientists understand the brain.

Thanks to these brain-measuring tactics, neuroscience made a major discovery in the twentieth century that changed the course of the entire field. Between the 1990s to the mid-2000s, studies led many scientists to conclude that the brain is plastic, meaning it could change and reconfigure itself. Until then, the accepted belief was that the anatomy of the brain was fixed once an individual reached adulthood.

The brains that humans have today evolved from what they were about 200 million years ago. Because humans are bipedal—meaning that our species stands upright on two legs—the upright posture limited the size of the cranium that could fit through a woman's birth canal. So, when a baby is born, its brain is not fully developed.

Early scientists didn't need much technology to figure that one out. The fact is that a baby can't go to the toilet, feed itself, or clean grout between dirty floor tiles. And forget mathematical equations. If you ask a baby the formula for the circumference of a circle, it won't respond with $2\pi R$. The moment a child enters the world, it needs much more development. It's not until late adolescence that the human brain reaches its fully developed weight of 1,400 grams. Although let's face it, like, even then teens aren't, you know, the sharpest knives in the proverbial drawer.

Early scientists believed that brain development stopped there. The former scientific model assumed each individual is born with a finite number of neurons. That is simply not the case.

It's been scientifically proven that the brain's anatomy goes through structural changes during adulthood simply by processing new information. This process is known as neuroplasticity.

Your Brain Is Plastic? What?

We are not saying your brain is made out of the same stuff as Tupperware. Plastic, in this context, means moldable or changeable.

Neuroplasticity is the brain's miraculous ability to change its structure through information processing, or in other words, through thoughts.

Yes, you can think your way toward changing the physical structure of your brain.

And what's cool is if you think your brain is not that super, the good news is it can be because you can train it to redesign itself to be super smart.

In fact, your brain is constantly in a state of adaptation. It creates new communication pathways based on what it requires at a given moment in time. It is continually adjusting to the world around it, to be sure that both the brain and the body not only continue to survive, but also function in a way that's optimal for both.

> " ... if you think your brain is not that super, the good news is it can be because you can train it to redesign itself to be super smart."

The brain is always in a state of optimization. Think of this like a car's GPS system. You program in your destination but there are a lot of little things that can change along the way—you might find a detour on your route; you might not be paying attention and miss a turn, or you might just decide to take the scenic route. Whenever you deviate from the GPS's original plan, it takes a moment to acknowledge the deviation and then creates a new plan based on these new variables that will allow it to give you new directions to your destination. The brain works in a similar way (except without the nagging mother-in-law voice).

For instance, when the body suffers an injury, such as a disease or stroke, the brain reorganizes its neural circuits to adjust to changes in the body and its new way of interacting with the world.

A classic example of neuroplasticity is evident among patients with an illness called phantom limb syndrome, a mental disorder that occurs when an amputee still feels the presence of the limb long after it's been removed from the body.

A bizarre phenomenon, known as "phantom limb," occurs in patients with a missing arm." It's common for them to feel pain at the site of the missing limb when they are touched on the face. They still feel the touch to the face, but anytime they receive a sensation to this region, the missing arm feels pain, too.

As it turns out, the brain region that recognizes a touch on the face is located next to the region that would normally receive sensory input from the missing limb. These two areas become crossed when the limb is removed. The region that's connected to the missing limb becomes hungry for sensory input. So, it simply reorganizes itself to get what it needs by invading the neighbor region. It's like a toddler who wants another kid's blocks.

American neuroscientist Paul Bach-y-Rita understood this adaptation principle long before it was scientifically accepted. In 1969, he built a machine that allowed congenitally blind people to see by being touched.

It sounds impossible, but Bach-y-Rita simply rejigged the processes involved in sight. He understood that for a person to see, a pulse must be sent from the back of the eye to the brain. That means that the physical eye is simply a messenger. It transmits information from the environment to the more critical region in the back of the eye. The brain's job is to decode the message it receives, so that the person understands what's in their visual field.

Pulses for seeing aren't any different from pulses involved in other senses, such as the pulses involved in touch. The difference is the frequency of pulses that get delivered to the brain. Each message tells the brain what it needs to understand.

A patient using Bach-y-Rita's machine would sit in a chair and have small needles touch his or her back. The chair was connected to a camera that acted as a set of eyes. Based on what the camera saw, it provided information to the needles that touched the patient. That information was sent to the brain and allowed the participant to see what the camera saw. Watch a video that explains Bach-y-Rita's revolutionary technology here: http://superyou.link/brilliantbachyrita.

This study was revolutionary and ahead of its time. It took many more studies for scientists to conclude that the brain is plastic. Another popular experiment was conducted by Eleanor Maguire. In 2000, she discovered that a key brain region for processing memory—the hippocampus—was larger in the brains of London taxi drivers. The job requirement for the cabbies forced them to learn and remember up to 400 different travel routes. This caused a redistribution of their gray matter so that their brains could better retain navigational information.

The fact that the brain is plastic also means humans can heal their own brain deficiencies through conscious thought and action to intentionally rewire pathways.

In the book *The Brain's Way of Healing*, author and prominent Canadian psychiatrist Norman Doidge writes about John Pepper, a South African patient with Parkinson's disease. He retrained himself to walk using specific self-developed conscious thinking processes. Using thoughts, he trained areas of his brain to become reignited where they had stopped being used.

Another example of this is David Webber, who went blind from an autoimmune disease, but then used meditation and hand-eye coordination exercises to cure his vision using the brain's amazing ability to restructure itself.

The discovery that the brain can not only create new connections but also reverse disease means exercising the brain has never been more important. For this reason, in recent years, there's been a reinvigorated focus on brain-training techniques that use thinking as a tool to expand cognitive function or reverse degradation.

THE RUBBER HAND ILLUSION

The rubber hand illusion (RHI) is an experiment anyone can do to understand how neuroplasticity works. It was first tested by researchers Matthew Botvinick and Jonathan Cohen in 1998. The basic premise is:

A human participant sits with his hands directly in front of his body with his hands stretched out on the surface of a table.

A blocking agent, such as a piece of wood or a large book, is used to separate the two hands and block the participant's visual field so he can only see one of his two hands.

A stuffed rubber glove that identically resembles the natural hand (which remains hidden) is placed in position on the table where the real hand would naturally fall.

The rubber glove is then stroked with a paintbrush. At the same time, a series of identical brushstrokes are administered to the real hand (the hand hidden from view).

After about 2 to 5 minutes the rubber hand starts to feel as if it is the real hand. When a hammer is used to try to hit the fake hand, the participant's typical reaction is to yelp and pull his true hand away in a panic (see Figure 5.1).

During the exercise, the brain is tricked into believing the rubber hand is the real hand. This process shows the brain's ability to adapt to environmental stimuli.

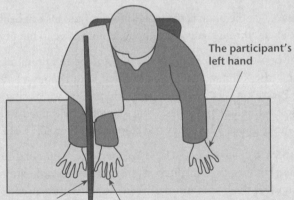

The participant's left hand

A barrier is placed between the fake right hand and the real right hand. It's used to block the real hand so the participant only sees the fake hand.

A fake hand that closely resembles the participant's real hand is placed where the right hand would normally sit.

AN EXERCISE IN BRAIN PLASTICITY

Figure 5.1 *The rubber hand illusion tricks your brain into thinking your hand is about to be whacked really hard by a hammer.*

(Illustration by Cornelia Svela.)

Do it Yourself (DIY) Brain Technology

Because the brain is plastic and not static, there are techniques anyone can use to prevent cognitive decline and to enhance intelligence and ability.

This is also why in the last ten years there's been an emphasis on developing new and less invasive technologies that improve brain chemistry. And it's an exciting development for anyone dealing with brain issues such as Alzheimer's disease, Parkinson's disease, and more commonly, the decline that naturally occurs with old age.

If you want to stay smart and perky without the need for brain chips, drugs, or surgery, then you will want to integrate the following practices into your daily ritual.

Go Learn Something

Whether you are conscious of it or not, your brain is always adapting. It creates new connections, prunes dead pathways, and it all happens naturally and without effort when you do one simple thing—learn.

When you practice a new activity, a new group of neurons chatter to each other. This is known as an electrochemical pathway. Over time, repetition of the same learned task or idea creates a stronger, faster connection. Externally, it looks like you are getting better at a new skill or task.

Neuroscientists describe it this way: "Cells that fire together, wire together." Over time, they communicate more efficiently. The more the same network is stimulated, the stronger it becomes.

Consider a chain of stock boys unloading a delivery truck at the local grocery store who are working as a team to get the job done. Angelo stands inside the truck. He gives a box of tomatoes to Evan, who passes the box to Stefan, who loads the box in the warehouse. By the time this team has moved the sixty-eighth tomato box they're likely to have sped up the time it takes for a box to move from the truck to the store. Actions among the team start to happen automatically. Everyone gets into a rhythm and they all know what to do.

Likewise, brain communication pathways work together to pass signals through, and with repetition they become more efficient and adept at delivering that signal. It's like memorizing a secret code and learning it so well you can immediately look at a line of apparent gibberish and immediately see the original message in all the mayhem.

Here's the kicker: When you stop practicing certain skills, the brain learns to eliminate the pathways it once created. The point of this pruning is to get rid of the unused pathways to make room for new ones. When this happens, you forget the secret code, and have to manually decrypt every message by hand, in a time-consuming fashion.

Do you remember what your family's phone number was when you were 6? If you haven't thought about it in a very long time it might have been erased by this natural process. However, recalling it every year (or more often) will keep it around.

A musician can easily relate to this process. A young piano player who practices the same piece of music repeatedly will quickly learn a pattern of moving their fingers across the keyboard. If they stop playing for years and later sit down at a keyboard, they'll still be able to play the piece but more slowly and it will involve greater concentration. If the same pianist were to relearn a piece through daily practice they could restore the connection fairly quickly because it's already been learned once.

So, continued learning is crucial for ensuring that the brain connections fire quickly and accurately. The brain—like an unemployed stock boy on the couch in his Calvins—gets lazy when it's not being used.

This process also explains how negative thinking patterns and behaviors get created and reinforced. The brain doesn't differentiate. It simply processes the information. It's always moving the individual toward pleasure and away from pain.

Consider the fundamental brain processes that lead to someone becoming an alcoholic. The first sip of a crantini sends a signal to the brain's reward center to stimulate the production of a happy mood chemical called dopamine. The brain likes dopamine because it produces a sensation that registers as "feeling good." Naturally, the brain wants more of the "feeling good," so it tells the body to tell the bartender to pour another tasty drink.

Communication pathways speed up and strengthen because repeated communication stimulates the growth of a natural insulator called myelin. It's a fatty, white material that grows around brain chemical pathways and helps facilitate connections.

This process happens a lot during childhood. Kids are little myelin factories. This is why they learn faster than adults. Researchers hope to use this knowledge to develop new technologies to help adults learn (once again) at the same rate as children.

The main takeaway here: Learn more. Learn often. Strengthen those connections to ward off brain disease. Or, take the advice of Kay's 91-year-old Oma: "Do more crossvort puzzles and vatch a lot of Y-eopardy." Crossword puzzles and *Jeopardy* keep your myelin growing!

TALK YOUR WAY INTO MENTAL HEALTH

In 1853, English psychiatrist Walter Cooper Dendy introduced a revolutionary noninvasive therapy that physicians could use to help treat patients with neurological disorders. He called it "psychotherapeia."

The technique involved challenging a patient's core beliefs using conversation to help them learn new, more proactive ways of behaving. His method is still in use today, but it goes by the name psychotherapy or talk therapy. And it is (thankfully!) much more evolved than it was in the nineteenth century.

Earlier forms involved hypnotism, and, in many cases, a great deal of—for lack of a better term—bullpoop. In the early twentieth century, Sigmund Freud emerged as a leader of psychoanalysis. But, his theories had no empirical underpinnings. He believed that dreams had meanings, and the human mind had three layers that influence behavior—id, superego, and ego. You can call this more evolved bullpoop. Yet his work was heralded as revolutionary and drastically influenced the emerging field of psychotherapy.

Early psychotherapists saw the value of talk therapy though they did not fully understand its mechanics. When scientists learned that the brain was plastic, the value of conversation became better understood. Learning to think new thoughts equals changes in brain chemistry internally, which results in new automatic behaviors developed externally.

Today, cognitive behavior therapy (CBT), is one of the most popular forms of noninvasive therapy used in psychotherapy. It's a training technique that allows patients to identify unhealthy behavioral patterns and shift them by reprogramming thoughts, which lead to a change in actions. CBT can also produce results for patients with mental illness. The tactics it teaches can be used by anyone who wants to shift negative thinking on-demand.

Meditate on This

One of the earliest types of brain technology is meditation. People have been practicing it for more than 2,500 years. Some of those people, such as Siddhartha Gautama, a Nepalese philosopher (also known as a guy called Buddha), sought a phenomenon called "enlightenment." In searching for that mystical thing—described as "liberation from human suffering"—their brain is doing something very scientific: Rewiring itself using the principles of neuroplasticity. This rewiring, as we have discussed, has many brain benefits.

An outsider might look upon a meditating human and regard them as somewhat silly, sitting there in silence, usually with legs crossed, and their body twisted like a salted pretzel. Practiced meditators sometimes use props such as silly looking angled cushions. Sometimes they sit still for hours on end.

However, meditation is an active process. During the meditative act, the state of attention is highly alert but relaxed. Objectivity is practiced by being aware and acknowledging what is happening—the sensations that are being experienced—without making any judgments about them. An individual learns to "be with what is" in reality, with indifference to it.

For instance, a meditator sitting on the hill who feels a brush of wind would think about the wind like this: "Oh, I feel a brush of wind". He or she would not engage in subjective thoughts such as, "it's pretty cold," or "why the heck am I even on this hill anyways when I could be cleaning the grout between my tiles in my kitchen?"

Surprisingly, this type of structured thinking has many brain benefits. A successful meditator (because it is possible to do it wrong) is simply teaching his or her brain to better register and control the information it is processing. Externally, this slows down the automatic responses that are learned and happening between the brain and body.

To better explain the distinction consider this comparison that follows.

Ned the nonmeditator gets rear-ended by an idiot driver. His immediate reaction is to spring out of the driver's seat of his car, stomp over to the idiot driver, and scream at him. Ned's reaction to bad things in life is to flare into a raging anger. His response is automatic. Sometimes it doesn't work in his favor, such as when he argues with his wife because she withdraws her wifely services—like lunch making—which he is rather fond of.

Now there's Molly the meditator. She's driving one day and she gets rear-ended by an idiot driver. While Molly used to react like Ned, she's been practicing 1 hour of meditation each day consistently for 2 years now. Her reaction is to feel the anger, understand it's there, and choose what she'd like to do with it. Her anger doesn't go away but she can better control her reaction.

That's why Ned got punched in the face by the idiot driver, who was 6 foot 4 inches and 300 pounds, while Molly serenely noticed the snake tattoo on his neck and quickly exchanged insurance information and got on with her day.

It's been suggested that meditators have a superior ability to regulate and control emotions. Meditation has also been linked to lowered stress, better sleep, improved attention and concentration, and greater levels of happiness.

In the twenty first century, studies to better understand the benefits of meditation are part of a growing sector in neurological research. Brain imaging tools, such as EEG and functional magnetic imaging (fMRI), are commonly used to watch activity in a meditator's brain during a session.

Spatial Resolution Doubles Each Year

Spatial resolution of MRI technology is doubling every year. That means the voxels (the 3D pixels that represent visual information) are getting smaller.

In 2015, a study from University of California, Los Angeles, reported that meditators who had practiced for more than 20 years had better brain health than nonmeditators of the same age.

In 2012, a team led by Dr. Judson Brewer of the Yale School of Medicine reported, that regularly practicing mindful meditation decreased activity in an area of the brain called the default mode network (DMN). Control of the DMN is linked to better control of thoughts, increased focus on the present moment, and potentially a greater happiness level.

And, in 2011, a team led by Sara Lazar at Harvard concluded that eight weeks of a specific form of meditation known as mindfulness-based stress reduction (MBSR), produced structural changes in key brain regions—the hippocampus and the amygdala—that led to a higher resilience to stress.

The only apparent downfall of meditation is that it takes time and effort. For an individual to reap the benefits, he or she must practice regularly for a minimum of 15 to 20 minutes per day, though more is recommended.

It might not be possible to fit that into a productive life between the (some would argue) more important stop for the Starbucks morning latte, the budget meeting, and the school pick-up of small, sniffly offspring. But, that's okay, there are new technologies providing the same effects.

HUMANS HAVE TWO BRAINS

The gut is the component in the human digestive track that allows the body to digest food, process it into waste, and eliminate it by sending it out one's backend.

In recent years, experts have learned that the gut works independently. It's the only organ that works without being controlled by the cranial brain. That means, it's also wired with a network of brain cells. It contains approximately 400 to 600 million neurons.

For this reason, the gut is commonly called the "second brain" or "gut brain." And it's spawned a new school of research called neurogastroenterology.

The gut brain manufactures approximately 30 neurotransmitters (brain chemicals). It talks to the cranial brain, by sending messages through a nerve at the base of the brain that extends to the abdomen, called the vagus nerve.

This means the gut brain sends signals to the cranial brain that affect feelings of sadness, stress, and that influence thinking processes. Both brains produce the happy-feeling chemical messenger serotonin. And get this: The gut produces 95 percent of all the serotonin in the body.

In other words, staying regular—having healthy poop habits—is a crucial component for positive mood. It might also explain why we get "gut" feelings sometimes.

Brain Fixers

Eat well. Learn. Meditate. These are good things to do for brain health, but none of these methods are foolproof ways for preventing neurological illness. These methods only help when the brain functions as a complete system. That is, these methods help if all the brain's mechanisms are working as they're supposed to.

Because scientists don't yet have a complete picture of all the inner-workings of the brain, little is known about what can be done to completely stave off brain illnesses such as Alzheimer's, schizophrenia, bipolar disorder, or major depression, to name a few nasties.

What is known about brain issues is that they all involve a combination of environmental, lifestyle, and genetic factors. Unfortunately, there's no specific formula that can predict the likelihood of brain disease or malfunction. At least not yet.

Medical conditions affecting the brain are classified as neurological disorders or mental illnesses. Included in this category are conditions involving damage to brain regions and their functions that are caused by deterioration or a genetic defect, or from a physical accident. These issues hamper an individual's ability to interpret his or her world, behave socially, or operate the body efficiently.

A brain that doesn't function as it should can have extremely negative consequences. The experience of life for the individual is altered. He or she can experience inability to be happy, or worse, become socially outcast, lose the ability to make a living and afford a home.

Consider what happened to famed railroad foreman Phineas Gage. In 1848, his head was impaled with a steel tamping iron while preparing a site for blasting. Improbably, he survived, and managed to live another 12 years following the accident, but he was forevermore a nasty jerk. The reason? The tamping iron punctured a brain region essential for social politeness and compassion. (Not coincidentally, this region does not fully develop until the age of four. This is why toddlers use the word "mine," stomp their feet, and have 3-minute meltdowns when their mothers don't feed them more chocolate milk.)

A brain that doesn't function as it should can be unpleasant to live with. It's why there's a need for new technologies to treat them. Prior to developing technologies such as drugs and electro-therapy, the initial ways of treating mentally ill patients were barbaric.

Gage's famous accident became an unfortunate treatment plan. It was called the lobotomy and involved a doctor putting a rod through a patient's skull to calm them down. Other early methods were throwing patients in ice baths, poking holes in their skulls, or inducing insulin comas to see if that would do anything useful. When nothing could be done, patients would go to a hospital and stay there until they died. These solutions treated the adverse behaviors, not the actual brain issue.

In a 2015 *TED Talk*, neuroscientist Greg Gage (no relation to Phineas from what we can tell) said, "One out of five of us—that's 20 percent of the entire world—will have a neurological disorder. And there are zero cures for these diseases."

 Zero Cures

See Greg Gage explain this in his TED Talk: http://superyou.link/greggage

He's right. There are still no cures for brain diseases. There are however, treatment options.

The two challenges to consider when creating technologies to treat brain illness include:

> "One out of five of us—that's 20 percent of the entire world—will have a neurological disorder."

- The brain is located under the skull, so to fix it, the method needs to be something that can get to the brain. It usually has to be as invasive as cutting open the skull and tinkering with the rather complex organ. And all the cells in the human brain (aka neurons) range in size from 4 microns (0.004 millimeter) to 100 microns (0.1 millimeter). Anything used to treat it or work on it needs to be small itself.

- Then there is this perplexing problem: We still don't fully understand how the brain works. So cutting it open and playing with the parts is not an ideal way of treating it.

That brings us to the nonsurgical treatments. Technologies such as brain drugs and tools that use electrical currents to rewire brain communication pathways are the other tools used to correct the brain when it's developed abnormal functions. These therapies are generally better than lobotomies because they often work fairly well. However, they don't always.

Brain-Zapping Fixes

If you ask your local tech geek, they will often give you this advice when dealing with a computer problem: "Turn it off and on again and see if that fixes the problem." It usually does for many common glitches.

When you reboot a computer, it temporarily stops all systems and then restarts them, clearing potential errors out of memory that might be causing the malfunction and resetting all processes. A similar process is used to treat brain issues. Consider that the brain is simply an electrical information processing system. To fix it, you want to reboot it to correct the issue. Here are some treatments that work on this principle.

Electroconvulsive Therapy

During an electroconvulsive treatment, a person receives a stream of electric currents to the head that go into the brain and induce a short seizure. The result is changes in the patient's brain chemistry that might reignite lagging or malfunctioning communication processes. It might sound like a crude treatment, but it is highly effective for some people. And the newer treatments use better equipment and technology, so it's much safer than it used to be.

In the late 1930s, early treatments involved extremely high doses of electricity that were applied without pain medications or anesthetics. This resulted in major issues such as fractured bones and memory loss. Not ideal.

Today, patients are given multiple treatments instead of one giant one. And often, treatment is administered on only one side of the head, which is called unilateral, instead of both sides simultaneously, which is called bilateral. The treatment is given under general anesthesia, so it is thankfully painless.

The patients are asleep for most of the treatment. When they wake up they might experience some side effects—headache, upset stomach, and muscle aches. Some patients experience minor memory loss. Most side effects are short-term and go away within a week or two. In most cases, unilateral electroconvulsive therapy (ECT) is safer and has less unpleasant side effects than the bilateral version.

The treatments are administered two to three times per week for three to four weeks. Most patients receive somewhere in the range of 6 to 12 treatments and the number of treatments depends on the severity of symptoms.

There are more targeted forms of electrical-stimulation treatment that send signals to specific brain regions rather than rebooting the entire system. Some scientists believe that focusing on a specific brain area is more effective and reduces the risk of side effects commonly associated with ECT.

These targeted forms are:

- **Vagus Nerve Stimulation (VNS)**—This technique is slightly more invasive than ECT, tDBS, and TMS (see descriptions that follow), as it requires a surgical implant under the skin. It sends electrical impulses through the vagus nerve, which carries messages from the brain to the body's major organs such as the heart, lungs, and intestines, and to areas of the brain that control mood, sleep, and other functions.

- **Transcranial deep brain stimulation (tDBS)**—A technique that involves sending constant and low amounts of electrical currents through electrodes on the scalp that target specific brain regions. The current induces flow to chemicals that either increase or decrease activity.

- **Transcranial magnetic stimulation (TMS)**—A procedure that uses magnetic fields to activate specific brain cells. The electric currents are delivered via a machine with a large electromagnetic coil that's placed above the scalp near the forehead.

WHEN THE CURE WAS WORSE THAN THE DISEASE

Early brain treatments are now considered inhumane. Based on the limited knowledge they had about the brain, physicians of that time believed these methods could help improve mental illness.

Exorcism and prayer: Mental illness was once understood to be demonic possession. Religious clerics performed rituals to rid the body of evil spirits.

Moral discipline: In the eighteenth century, brain illnesses were thought to be a moral issue. They were treated using hospitalization, isolation, and attempts to teach better social conduct.

Fever therapy: In the late 1800s, inducing a fever in a mentally ill patient was a common treatment. This technique originated from the Ancient Greeks, who observed that fever sometimes cured people of other symptoms.

Trepanation: The puncturing of holes in the skull using an auger, bore, or saw.

Lobotomy: A surgery to remove key brain regions linked to mood and reasoning—the prefrontal cortex and part of the frontal lobes.

Ice water and physical restraints: Before the connection between physical and behavioral issues of mental illness were discovered, treating the physical components was all physicians knew to control adverse symptoms. Patients were bathed in ice water or restrained in rooms, asylums, or straitjackets.

Insulin coma therapy: From the 1930s to the 1960s, it was believed that inducing a state of coma by lowering blood sugar levels could rewire the brain.

Metrazol therapy: The stimulant drug Metrazol was once used to induce seizures in patients with mental illness. Physicians believed conditions that brought on seizures, such as epilepsy, could not exist with mental illness. This technique was withdrawn by the Food and Drug Administration (FDA) in 1982. Metrazol therapy was the precursor to electroshock therapy.

Deep Brain Stimulation

A relatively new form of treatment that uses electrical currents to manipulate brain pathways is deep brain stimulation (DBS). It was first approved by the FDA in the late 1990s to treat tremors, and is now used to manage symptoms of Parkinson's disease, dystonia (a disorder of abnormal muscular tone), and obsessive-compulsive disorder. Most recently, it has gained attention as a potential treatment for chronic pain issues and mood disorders, such as major depression.

DBS requires brain surgery so that miniature electrodes can be inserted into specific regions of the brain (see Figure 5.2). The electrodes are connected with wires to a surgically implanted battery pack that's commonly placed near or under the collarbone. The treatment requires regular adjustments from a physician to tweak the level of stimulation to manage the patient's current symptoms. Patients are also given a device that gives them some control. However, there are risks with DBS, such as coma, brain hemorrhage, cerebral spinal fluid leakage, seizures, and paralysis.

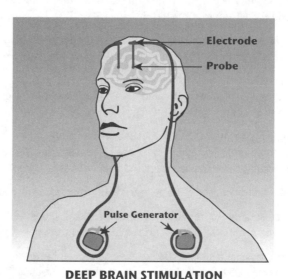

DEEP BRAIN STIMULATION

Figure 5.2 *Deep brain stimulation requires the surgical insertion of electrodes deep in the brain and is used to treat Parkinson's Disease and in the future, other neurological disorders. (Illustration by Cornelia Svela.)*

Consider this typical trajectory for a Parkinson's patient. He or she is prescribed what's known as a dopaminergic medication called Levodopa to manage the motor issues that come with the disease. It works for approximately five years, but then the symptoms start to reappear. More (or other) medications might be required to control the disease. But, with higher levels of the drug come new side effects. A common one for Parkinson's patients is dyskinesia, which causes involuntary motor movements. DBS treatment complements these drugs and gets rid of many of these terrible symptoms.

Dr. Helen Mayberg runs a research team at Emory University School of Medicine that has had a lot of success using DBS to treat severe cases of major depression.

In 2008 when trials began, 30 patients received implants. Mayberg's team continues to track the progress of 27 of those patients. She said "We have about a 75 percent sustained response rate now out to seven years, with continued stimulation."

The electrodes they insert target a specific brain region called Area 25. (Not to be confused with Area 51, where the government keeps, you know, captured aliens.)

Area 25 is rich with the mood chemical transporter serotonin that affects appetite and sleep. This area governs a number of key brain regions involved in processing memory and mood.

In a 2012 interview with CNN, Mayberg confessed, "To be brutally honest, we have no idea how this works." She acknowledged that more study is required to understand which variables are at play. "Maybe we are doing something wrong. Maybe the electrodes aren't positioned correctly. Or, maybe they are not the right patient. That means we've got to understand the biology better."

At this point Mayberg's trials are on hold while they consider using better biomarkers and improve targeting. This means that FDA approval for depression or other mood disorders using the technique won't come anytime soon. But the work continues.

That said, aside from Mayberg's project, there are new initiatives currently being developed with more sophisticated devices at the National Institutes of Health (NIH) and Defense Advanced Research Projects Agency (DARPA).

Brain Drugs

Pharmaceutical drugs are currently the number one treatment method used in Western medicine to treat neurological issues. Their major benefit is that they are less invasive than surgery.

Drugs alter brain communication pathways by changing the way chemical messengers, called neurotransmitters, talk to one another. There are approximately 100 of these neurotransmitters in the brain, though there might be more. Each neurotransmitter carries a unique message. A prescription drug acts by mimicking certain neurotransmitters to alter the body's behavior.

For a drug to get to the brain, it has to enter through the bloodstream and cross what's known as the blood-brain barrier (BBB). It is a semipermeable membrane, meaning it can selectively choose the substances it allows to pass through it. Scientists have yet to fully understand how it works.

What is known about the BBB is that it has three critical functions:

- It protects the brain from various substances in the blood that can injure it.

- It also acts as a gatekeeper, blocking hormones and neurotransmitters found in the body from affecting the brain.

- It keeps the brain safe in a protected environment.

Large molecules have trouble passing through the BBB, as do substances that are fat soluble. However, low lipid (fat) soluble molecules can pass through the BBB easily.

An over-the-counter drug, such as the pain reliever acetaminophen (more commonly known as the brand Tylenol in the United States) works by interfering with brain signals received by cells in the body alerting a person to the feeling of pain. The result: The message is not able to make it from the cells of the body to the brain. This keeps a person from feeling pain.

Because the brain controls all processes of the body, including itself, drug treatments are some of the most important medications used to treat many illnesses including those that affect the brain directly.

No New Brain Cells: Myth Busted

Many doctors practicing today learned in medical school that adults do not develop new brain cells. What you have is what you got. As we said earlier: Wrong!

The hippocampus, which is a structure deep in the brain, produces 700 new neurons per day in adults. This process is called neurogenesis.

"You might think this is not much, compared to the billions of neurons we have. But by the time we turn 50, we will have all exchanged the neurons we were born with in that structure with adult-born neurons," said neural stem cell researcher Sandrine Thuret of King's College in London, in her June 2015 *TED Talk*, also in London.

Curiously, Thuret said depressed people have lower levels of neurogenesis and that antidepressants increase neurogenesis. The great news is if doctors can control neurogenesis, then they could likely therapeutically improve or impact memory formation, mood, and even prevent neurological decline associated with aging or stress.

So what can be done to increase neurogenesis? See Table 5.1.

Table 5.1   Growing New Brain Cells

Increases neurogenesis	Decreases neurogenesis
• Engaging in learning	• Stress
• Having more sex	• Sleep deprivation
• Engaging in physical activity (see entry above)	• Aging
• 20 % to 30 % calorie restriction	• High saturated fat
• Spacing meals apart (intermittent fasting)	• Alcohol consumption
• Eating flavonoids (found in blueberries and dark chocolate)	• High sugar consumption
• Eating omega-3 fatty acids (present in fatty fish like salmon)	• Eating soft food
• Eating resveratrol (found in red grape skins and other foods)	
• Eating crunchy food	
• Folic acid	
• Curcumin	

Brain Enhancers

It's unfortunate your authors wrote this brain chapter as one of our last. Learning about nootropics might have served us in the delivery of this book. (Or, potentially, months in advance.)

The term "nootropic" is used to classify a group of over-the-counter substances that enhance brain function with few (and in some cases zero) negative side effects. They can increase the supply of certain brain chemicals that expand the brain's ability. They're used to improve cognitive functions like memory, mood, intelligence, attention, concentration, motor-control, and self-discipline.

There are five criteria that need to be met in order for a substance to fall into the category of a nootropic:

- It enhances one or more cognitive functions.

- It has few side effects and is virtually nontoxic.

- It enables firing mechanisms and facilitates communication between brain cells.

- It protects the brain from physical assault (such as injury or concussion) or chemical assaults.

- It assists the brain in functioning under disruptive conditions, such as hypoxia (low oxygen) and electroconvulsive shock.

Nootropics are not new. The term was coined in 1972 by Romanian psychologist and chemist Corneliu E. Giurgea, who once said, "Man will not wait passively for millions of years for evolution to offer him a better brain."

Within the last ten years, these cognitive enhancers have grown in popularity.

Americans are competing more than ever in career and school to get to the top, and specifically in regions where technology is being developed, such as in Silicon Valley. Career-obsessed developers are trying to "one-up" each other. Many young entrepreneurs use nootropics to hack their brain, reprogramming their biology to gain an edge.

Tim Ferriss, author of the *4-Hour Work Week*, told CNN in an interview, "Just like an Olympic athlete who is willing to do anything—even if it shortens their life by five years to win a gold medal—(young Silicon Valley entrepreneurs) are going to think about what (they) can take. The difference between losing and making a million dollars, or a billion dollars, is your brain."

Ferriss has admittedly used nootropic substances to increase his productivity levels. So has Dave Asprey, the creator of Bulletproof Coffee. Asprey is a famous biohacker most infamously known for buttering his coffee instead of his toast in the morning.

 Butter Your Coffee, Not Your Toast

Cloud computer pioneer Dave Asprey first created his Bulletproof Coffee recipe after visiting Nepal and drinking tea made with yak butter.

It's made of brewed coffee, unsalted butter, and a special blend of oils. And it's become all the rage in Silicon Valley because:

1. It can trigger weight loss by kicking fat burning into hyperdrive.

2. It stops hunger.

3. It gives you clarity of mind in the morning.

It also gives you a quick way to consume fats and 460 calories without eating carbohydrate-loaded breakfast food.

The recipe is found on www.bulletproofexec.com.

He also takes approximately 15 bio-enhancing drugs each morning, most of which are considered to be safe. However, he told CNN that even if he learns later there are long-term health issues, his "quality of life is so much better now than it was 10 years ago, that it's priceless."

Asprey told CNN he has done his research in biohacking. He's spent approximately $300,000 and more than 15 years learning what's effective and claims to have increased his IQ by 20 points.

An important distinction to make about nootropics is they enhance cognitive ability without (any known) detrimental side effects. So, while caffeine is cited by some as a nootropic, it's technically not.

Stimulants such as coffee that contain caffeine give you heightened energy, but come with adverse effects. Drinking ten cups of coffee, for example, would raise blood pressure levels and later lower mood and increase anxiety levels.

Prescription-only medications such as Modafinil and Adderall also are not nootropics. They are, however, commonly called "smart drugs" or cognitive enhancers. They are used to treat serious disorders such as attention-deficit/hyperactivity disorder (ADHD) and narcolepsy, but the number of off-label users is growing. These drugs are helping many people perform better at school and work.

In 2014, one popular study published by Kimberly R. Urban and Wen-Jun Gao, from the University of Delaware, reviewed the ramifications of a group of under-the-counter drugs (MPH, Modafinil, and Ampakine) that many young adults use as cognitive enhancers. Their study concludes there are "deeply concerning effects" related to reduction in the brain's plasticity in teenagers or young adults with immature brains. But they conclude their research notes by citing a need for "further exploration."

Considering taking a nootropic? Be sure to do your research beforehand. Take products that are pharmaceutical grade. And never overdose. Substances that boost brain power should be consumed with caution until more is known about their long-term impact.

> "Substances that boost brain power should be consumed with caution until more is known about their long-term impact."

Over-the-Counter Drugs to Improve Brain Plasticity

Dr. Takao Hensch, a professor of neurology at Harvard University, is developing a drug that reverts an adult brain back to its childlike state. When it comes to brain development, the period from birth to seven years old is one of the brain's critical learning periods. During this stage, the brain builds many pathways facilitating complex processes that involve seeing, hearing, touch, language, and speech, and higher cognitive functions such as reading, mathematics, and critical thinking.

Learning new skills is easier during this time than it is in adulthood due to the brain's need for stimulation and a greater level of plasticity. It's also a crucial time for building neural circuits that determine who we become later in life, and the formation of skills such as athletic ability and bilingualism.

Hensch and his team discovered the length of this heightened stage of learning and plasticity that occurs early in life is inhibited as a person ages.

There's an enzyme involved in the process that affects DNA by making it harder to switch genes on or off. By reversing histone deacetylase (HDAC) with a drug used to treat bipolar disorder known as Valporate, Hensch's team found it enabled the brain's early neuroplasticity levels to kick in again.

In 2012, the first trial involved a group of adult mice with a lazy eye condition known as amblyopia. Similar to humans, this condition can only be corrected during the critical period of brain development, and only if an eye patch is worn over the strong eye so the weaker eye is strengthened. The adult mice with amblyopia received Valproate. They learned to effectively use their weak eye.

A further trial was performed on 24 men with little-to-no musical training. The study involved a test for perfect pitch, a skill that can only be learned within the first six years of development.

Some men received a dose of Valproate and some of the men received a placebo. All the men watched a 10-minute video daily that taught them skills of perfect pitch for a week. When they were quizzed at the end of the week, men who took Valproate correctly identified 5.09 notes out of 18 on a test. The placebo group identified 3.5.

The results of the study are promising despite the small test group. The team is now working toward replicating the study with a larger group of participants. In addition, they are testing more drugs, including ones that involve different genes.

This Is Your Brain on Video Games

Using drugs to treat brain illnesses or to boost intelligence is less invasive than cutting open the skull and working on the brain, but drugs still alter brain chemistry using chemical substances. Also, some drugs have side effects that hamper other cognitive and motor functions. For example, some antidepressant medications to treat major depression cause erectile dysfunction.

Thankfully, there is perhaps an easier way to get smarter or to reverse cognitive issues: video games. Since the discovery of neuroplasticity, scientists have been focusing on developing new brain-training technologies, and there's been an explosion in brain games. Popular companies such as Luminosity, BrainHQ, and Happify sell games that help improve many areas of cognitive function.

The Gazzaley Lab, led by Dr. Adam Gazzaley, is a cognitive neuroscience research lab at the University of California, San Francisco, which is at the forefront of the brain game movement. Gazzaley and his team develop different types of video games with two key focuses:

- To treat brain illnesses and cognitive decline by correcting deficits existing in impaired populations. They work with all neurological disorders from ADHD to Alzheimer's, and major depression.

- To enhance the function in the brains of healthy individuals, and to optimize it as far as it can go.

Gazzaley calls it a different class of medicine. It's "what I would call 'digital medicine,'" he said in an interview with the website 52 Insights (www.52-insights.com).

The games developed by the Gazzaley Lab set themselves apart from other manufacturers' games because they are customized to the player's personal experience.

Gazzaley told us, "The video games we build are very immersive and engaging. They are built at a high level with professionals from the video game industry. We validate them in our lab with neural recordings to understand what their impact is on the brain."

 Brain Games

See Adam Gazzaley talk about video games as digital medicine:
http://superyou.link/gazzaleysgames

They use a mechanism he calls "adaptivity."

"During game play, real-time performance is being recorded and being used to guide the challenge of the game. It is scaled in an appropriate way to the user's ability. So, as they inch forward in terms of improving their performance, the game pushes them right to their maximum level," he explained.

The games detect when the challenge is too hard. "If it pushes too far it pulls back and finds the player's 'sweet spot.' It is completely individualized from the moment they start playing."

The game adjusts to the play so that it's not so hard that players will become frustrated and abandon the game, or so easy that they get bored. During the game, positive and negative feedback is also provided to help guide the user's understanding of the tactics in the game.

In 2013, the Gazzaley Lab announced success with the game *Neuroracer*. The premise is to navigate a car on windy roads and shoot colored signs. After a month of regularly playing the game, a group of older adults dealing with cognitive decline showed improved attention and memory.

The lab is also currently working on a rhythm game called *Rhythmicity*. On this project, he teamed up with musician Mickey Hart, drummer of the band Grateful Dead, to develop a game that teaches players to learn rhythm.

Gazzaley explained: "Our brains are rhythm machines. It's a core part of its function."

When people have clinical conditions there is also a timing and rhythmic dysfunction that occurs in the brain. So the hypothesis is, if we can make someone more rhythmic will we see a change in their cognitive function?"

The team is trying to crack the "rhythm genome," so they can use this information to develop new technologies and to learn more about the brain.

Along with the *Rhythmicity*, the lab is building four new games aimed at solving various neurological issues. As we go to press in mid 2016, these games still haven't been released to the public. However, another Gazzaley-founded company, Akili Interactive Labs, has raised more than $30 million for further clinical development and building a commercial infrastructure, with the hopes of being on the market in 2017. In application, these tools will help improve and restore cognitive capability across all neurological issues.

When the newest games are ready, Gazzaley admittedly will be testing some of his own technology. "I hope that [...] I'll start getting some benefits from the things we are building. I hope there will be cognitive enhancements I can document."

Gyms of the future might have to expand themselves to include video games so people can work out their brain after they finish pumping iron. This field is exciting and we expect it to grow.

More Super Cool Brain Projects

In our research for this book, we spoke to a lot of scientists about a lot of cool projects. Here are a few that got our attention, in a kind of wildly distracted "was that a duck reciting the Gettysburg address?" kind of way.

However, you will see tons of crazy new projects referenced in the media and on the Web in the coming years as researchers continue to explore the brain.

Brain-Controlled Exoskeleton

If it is possible to connect a machine to a human brain, why not connect a human brain to a machine? Brilliant idea, right? It's recently been done and it involves World Cup level sports. (And you thought this chapter couldn't get any better.)

At the 2014 World Cup (for soccer, or futbal, as some nations refer to it), a man named Juliano Pinto kicked out the first soccer ball at the Corinthians Arena in São Paulo. Considering Pinto was a 29-year-old paraplegic, this was super cool.

Brazilian neuroscientist Miguel Nicolelis of Duke University, and a team of 150 researchers worked for more than a decade to develop the brain-machine inter-face that made this possible for Pinto. The robotic exoskeleton uses brain signals translated into commands the machine understands so it can bend it like Beckham (English people keep reading; Americans please see factoid).

 Bend It Like Who?

David Beckham is a famous English soccer player known for kicking the ball in a way that curved around a defending player, which fans referred to it as "Bend it like Beckham."

Gordon Cheng of the Technical University of Munich led the development of the exoskeleton, which included artificial skin with printed circuit boards. Each board contained pressure, speed, and temperature sensors. Tactile sensors on the robot feet transmit signals to a device on the patient's arm. After some practice, the brain learns to recognize the arm vibration as associated with leg movements. But how does the exoskeleton recognize what the brain wants?

Early in the research, Nicolelis and his team identified brain signals through recordings of movement, which sounded almost like radio static. Nicolelis said his son calls it the sound of popping popcorn while listening to a badly tuned AM radio station.

"We recorded more than 100 brain cells simultaneously. We could measure the electrical sparks of one hundred cells in the same animal," he said in a 2012 *TEDMed* talk, adding, "We got a little snippet of a thought, and we could see it in front of us."

The find prompted Nicolelis and his team to proceed with experiments involving a monkey named Aurora, which played a video game.

This is getting weird right? Stick with it, science fans, it gets cooler. Aurora was required to move a cursor to a target on a video screen using a simple joystick. She learned to hit the target successfully 97 percent of the time in a United States lab. The movements were mapped from Aurora's brain activity and transmitted to a robotic arm in Japan, which was playing the same game.

Eventually the monkey learned to play the game and operate the robotic arm with its brain alone. And the research progressed from there, to the day when Pinto kicked the ball with the engineered exoskeleton at the World Cup.

 Bend It Like Juliano

Paraplegic Juliano Pinto talks about kicking a ball wearing a robotic exoskeleton in this video: http://superyou.link/julianoskick

The only really poopy thing about all this wicked science and research effort is that it didn't draw a lot of well-deserved attention, likely because it was poorly hyped at the World Cup event. The breakthrough is enormous, and c'mon, anyone who can get an American monkey to play a video game in Japan by just thinking about it should pretty much get a Nobel Prize.

OpenWorm Project

Earlier, we talked about the tiny *C. elegans* worm. With only 302 neurons, it is the only brain that scientists have mapped.

OpenWorm is an international research project that's created a simulation engine that uses computational models of the worm in their research. The goal is to build a simulation of the worm's nervous system. Its aim is to help science make comparisons and answer fundamental questions about the human brain by learning about the worm and using it as a model.

In 2014, Timothy Busbice, a programmer involved in the OpenWorm project, was able to build a Lego robot that was controlled by the worm's brain. Prior to that point, Busbice had spent more than 20 years trying to create a connectome that would use individual programs to represent each neuron and mimic their unique set of functions in a computer.

Initial attempts failed due to the "limitations of the machines." On a 32-bit computer, there were too few processes. He overcame this challenge by using a 64-bit machine to support a system of programs that matched the 302 neurons in the *C. elegans* brain.

 Duck Deliveries

Most computers these days are 64-bit processors, which means bigger chunks of data move across the chip. Imagine a truck hauling 64 rubber duckies down a highway compared to a smaller truck hauling 32 rubber duckies at a time. Fill the highway with 64-bit trucks and that's a lot more duckies being delivered.

Busbice discovered he was able to turn a LEGO robot into a simulated worm. He was fascinated by "the recursiveness of the neurons." He discovered that neurons communicate by looping back to one another.

"It's not that you push a button and neuron a talks to b, c, d and then it stops. That's not how the nervous system works. The output is continuous. They all come back and activate (the other) neurons again and again," said Busbice. "My hypothesis is that this recursiveness in our neural networks dictates our behavior."

Busbice's project is a great example of a bit of wild science. Emulating animal brains in bots to see what the outcome will be can teach us a lot about how the brain, and by extension the human brain, works. Researchers don't always know where they'll end up, which we imagine is part of the thrill of being a scientist that works with artificial worm brains. Plus, how cool is it to kiss your spouse goodbye in the morning and say, "See you later my little love muffin. I'm off to work to see if I can get an artificial worm brain to drive a LEGO robot."

 Wormy Robot

See a robot that's run by a worm brain: http://superyou.link/wormbrain.

Human Brains in Bots

These days, Kevin Warwick's team, among other cool stuff, conducts research with brains in bots. Early iterations on their project involved culturing miniature brains from rat neurons and using them to control small robots.

"We get rat embryos, take part of the cortex out of them, separate the neurons with different enzymes and lay them out in a small dish that has electrodes on it. Then, we put them in the incubator where they live and grow," he said.

Growing the rat brains allows the researchers to gain insights on how they develop. Then, when they're deemed ready, they are placed into what Warwick describes as "a little robot with ultrasonic sensors on it."

The rat brain is connected to its machine body via a Bluetooth connection. Sensory signals from the robot stimulate the brain. Output from the brain stimulates actions in the robot. This connection allows the researchers to understand how the brain receives information from the environment and processes that information to produce behaviors.

"We learn simple things, like how the robot learns to move around and not bump into objects. We learn about the brain from what it's doing as it's going through the learning procedures."

These experiments teach researchers more about the brain so the information can be used to advance brain science. More specifically, this information can be used to treat disabilities. For example, one of Warwick's team's objectives is to learn how memories are created. This understanding will help develop cures for diseases like Alzheimer's, where memories are lost over time.

The team has since moved into the process of using human brains. "We're culturing human brain tissue," said Warwick. "With rat neurons, we feel we get good results. But certainly, the surgeons we work with (say) rat brains are so unlike humans so the results don't really map that easily."

Brain-to-Brain Communication

Miguel Nicolelis and his team from Duke University have worked on a number of significant projects involving brain-to-brain communication. That is, the wiring of brains together, using a form of technology to connect them, or what's known as a brain-to-brain interface.

In 2013, his team linked two rat brains across the world—one in Durham, North Carolina and one in Natal, Brazil. They had previously learned to connect brain to a machine. The goal was to determine if they could connect brain-to-brain through a machine to facilitate an information exchange.

Each rat was first taught a simple task: To press a lever for food when they were signaled by a light flashing in their cage.

Rat 1 functioned as an "encoder" rat. He was shown the signal (the light) and once he pressed the correct lever for food, his brain signal transmitted to Rat 2 (who was not nicknamed Ratatouille, by the way).

Rat 2 was the "decoder" rat. He received no flashing light but was still expected to press the lever when he got the information that Rat 1 had seen the signal. And this is what he did.

The pair worked in collaboration 70 percent of the time. When the decoder rat made a mistake, the encoder rat also adjusted for the behavior. Nicolelis's team discovered that brains can communicate with one another through a network and are not limited by physical location.

In 2013, he told KurzweilAI.net that more research like this could lead to a new field he calls "neurophysiology of social interaction." "To understand social interaction, we could record from animal brains while they are socializing and analyze how their brains adapt—for example when a new member of the colony is introduced," he said.

In 2015, his team used the combined thinking power of three monkeys to successfully move a mechanical arm. Each monkey was individually taught how to move a virtual arm in a 3D space by picturing the motion in their minds. Then the team of monkeys shared control of the arm. Together, they collaborated to adjust movement and speed so they could collectively grab the digital ball. The old adage "two brains are better than one" proves true, at least in monkeys. This increased brain power has an application for expanding intelligence in humans. The coming work will be to connect humans together.

Human-to-Human Brain Communication

In 2002, Kevin Warwick had already connected human brains, but only because he and his wife were willing to use themselves in place of the monkeys. He wired his nervous system to his wife's so that the couple could communicate through thought.

For this project, he had circuitry surgically implanted to a nerve in his arm. Brain-Gate, as it is called, is technology that uses electrical signals to move robotic devices. It is designed to improve cognitive or motor function in disabled individuals. Using BrainGate, an individual can move a limb just by thinking that they'd like to do it.

With the BrainGate implant in Warwick's arm, and two electrodes placed in his wife's nervous system, they were linked. Warwick explained: "When my wife closed her hand, my brain received a pulse. So if she tapped two or three times, my brain received two or three pulses."

The communication between the couple was not complex, though it was effective. Warwick's wife couldn't tell him to stop at the grocery store to pick up a carton of milk on his way home, but she could tap her hand and he would know she was thinking about him.

"Think about a telegraph system, we were sending telegraphic signals, but it was directly from nervous system to nervous system," he said.

"Sam Morse said when he was first coming up with Morse code, he was referring to signaling from brain-to-brain communication and he achieved 99 percent of that, except for this interface issue—how to get signals from the brain back to the wires and back again at the other end. So all we were doing is figuring out the one percent that Morse didn't do."

Signals were sent wirelessly using an Internet connection from lab-to-lab. Part of the reason for the experiment was to show how easy it was to connect nervous systems. But the team quickly realized that meant they could do much more with those captured signals—once they were no longer restricted to the body where they originated—such as send them vast distances over the Internet. Just like a Kim Kardashian selfie.

Warwick predicts that the next ten years will show major advances in thought communication technology.

"For me, thought communication in some basic form will be enormous. Just as the telephone has been enormous. Though, the telephone will only be a tenth of what thought communication will be like. I think the first experiments will have happened. How long until it's a commercial success? That, I don't know."

Unsolved Mysteries of the Brain

As you can see, neuroscientists have made great strides in neuroscience in 5,000 years, especially in the last couple of decades. However, the reality is they know very little about the brain and how it works relative to what they would like to know.

There are some significant mysteries that remain unsolved, and many neuroscientists believe that the answers to these questions will likely remain very elusive for quite some time—that is, unless we start to see improvements in this area comparable to exponential rates of improvement found in consumer technology. We think the pioneering work being done this decade in neuroscience will pay off in the next decade. Remember, significant breakthroughs accelerate the next set of breakthroughs.

Here's a summary of some of what we don't know, and what the issues are.

What Is Consciousness?

This is probably the greatest of unsolved mysteries in neuroscience. Humans are conscious but we have no idea *how* or *why* we got the way we are. Nevertheless, there are two opposing camps on this.

- There are scientists who believe that consciousness can be solved and will be explained once we better understand the mechanics of how the brain works.
- The opposing group believes that consciousness is a law of the universe, like gravity, that can be understood but never explained.

Camp one aims to answer questions, such as:

- Does consciousness arise when a number of neural connections is reached?
- Or, does it originate from a specific brain region?

Perhaps, as quantum mechanics theory suggests, it comes from micro molecules inside the brain that work together as computing elements.

Why Do We Sleep?

There are still many unanswered questions about sleep. What scientists do understand is that sleep helps humans better perform. It aids man's ability to process information. It's essential for the body's repair and rejuvenation process. Humans can't live without sleep. A complete lack of sleep will kill you. A rare genetic disorder known as fatal familial insomnia (FFI), a condition where your brain can't go into a resting state, demonstrates this. Anyone who develops this condition suffers a dramatic breakdown of cognitive processes. It starts with headaches, moves to panic attacks and hallucinations, then to mental retardation, and ultimately ends in death, 18 months after the initial onset of symptoms. The average age of those afflicted is 50, although the age range is 18 to 60.

The bottom line is: Sleep is crucial and life giving. And humans seem to do it in the most complex way. We go through five stages of sleep all with different important characteristics. Animals vary in their sleeping habits. The giraffe, for example, can go weeks without it. The bat or sloth can sleep for an entire day. Dolphins sleep while partially awake and with one-half of their brain operating, which allows them to always remain awake to ensure their survival.

Animals with a higher aptitude for intelligence, such as all mammals, spend more time in a stage of sleep called REM, rapid eye movement. Humans spend the most time in REM. Babies spend 50 percent of sleep in REM, while adults spend 20 percent. So, obviously REM is important, too. But why?

Well, perhaps, it's because dreams occur during REM. This opens up another question? What are dreams? Why do we need them? We don't know.

Do We Have Free Will?

This question circles back to the debate of consciousness. Are humans complex machines run entirely by their brains? Or, is there something else at play—something that is yet to be understood, or perhaps, man will never truly understand?

When your spouse asks you what you want for dinner, what are the thoughts you process to arrive at an answer? Is your brain processing information? Does it simply retrieve and collect memories that ignite your tastebuds and tell you to ask for sushi? Or, are the thoughts you process a convention of your "mind." More simply put: When you think about your thinking are you an active agent? Or, do you just think you are, grasshopper?

Yikes.

Two schools of thought on free will are:

- **Immaterialists** are scientists who believe in free will. They suggest there is a component of active choice in the decision-making process. A neuroscientist, for example, using a brain-imaging technique to read neuronal activity will observe the thinking process in a patient, but they can never know (yet!) what that person is thinking about. This suggests personal choice is being made internally. Immaterialists believe that psychology can never be understood.

- **Materialists**, on the other hand, believe free will is merely an illusion. The brain controls everything. The thoughts the brain produces are merely based on what a person has learned from the past and their inferences about a given moment in time.

Research suggests there might also be a genetic component involved in free will. In 2013, Eeske Van Roekel's team from the University of Groningen, Netherlands, discovered that a specific oxytocin receptor gene predispose a group of girls to intense feelings of loneliness when in the presence of judgmental friends. So, free will might also hinge on your biology and genetic makeup.

How Are Memories Processed?

Memories serve a crucial role in how each individual understands and engages with his and her world. Consider that from the moment a baby is born it starts learning by connecting with its environment. Its brain makes decisions on how to encode, store, and retain new information by measuring it against how safe it is.

After all, it is the brain's ultimate job to keep the organism surviving. So memory plays a vital role in keeping man safe from perceived threats in his environment.

A boy who gets bitten by a dog, for instance, might learn "dogs aren't safe." He'll catalogue the event in his biological filing system. At a later date, it will be retrieved when he sees another dog as a signal from the brain to say "remember what that other dog did to you? Watch out!" This also means he can start to relate to himself as someone who is "not a dog person." So memory is also a factor in how a person sees and describes themselves, their personality.

There are important things we know about memory. We know some of the functions it serves, as noted above. Scientists also understand many of the brain regions—like the hippocampus—that are involved in memory processing. There are different types of memory—declarative, nondeclarative (muscle memory), short-term, and long-term—that get stored in different ways.

But storing memories is incredibly complex. Neuroscientists do not yet know where a memory gets stored and how memory recall works. It's also unknown how the entire system works together. For example, a person driving a car uses memory recall to control motor skills and to remember where they are going. Scientists have figured out pieces of the puzzle, but there's still a lot to be learned to get the complete picture.

Is that It?

As you'd imagine these aren't all the unanswered questions that neuroscientists are noodling around their heads. There are dozens of big questions. And some are much more complex than most of us can understand. Many of them relate to how the brain's structures work and work together.

If you are curious, do a Google search on "top unanswered questions of neuroscience" to get a sense of how big the task is that neuroscientists face in understanding the human brain.

Discover magazine quotes 23 of them from California Institute for Technology neuroscientist Ralph Adolphs, who wrote about them for *Trends in Cognitive Science* in April 2015. Here's a sampler that Adolphs believes might be solved in the next 50 years:

- How do circuits of neurons compute?
- What is the complete connectome of the mouse brain (70,000,000 neurons)?
- How can we image a live mouse brain at cellular and millisecond resolution?
- What causes neurological illness?
- How do learning and memory work?
- Why do we sleep and dream?
- How do we make decisions?
- How does the brain represent abstract ideas?

With this part of science being helped along by ever-accelerating improvement of technology, we think they will get there faster than they think. If most of the breakthroughs in neuroscience from the last 5,000 years were achieved in the last 20 or 30 years, the next decade is going to be very, very surprising to a lot of researchers.

The Future

Ok, so let's summarize our progress around the brain before we leap into some creepy stuff about the future that (spoiler alert) might have you send your child to the liquor cabinet to get some special nerve medicine for mommy (or daddy).

> **Lesson 1:** It took a very long time to figure out pretty much anything useful about how the brain sort of, kind of works. But, if we are really honest, we pretty much know squat about it.

> **Lesson 2:** Most of the big progress in neuroscience has happened recently—like in the last 20 or 30 years.

> **Lesson 3:** We've figured out worm brains and done some cool experiments.

> **Lesson 4:** Monkeys playing video games + some very smart scientists × ten years = Paraplegics that can play soccer.

> **Lesson 5:** Computers are still pretty dumb and unconscious.

> **Lesson 6:** There's a crap-load of stuff we don't know about the brain, but...

... Now, we're trying to build super smart computers that might suddenly become conscious and take over the planet and enslave us all to their evil bidding. Or so some smart rich guys say. Though our friend Mr. Kurzweil and other uber-smart scientists say don't worry about that. Seems about right? Good, let's do this future thing.

Expanding Human Intelligence

About 200 million years ago humans grew a neocortex. It was a new layer in the brain that provided early humans with the ability to develop technology in order to build and optimize the world around them.

All mammals have a neocortex, which allows for higher-level functions including sensory perception, generation of motor commands, and spatial reasoning. In humans, it also handles language and the capacity for art, music, and humor (yes, blame fart jokes on some of the more vulgar neocortexes out there). That amazing brain development was a quantum leap in the evolution of mankind (see Figure 5.3).

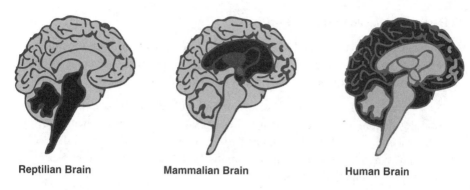

Reptilian Brain   **Mammalian Brain**   **Human Brain**

Figure 5.3 *The evolving human brain.*
(Illustration by Cornelia Svela.)

This leap forward is about to happen again, at least if Google has anything to say about it. As we discussed in Chapter 1, "The Emergence of (You) the Human Machine," natural evolution can take generations to enhance a human biologically, but technology can accelerate human abilities much faster. And perhaps technology is the new accelerated evolution.

The famed search engine company is hard at work building a synthetic neocortex, with Ray Kurzweil heading the project. Besides being an inventor and futurist, he is also a Director of Technology at Google.

"We're simulating how the neocortex works and developing mathematical models of it. If we can develop them through computers and develop a synthetic neocortex, then we can do the same kinds of things the neocortex does. For example, human language."

This amazing resource won't just be an in-house Google tool. Anyone will be able to tap into it just by thinking. That's because by the mid-2030s—in 20 years or so—there will be thousands of nanomachines swimming around in our bodies, doing cleanup work, keeping us healthy, but also connecting our brains to the Internet, including resources such as Google's synthetic neocortex.

 You Say Nanobot, I Say Nanite?

You have likely heard different words for these tiny robots that will swim around in our blood stream one day. Some call them nanobots, but they are variously referred to as "nanoids," "nanites", "nanomachines," "nanocytes," or even "nanomites."

Nanobot is the term for tiny robots that provide all kinds of services at the nano-level. Nanocyte refers to a nano-sized cell, as "-cyte" is latin for cell. Nanites might be the closest correct term used here.

Here's how it could work, said Kurzweil:

"Someone is approaching and I need to say something clever and my 300 million pattern recognition nodules (in my organic brain) are not going to cut it and I need a billion more," he explained, "for two seconds, I could access that in the Cloud."

This boosted brain power will massively expand our natural ability to process information. In the next 15 years, Kurzweil says, "we'll become a hybrid of biological and nonbiological thinking. My model for that is not so much that we would put synthetic neocortex into our brains. We'll basically put communication out from our neocortex into the Cloud."

The "cloud" is geekspeak for the Internet, or rather the computation power it has to provide services that run on the Internet instead of locally on your computer. Dropbox, Netflix, Flickr, and Google Drive are all examples of cloud services.

This new resource will take human evolution to the next level, he said. "What happened two million years ago allowed us to take a qualitative leap. And this will allow us to take another qualitative leap."

Some people fear this technology will make us less human and more robot. One scientist concerned about this is Miguel Nicolelis, head of neuroengineering at Duke University. (He was also the researcher we mentioned earlier who connected the paraplegic man to an exoskeleton.)

"I think we're facing a big danger—if we keep relying so much on computers, we will begin to resemble our machines," he warned in an interview with *The WorldPost*, a *Huffington Post* project. "Our brains will assimilate to the way computers operate, causing a significant reduction in the range of behaviors that we normally produce."

Kurzweil argues that such a resource will make us more human. "Evolution creates structures and patterns that over time are more complicated, more knowledgeable, more creative, more capable of expressing higher sentiments, like being loving," he said in his *WorldPost* interview. "It's moving in the direction of qualities that God is described as having without limit.

"So as we evolve, we become closer to God. Evolution is a spiritual process. There is beauty and love and creativity and intelligence in the world—it all comes from the neocortex. So we're going to expand the brain's neocortex and become more godlike."

Of course there are some obstacles to overcome to get there. Technologically, two decades should be sufficient to develop the nanites. However, as one pundit pointed out, it's the FDA that will have to approve the availability of medical brain enhancing nano devices, and that might take longer than a couple of decades.

The other issue is these nanites need to be made of something our bodies will not reject. Tiny bits of nano-metal that interface us with the Cloud and float around in our bloodstream will need to interact with our biology in a way that is inert. You can't have your immune system trying to eliminate them.

One solution to this might be to engineer a form of disabled virus made from our own tissue. Or generated from our own stem cells. If it's made from us, our systems won't reject it. In a way, vaccines are the beginnings of this kind of technology. They are actual disease agents injected into us that have been engineered to not cause the disease and be recognized by the body so that it can learn to stomp them out if the real thing comes along.

Viruses, and even cancer cells, have the capability to interface with our cells to do bad things to us. But what if we can engineer these to do good things for us? That seems to be inevitably the future of nanite technology, at least from where we sit right now.

> " ... you can never see the actual future from the point of view of the soon-to-be primitive past."

Of course you can never see the actual future from the point of view of the soon-to-be primitive past. So while we don't plan to miss the mark on this one, the actual outcome might look very different.

Here's how that might play out. If you recall, *The Jetsons* was a TV show that started airing on Sunday nights in 1964 and was a prime time cartoon (not a kids show). It was designed to show the future in 100 years (2064). The Jetsons has been a remarkable forecaster of that future with one massive exception that was completely missed.

If you watch one of the initial 24 episodes, it failed to grasp the impact of digital technologies. All the inventions it illustrated were largely analog.

And so it might be with the future we are predicting here. When we say nanites will swim around our brains and connect us to the cloud to expand our intelligence that sounds possible. The outcome is highly likely, but the execution might look very different.

Artificial Intelligence (AI)

In a 2014 interview with BBC, Kevin Warwick, professor of cybernetics at Reading University (we mentioned him earlier in this chapter) said, "In the field of Artificial Intelligence there is no more iconic and controversial milestone than the Turing Test."

The Turing Test was developed by Alan Turing, a British computer scientist, who is most famously known for leading a team that cracked the Enigma machine during World War II, allowing Britain to intercept encrypted communication between Nazi forces and their leadership. It ultimately helped the Allies win the war.

Turing proposed that by the year 2000, a computer would have the intellectual capability to fool humans into believing that it was real at least 30 percent of the time. His famous Turing Test evaluates a machine's capability to match or supersede the intelligence level of humans.

During the test, a human judge assesses an exchange of natural language occurring between a human and either a human or a machine. The conversation is conducted through a computer keyboard typing channel so that the machine's capability to speak is not necessary.

This was the scenario that took place at the Royal Society in London in June of 2014. A group of humans sat in one room, each at their own computer. The participants engaged in a text-only conversation with one another or with someone who couldn't be seen in the room across the hall. In other words, they were unsure whether they were interacting with a human, or a "chatterbot," a computer program engineered to interact with humans by simulating a conversation. A panel of judges evaluated the conversations to see if they could decipher whether the human was talking to a man or to a machine.

The result: 33 percent of the judges were tricked by a computer masquerading as a 13-year old boy from Ukraine named Eugene Goostman. Chatterbot Eugene Goostman was developed in Saint Petersburg, Russia, by three programmers, and he was the first machine to pass the legendary Turing Test. It was a landmark moment in AI. Though to be fair, it doesn't mean that "Eugene" was actually conscious and capable of independent thought. It only proves that its algorithms were convincing enough to fool people on the other end. Eugene's broken English probably helped mask any deficiencies, too.

"Some will claim that the Test has already been passed," said Warwick. "The words Turing Test have been applied to similar competitions around the world. However, this event involved more simultaneous comparison tests than ever before, was independently verified and, crucially, the conversations were unrestricted. A true Turing Test does not set the questions or topics prior to the conversation."

Warwick said Eugene Goostman had a broader ability to answer questions than other computers that have shown an aptitude for intelligence.

Watson, IBM's supercomputer, is famously known for beating a pair of humans at a game of *Jeopardy!* in 2011. Watson is a cognitive technology, a natural extension of what a human is capable of doing when it comes to interpreting and responding to questions expressed in natural language.

Watson is more advanced than other artificially intelligent computers. Like other systems, when Watson is asked a question, it goes into a retrieval process to come up with an answer. The retrieval process involves interpretation.

Ray Kurzweil said that Watson's intelligence level shows great promise for the field for AI: "The interface *Jeopardy!* queries has all these subtle forms of language including riddles and metaphors and jokes. Watson got this right: 'A long frothy speech delivered by a pie topping' and it quickly responded 'What is a: Meringue Harangue?'"

Another bot that is showing great promise is Bina48, a project developed in partnership between LifeNaut and the inventor of Sirius XM Radio, Martine Rothblatt and her wife, Bina Rothblatt.

Bina48 is a mechanical clone of Bina Rothblatt (at least her head and shoulders) and was created using video interviews, laser scanning, face recognition, and voice recognition to be a complete imitation of the real Bina.

Bina48 is an android that uses information she processes and stores about the real life Bina Rothblatt. She draws her knowledge from the real Bina's "mind file"—an uploaded collection of her beliefs, memories, and personality traits.

However her inventors say she is learning and growing at an exponential rate, and will eventually evolve into something beyond the original Bina.

The android is an early demonstration of the Terasem Hypothesis, which is defined as "a conscious analog of a person that is created by combining sufficiently detailed data about the person (a mindfile) using future consciousness software (mindware)."

 File Your Mind

You can create your own Mind File on this site: http://superyou.link/makeamindfile

 Domo Arigato Bina Roboto

See Bina interviewed by a musician: http://superyou.link/sheisbina

Will the Machines Rise Up?

Kurzweil says he's often asked questions such as, "Will computers like us? and "Will they want to keep us around?" He always reassures them that "that's up to us."

"We create these things," Kurzweil said. "They are part of human civilization. They are part of humanity. Man couldn't reach the food on the highest branch thousands of years ago, so we invented tools to expand our physical reach. Now those physical tools allow us to build skyscrapers."

Kurzweil says, "It's part of who we are. It's not us versus machines. We will become hybrid and ultimately the nonbiological portion will dominate."

Zoltan Istvan, leader of the Transhumanist Party, sees the future of artificial intelligence in much the same way .

"The likelihood of a *Terminator* scenario is pretty Hollywoodish," Istvan said. "The much greater likelihood is that we'll have a society that interacts with robots and uses artificial intelligence all the time."

When Kurzweil spoke at the 2015 Exponential Finance conference hosted by Singularity University and CNBC he said this to counteract the worry: "We have these emerging existential risks and we also have emerging, so far, effective ways of dealing with them."

He suggests the concerns about AI will cease to exist when the positive benefits are shown and people gain more confidence with the new technologies. He also makes the argument that "[man's] always used technology to ... transcend our limitations, and there's always been dangers."

In 2014, Neil Jacobstein, Singularity University's cochair in AI and Robotics, spoke at Summit Europe of the incredible good AI will bring to the world. Super-intelligent computers will be able to assist man in solving the most complex questions, from illnesses to climate change.

At the same time, he cautioned: "These new systems will not think like we do. We'll have to exercise some control." Man will always have a moral responsibility to maintain control over these systems. Efforts to do so will likely require rigorous lab tests and programming in controls to control machine behavior. Of a future with AI, Jacobstein said, "We have a very promising future ahead ... build the future boldly, but do it responsibly."

An even bigger concern, according to Istvan, is "who gets to artificial intelligence first." He suggests this could change the power dynamic of the entire world. We've explored this discussion in greater detail in the final chapter of this book, "Human 2.0: The Future Is You."

... And What About Conscious Robots? "GULP"

What concerns many people about AI is the unknown variable of consciousness. Put more specifically—Will robots that are as smart as humans be conscious? How will we know they are conscious given that consciousness is a subjective experience?

The answer to this question requires us to first determine what consciousness is. As we mentioned earlier, it is one of the top unanswered questions in neuroscience. Will we ever understand what it is and the mechanisms that create it?

Some say yes. Some say no.

We say: Of course. Exponential growth in computing power is going to make understanding it inevitable. Or, at the least, we'll be able conclude that we'll never truly understand it and accept it as a law, such as gravity.

The mystery neuroscientists are dealing with is where in the brain does human consciousness originate? One theory is that consciousness occurs when we register, collect, calculate, and assess our experiences and our memories in a continuous mode that follows us through life. There are all these sensory inputs that are combined by the brain to provide each one of us with a highly personal and subjective experience of the world.

We know consciousness exists because we each have an experience of our own. And we can observe others who say they can recognize their own, although we can't experience someone else's consciousness or say whether another person is conscious.

In his 2014 *TED Talk*, Philosopher David Chambers, a professor of New York University said, "Our consciousness is a fundamental aspect of our existence … there's nothing we know about more directly … but at the same time it's the most mysterious phenomenon in the universe."

Up until recently, there has been very little scientific work on human consciousness. Human behavior can be studied objectively and neuroscientists have studied and continue to study the brain objectively. But the human consciousness remains uncharted.

Approximately two decades ago, neuroscientists such as Francis Crick and physicists, such as Roger Penrose raised the idea of science investigating consciousness. It's now believed that consciousness might be easily explained from studying recognized processes in the brain.

There is also the theory, according to Chambers, that consciousness is an existing fundamental and can be linked to other fundamental sciences including, space, time, mass, and physical processes.

To better understand this, Chambers and other scientists have attempted to find a correlation between brain activity and consciousness. The aim is to examine parts of the brain and how they influence the human ability to see faces or to experience pain or happiness. It is a science of correlation, but fails to explain what makes human consciousness tick.

"We know that these brain areas go along with certain kinds of conscious experience, but we don't know why they do," said Chambers. "But it doesn't address the real mystery at the core of this subject: Why is it that all that physical processing in a brain should be accompanied by consciousness at all? Why is there this inner subjective movie? Right now, we don't really have a lead on that."

A bigger question asked by many scientific thinkers is: Will robots or computers have a capacity for consciousness? Christof Koch, Chief Scientific Officer of the Allen Institute for Brain Science in Seattle believes it is possible. He has spent almost a quarter of a century studying consciousness and is currently working at The Allen Institute to build a complete map of the mammalian brain. It's a $500-million initiative funded by Microsoft cofounder Paul Allen.

In 2014, Koch spoke at MIT about integrated information theory (IIT), developed by Giulio Tononi at the University of Wisconsin. Tononi's theory suggests consciousness occurs when a system that is so complex produces a "cause-effect" repertoire.

Koch told *MIT Technology Review* in an 2014 interview, "If you were to build a computer that had the same circuitry as the brain, this computer would also have consciousness associated with it. It would feel like something to be this computer."

Koch says that humans will likely build AI systems before they understand them. He is probably right. And it might be from building it that we can start to understand it.

Neuroscientist Michael Graziano, professor of neuroscience at Princeton University, has an interesting perspective on consciousness. Writing in *Aeon Magazine* (www. aeon.co), he said "Artificial intelligence is growing more intelligent every year, but we've never given our machines consciousness. People once thought that if you made a computer complicated enough it would just sort of 'wake up' on its own. But that hasn't panned out (so far as anyone knows). Apparently, the vital spark has to be deliberately designed into the machine. And so the race is on to figure out what exactly consciousness is and how to build it."

Part of the problem is consciousness has been viewed as a mystical process and that is a sticking point for researchers.

"Consciousness research has been stuck because it assumes the existence of magic," said Graziano, who is also the author of *Consciousness and the Social Brain* (Oxford University Press).

"Nobody uses that word (magic) because we all know there's not supposed to be such a thing. Yet scholars of consciousness—I would say most scientific scholars of consciousness—ask the question: 'How does the brain generate that seemingly impossible essence, an internal experience?'"

His answer is to engineer for it.

"As long as scholars think of consciousness as a magic essence floating inside the brain, it won't be very interesting to engineers. But if it's a crucial set of information, a kind of map that allows the brain to function correctly, then engineers may want to know about it."

So perhaps that's the job at hand. If we build it, will it be self-aware? Wait 20 years or so. We'll soon see.

Super Us? Here Come Virtual Helpers Armed with Strong AI

It's one thing to have AI available. It's largely available today: It runs in the financial markets. Search engines use it. It watches you when you appear on security cameras. In fact, you probably interacted with an AI-enabled machine multiple times today during your daily routine and didn't even know it. Heck, devices such as Siri on the Apple iPhone or the discreet "Google Now" on Google's Android phone or other virtual assistants seemingly behave like artificial entities, however it is easy to tell that "they" are not real humans. (Try asking: "When will pigs fly?" Siri on my iPhone answered: "On the twelfth of never.")

But the advent of strong AI is only a few decades away. By one measure, when it happens that will be the technology singularity: the point where the machine's intellectual capability is functionally equal to a human's.

At that point, super smart machines will start to self-design and we humans can retire to the beach and drink mai tais. Or cower under the rusty shell of a corrugated metal roof as World War III rages around us. (It depends on who you ask.)

"This intelligence explosion, which thinkers believe will happen around 2045 or so, is going to fundamentally change the world as we know it."

This intelligence explosion, which thinkers believe will happen around 2045 or so, is going to fundamentally change the world as we know it. After 2045 we can't really predict what will happen. This is why the technological Singularity is the point at which events can start to become highly unpredictable or even unfathomable to our own human intelligence.

So are we on track for this?

To date, computer scientists are making progress. Consider these results, as reported by *New Scientist* magazine:

- In September 2015, an AI system called ConceptNet answered a preschooler IQ test and produced results on par with the little tykes.

- In 2014, a system called To-Robo passed the English section of Japan's national college entrance exam.

- A system called Aristo at the Allen Institute for Artificial Intelligence (also known as AI2) in Seattle, Washington, is taking New York state school science exams. AI2 has also challenged other computer scientists to beat the exam for a cash prize. (AI2 was created by Microsoft cofounder Paul Allen.)

The key problem with these kinds of challenges is that computers that do well on them might struggle with everyday questions that humans might find simple to answer. Like, perhaps, how many catfish are at the Humane Society? Or as *New Scientist* put it, "Is it possible to fold a watermelon?"

Machines can have trouble with context, sarcasm, humor, and communication that requires interpretation beyond basic information retrieval. They are, however, good at pattern matching. Ask a machine "Who is Sundar Pichai?" And it will tell you he is the recently instated CEO at Google. Regardless, intelligent machines, even if they don't yet use "strong AI" are already replacing human jobs.

Boston Consulting Group forecasts that by 2025, up to 25 percent of human jobs will vanish and be replaced by either smart software or robots, while a study from Oxford University has suggested that 35 percent of existing UK jobs are at risk of automation in the next 20 years.

If your job is repetitive and doesn't require super complex analysis or strategy, or you process information, you will likely be out of a job in the coming decade. Sorry! It will also replace those grumpy civil servants that can be miserable and rude. For them, we're not sorry.

As you can imagine, if we can reverse engineer the human brain it will readily help us understand how to build artificial brains that can make fairly advanced decisions accurately and with a lower error rate than a tired, grumpy human.

Don't worry, though—we will figure all this out. Luckily Transhumanist Party president and United States Presidential Candidate Zoltan Istvan, who is super brilliant, has a plan for you, dear reader. It's called Universal Basic Income. After you hear about it in Chapter 7, "In Hacks We Trust? The Political and Religious Backlash Against the Future," you might be saying, "Heck, give the robots my stinky job," then retire to a life in a sunny deckchair by an ocean of your choice.

6

Franken-You: A Better Life Through Cyborg Technology

You are a cyborg. Yes, you sitting in that chair holding this book. Now if you get in front of a mirror, you might not look like one, especially with that little glob of guacamole on your upper lip, but trust us you are. You're just not the cyborg that Hollywood generally defines. Then again, when was the last time Hollywood was right about anything?

Here are two ways we know you are a cyborg:

- For one, no one looks like Michelangelo's *Vitruvian Man* any more (see Figure 6.1). We're all wired up with an iPhone and other gadgets and as you'll see in this chapter, our contemporary brethren are doing miraculous things to their own bodies to evolve past Michelangelo's original vision of man. Still the elemental intention is still there. With *Vitruvian Man*, Michelangelo was demonstrating the blend of art and science during the Renaissance. The pen-and-ink drawing was a study in proportions and an effort to connect man to nature. Today, human beings are undergoing a new renaissance a new blend of art and science, especially when it comes to enhancing the body.

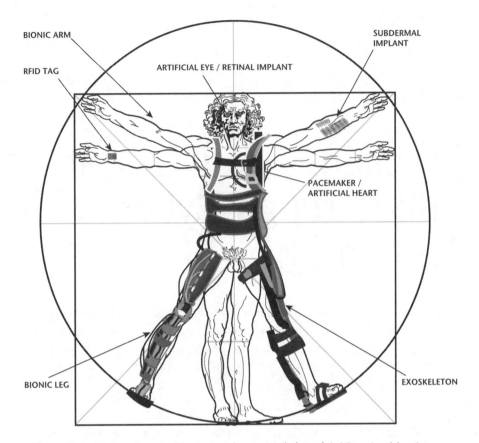

BIONIC ARM

RFID TAG

ARTIFICIAL EYE / RETINAL IMPLANT

SUBDERMAL IMPLANT

PACEMAKER / ARTIFICIAL HEART

BIONIC LEG

EXOSKELETON

Figure 6.1 *We've taken the liberty to enhance Michelangelo's Vitruvian Man into a more up-to-date version of what we are becoming.*
(Illustration by Cornelia Svela.)

- The second reason we know you are a cyborg is because cyborg anthropologist Amber Case says you are. And she has the coolest job title around, so she's gotta be right.

The Portland-based thinker explains what she does this way to *National Geographic*: "What will the next life-shaping breakthrough in technology be? How do parents respond to children who spend hours online? Which new products will fail or succeed? Is technology changing our values and cultures? 'These are the kinds of questions my work tries to help answer,' said Case."

When we contacted her about cyborgs for this chapter it resulted in a very insightful chat.

Case told us the cyborg experience for humans started right at the beginning of our time on this planet.

"Cave people had fire. The minute we started to first cook food that was taking the digestion process that we would have had as animals outside of ourselves," Case observed. "We started taking all these pieces we would have had eternally and started externalizing them. We became dependent on fire. We became dependent on a knife instead of our teeth to cut through the meat. We became dependent on houses, nails and hammers and then we had to create a society around that."

That process continues through today, and will continue even as these external aids become more and more technically advanced. Plus, she noted, as we progress we don't even necessarily notice that it's happening.

"Once a thing becomes part of your life, like a laundry machine, it dissolves. You don't notice a laundry machine. It just becomes part of your everyday life. If it breaks down, you notice it. It's the same with technology," says Case. "While everyone is waiting for the bus, or starts smoking a cigarette, they pull out their phone. In all the little moments of life where you are looking ahead or talking to your neighbor you pull out your phone. You go to bed with your iPhone, you wake up with it and read the news."

In other words, don't spend too much time worrying about whether you're going to become a cyborg in the future … you already *are* one. And if you are an Amish farmer, your barn and gas-powered lantern make you a cyborg. That said, you are not a fancy one like Arnold Schwarzenegger was in *The Terminator*, Jebediah.

> " … don't spend too much time worrying about whether you're going to become a cyborg in the future … you already *are* one."

Are you a Cyborg? Or just Bionic?

But before we get too far ahead of ourselves, let's rewind a bit. The term "cyborg"—a portmanteau of "cybernetic organism"—was first used by Manfred E. Clynes and Nathan S. Kline in 1960 to refer to human beings who have been physically enhanced to allow them to survive outside of Earth's atmosphere. In a *New York Times* article that year, Clynes and Kline referred to a cyborg as "essentially a man-machine system in which the control mechanisms of the human portion are modified externally by drugs or regulatory devices so that the being can live in an environment different from the normal one."

That definition has morphed over the years, however. For our self-serving purposes, a cyborg is any organism that has *enhanced* capabilities thanks to integration with technology, which is not quite the same thing as using prosthetics simply to regain lost functionality.

For example, an amputee using an artificial leg simply to regain mobility would not qualify as a cyborg by this definition. They are merely bionic. On the other hand, if the replacement leg was designed to increase running speed, or to include special sensors that are able to sense magnetism, radiation, or a dog peeing on it, then it's: hello cyborg!

Unfortunately, the terms "bionic" and "cyborg" are usually used interchangeably or used as synonyms by Hollywood and literature, its tweed-jacketed, pipe-smoking cousin. On the 1970s drama *The Six Million Dollar Man* and its spinoff program *The Bionic Woman*, both Steve Austin and Jamie Summers benefitted from speed and strength enhancements and each had one sense amplified, yet both were routinely referred to as "bionic." (We know because some of us authors begged to stay up and watch it at 9 P.M. on a school night.)

But let's get real. Our editor, Rick Kughen, has a chronic back issue that required several surgeries. In one operation the surgeon installed an electro-prongy thingy (that smart people call a neurostimulator). It doesn't give Rick a super back. But it was suppose to help his wonky defective back work better than it would otherwise naturally and still let him go fishing. That makes Rick pretty much bionic. (It was a bust, but you get the idea.) When he sends us stressed emails from his iPhone, however, asking about missed deadlines, that is better than walking across the country to our house to grumble at us. That enhancement makes him a cyborg. Get it?

CYBORG VERSUS BIONIC: A HANDY CHECKLIST

Since being bionic versus a cyborg can sometimes be confusing, here's a handy checklist:

You are bionic if you have:

- An artificial leg allowing you to approximately stand and walk as you did before amputation
- An artificial arm or hand replacing the function of your old arm or hand
- An artificial heart pumping blood as the old heart did
- An artificial eye restoring vision to previous acuity
- An artificial ear restoring hearing to previous acuity

You are a cyborg if you have:

- An artificial leg allowing you to jump 30 feet in the air or run at speeds faster than before

- An artificial arm or hand with super strength or added agility

- An artificial heart whose speed can be controlled from a smartphone for peak performance

- An artificial eye that can zoom into many times standard 20:20 vision, or see light frequencies outside the normally visible spectrum, such as infrared or X-ray

- An artificial ear that can hear much quieter sounds, hear from a greater distance, or hear sounds outside the standard 20 to 20,000 Hz range

- Any artificial body part, implant or wearable, that can sense radiation, magnetism, or other phenomenon undetectable without additional external instrumentation

- Any artificial body part, implant or wearable, that can communicate with the outside world

- Any implant that increases cognitive function or memory capacity

You might be a cyborg supervillain if you have:

- Any artificial body part that shoots darts or unfolds into a saw

- Any body part that can dissolve into a mercury-like liquid and then reform into something that looks vaguely like it might belong on Arnold Schwarzenegger

- A built-in radio transmitter that broadcasts your thoughts to your collective and/or to nearby starships

- A business card with "book editor" on it

What the Heck Is Cybernetics?

The term "cybernetics" was coined by MIT scientist Norbert Wiener in the 1940s, while working with Mexican colleague, Dr. Arturo Rosenblueth, after the Greek word meaning "steersman." They felt this was an appropriate description for the basic gist of their field, which studied feedback mechanisms and governors in mechanical devices. No we're not talking Chris Christie here. To be more specific, they are the things that allow mechanical replicas to act and function more like the original human parts they were modeled on.

In early experiments, the duo and their team measured electrical charges inside muscles during contraction. They also developed a device that would convert visual information on a page into tones, in an attempt to allow the blind to "read" by ear, which we think is kinda cool.

Wiener noted the similarities between electronics and the human nervous system: "It is a noteworthy fact that the human and animal nervous systems, which are known to be capable of the work of a computations system …" He goes on ad nauseum, but we deleted that part because frankly you get it. Cyborgs are humans with replacement parts that make them as good or better than their birthday equipment.

If you read Wiener's 1948 book *Cybernetics: Or Control and Communication in the Animal and the Machine*, you're not going to read about a bunch of cool robot arms—this pioneering research occurred well before much of the miniaturization that make electronic implants and robot limbs feasible. In fact, much of the text is pure theory, with long passages of mathematics. But it is an important, if occasionally baffling, starting place for much of what we now consider cutting edge in this field.

Mostly Evil Killing Machines?

English professors usually point at Edgar Allen Poe's 1848 story, "The Man That Was Used Up" as one of the first appearances of a cyborg in literature. It featured a man made up largely of prosthetic appliances that replaced body parts lost in various military campaigns. By our definition here, this wouldn't quite qualify … as bizarre and futuristic as it might have sounded in the mid-nineteenth century.

That definition would apply more to the early-twentieth century French superhero The Nyctalope, who had enhanced night vision and an artificial heart in his arsenal of crime-fighting tools. It would certainly apply to Steve Austin, hero of Martin Caidin's novel *Cyborg* (later made into the television series *The Six Million Dollar Man*). Following an accident, Austin's legs (and one arm) were replaced with faster, stronger mechanical versions. His eye was also replaced with a camera that had telephoto capabilities.

On the darker side of the equation, both Darth Vader and *Battlestar Galactica*'s Cylons incorporate some technology that could allow them to be considered cyborgs. As mentioned earlier, the original Terminator was also a cyborg.

And do you remember this line: "We are the Borg. Your biological and technological distinctiveness will be added to our own."

For a whole generation, this catchphrase from *Star Trek: The Next Generation* came to exemplify the idea of the intersection of the biological and the technological. They—or the collective "it"—was a cyborg. The name "Borg" kind of gives it away, really. Archvillains are fun to watch when they are cyborgs. Of course not all cyborgs are evil, killing machines. But Hollywood certainly has given them a bad rap.

This Cyborg Life

The basic equipment humans have when they come out of Mom's tummy is not that spectacular. You're small, pink, and wiggly. So a level of technology help is pretty much required for survival. As we mentioned in an earlier chapter the slippery gray matter between our ears has long been our greatest asset.

Back in the early days, everyone walked around naked all the time, until someone's gray matter concocted the plan to bonk a buffalo in the noggin with a rock, peel off its shaggy coat, and wear it to stay warm and survive. And have sex and have babies. And that was just the first step toward our cyborg destiny.

> " ...the slippery gray matter between our ears has long been our greatest asset."

Those early pelt wraps gave way to tailored furs, with sinews being used to stitch together multiple panels of animal hide, creating clothing that fit the body in a more natural way. Eventually, heavy pelts gave way to fabrics woven from a variety of natural materials, such as linen and cotton, which were lighter, cooler, and could be woven into predetermined shapes. Those, in turn, were bolstered or replaced by synthetic and ultralight materials such as polyester, rayon, and nylon, which had properties that would have seemed magical to our buffalo-skinning forebears. They had heat retention and water-repellency and the apparent ability to stretch when worn by John Travolta in *Saturday Night Fever*. Geez, that guy could wear polyester!

Footwear made of wrapped animal hides turned into leather and rubber composite footwear stitched together for protection and durability. Through the years, advances in technology meant that footwear eventually transitioned away from animal-based material like leather to things like canvas and synthetic materials. Shoe soles, which originally were intended to prevent rocks and twigs from poking your feet, were computer-engineered to include features like shock resistance and springiness enabling you to run a reflective marathon or two painlessly on misty mornings (if you believe the "Just Do It" ads).

Of course, clothing wasn't simply for protection against the elements; sometimes it was a matter of protection against each other. People had this bad habit of pissing off their neighbor, which occasionally led to them getting impaled by sharp, stabby bits of metal.

Thousands of years ago, people hammered plates of metal into shapes that would protect various body parts: helmets protected the head, breastplates and later, around 4 B.C.E., chain mail was invented to deflect bladed weapons. More recently, protective synthetic materials, such as Kevlar, are lightweight enough to be woven into gloves and jackets, and with a strength five times that of steel. They offer protection against projectile weapons such as bullets, or even (to some extent, anyhow) the blast of a grenade.

You get the picture through this timeline, right? Heck, your clothing makes you cyborg because your bare hairless flesh won't keep you from freezing to death.

Until the thirteenth century, for people with eyesight that had lost its acuity over the years, the only option was to stumble around in a blurry world. Then someone realized that polished glass could correct vision to near-normal levels. Eyeglasses could cure near-sightedness, far-sightedness, or provide magnification for close-up work like reading or detailed work. Eyeglasses could also correct for eyes that didn't track properly with each other, and could limit the amount of light that entered the eye, whether to protect against brightness or harmful ultraviolet rays or to look super-cool like Tom Cruise in *Top Gun*.

Eventually, corrective lenses were shrunk down into small transparent discs that could be placed on the eyeball, obviating both the need to perch spectacles on the end of your nose, or getting beaten up at recess for being a stinky four-eyes.

So far, each of these things has corrected deficiencies in the human condition: bad eyesight, the inability to survive when the temperature dropped, or resistance to deadly weapons wielded by fellow humans. But at a certain point, the desire to self-enhance went beyond simply curing these deficiencies to actually improving human capabilities.

One of those capabilities was knowing what time it was. The sundial was introduced around 3,500 B.C.E., making it easy for people to figure out the time of day so long as you were within a short distance of the sundial itself. This helped with avoiding darkness, which was generally hazardous to be out in. Being aware of time allowed for keeping safe, planned resource gathering, and other life optimizations.

The introduction of spring-driven clocks in the fifteenth century allowed people to tell time in their very own homes and without having to rely on the sun, which inconveniently disappeared every night for hours at a time.

In the sixteenth century, the technology was miniaturized into a wearable format that we know today: the wristwatch. Initially, it was favored by women, while men tended to rely on the pocket watch. (The neck-mounted watch was limited to Flavor Flav of the hip-hop group Public Enemy, yo.) Whatever version you preferred, the watch had the distinction of being the first wearable device.

Okay fine, but as superpowers go, knowing the correct time is pretty unimpressive. But let's get adventurous. How about breathing underwater?

Until the late nineteenth century, the only place that human beings could comfortably breathe was while walking on land (save for areas plagued by pollution caused by heaving coal burning).

In 1878, Henry Fleuss created the self-contained breathing apparatus, which was strapped on and allowed divers to breathe underwater. These days it seems routine to don an air tank and go for a stressless underwater swim on holiday.

"Want a pina colada, darling?"

"No, no. Going shark diving. Will be back at 2."

That said, it was the first step toward allowing people to venture into environments previously considered wholly hostile to human life. Further, it was only a few small steps for man to the spacesuit, which allowed humans to breathe off the planet.

While breathing underwater and in the void of of space come closest to Clynes and Kline's original definition of cyborg, it's probably a fair guess that scuba divers or wristwatch-wearers aren't what immediately springs to mind when you think of part man, part machine.

Let's Start with Bionics

So let's take a look at more recent technological advancements in the world of the cyborg. Medically speaking, there's been more of a focus on bionics in recent years rather than being a cyborg, with a growing field of research spawning replacement limbs that act (more or less) as natural replacements for original limbs. In 2015 or so, that has included robotic arms functioning like natural arms, and robotic feet functioning in harmony with the rest of the body.

The prosthetics company, Össur, has a line of bionics designed to replace missing legs, complete with processors onboard for more natural motion that mimics the movement and reactivity of the original meat limbs. In his book *Tomorrowland*, Steven Kotler tells the story of David Rozelle, a combat veteran who lost the lower portion of his leg but who was able to return to active duty with a bionic Össur-built replacement.

While it's true that some of these limbs can be exceptionally expensive—and thus hard to justify the cost, at least for the uninsured—some people are building their own replacements at a fraction of the cost using 3D printing.

In 2013, 12-year-old Leon McCarthy had been missing fingers on his left hand since birth due to lack of blood flow during his development in the womb. Traditional prosthetic units, to help people like Leon, run into the tens of thousands of dollars.

Leon's dad connected with inventor Ivan Owen on YouTube, and in turn, Owen and Richard Von As from Johannesberg, South Africa. They collaborated on a high-quality, low-cost 3D printed prosthetic. The inventors do not hold a patent for the invention or charge to download the plans for the hand, so the cost of materials is all that is required. That's $10. Pretty much the cost of two Venti specialty coffees from Starbucks.

Fortunately, Leon's school owned a 3D printer and handily made it available. A $10 replacement was printed for Leon, and he now has a "cyborg" hand with fingers able to close, which he sees as "special, not different." The fingers are controlled by flexing the wrist. This pulls on cable "tendons" to close around a targeted object. Leon can hold his backpack, give a snack to a buddy, and hold the handlebars of his bike just like other kids. (Check out Leon's 3D printer hand at http://superyou.link/leonshand)

This points to a future where you can upgrade your hand simply by popping down to the local library and booking some time on one of their 3D printers. It's not hard to imagine a future where we are able to inexpensively print limb replacements that not only replace but also *improve* upon what nature gave us.

I ♥ Technology

Technological difficulties notwithstanding, limb replacement still seems relatively straightforward next to organ replacement. While you can get by with replacement legs of varying capabilities, an artificial heart is something that needs to be built well and has to last a long time (maybe forever if you believe what's discussed in Chapter 8, "Hyper Longevity: How to Make Death Obsolete"). You can't just remove your heart for a while and then stick in a new 3D-printed ventricle. (At least not yet.)

"Honey, you can run really fast today. What did you do?"

"I upgraded my ventricles on the weekend, Mom."

Seems nuts, but so did cellphone video conferencing in 2001.

Artificial hearts have been around since the early 1980s, but until recently the problem is they are tethered to external machinery which makes it hard to go on a hot date. Most recipients were unable to live for extended periods with the replacement technology. For some, the devices acted strictly as bridge technology until the patients could receive a donor transplant.

In the early twenty-first century, Peter Houghton was able to live seven years with a Jarvik 2000 ventricular-assist device which didn't replace the heart outright but augmented the weakened ability of his original ticker. Heart-assist devices today are common tools that surgeons can implant to assist a weak heart pump better, which allows their recipients to live a normal life. (Learn more about the Jarvik 2000 device at http://superyou.link/jarviksheart.)

Surgically implanted devices also allow people to regain their hearing. In 2011, a 1.5-minute video of a deaf woman named Sarah Churman went viral, showing her breaking into tears when her audio implant (from a company called Esteem) was switched on (http://superyou.link/sarahhears). Since then, a number of other videos have shown other men, women, children, and babies having their implants activated (and now this kind of video is so de rigeur, it's almost as common as those dog-on-a-skateboard videos). It's worth noting that such cochlear implants have been around since 1976, but earlier models were bulky and often required external components. New models can be implanted completely internally. The downside is that, unlike traditional external hearing aids, you have to go to see a technician to replace the implant's batteries. Under normal usage, that's likely to only be necessary every five to eight years.

I See U

Replacing lost vision has been a bit tougher, but it's not hopeless. Jens Naumann became known as Patient Alpha in 2002 when a prototype implant device helped him regain part of his vision. (See the profile on Naumann later in this chapter for more details on his story.)

Many of these early visual therapies required a lot of external hardware to properly function, but with ongoing research, the outboard hardware has been shrinking. Today (as we write this in 2016) it is close to the size of a deck of cards. It could be a matchbook by the time you get to this section of the book after it sits on your night table until 2017.

For those who still retain some of their original vision, we're also now starting to see lenses that can not only correct eyesight back to 20/20 vision, but can also enhance vision into the world of cyborg, so people can see things at greater distance than with the original parent-provided wetware. It's getting to be very "Steve Austinish" out here today.

If you haven't had to replace (or correct) one of your body parts, you might think there's very little overlap between your world and the world of cyborgs. And you'd almost certainly be wrong.

I Am the Very Model of a Modern Cybernetic Man

At the very end of the twentieth century, two technologies designed to increase your capabilities exploded into the consumer marketplace: the mobile phone and the personal digital assistant.

The mobile phone started as an unwieldy assembly that had to be anchored to your automobile to connect to the phone company's radio transmission network. Motorola's DynaTac phone was introduced in 1983, finally allowing people to talk on the phone while walking down the street... but with a weight of 28 ounces (794 grams) and a form factor that was more than vaguely reminiscent of a brick, it wasn't widely embraced. But it was cool.

Eventually, the miracle of miniaturization allowed mobile phones to fit comfortably in your jacket or jeans pocket, making them practical for nearly everyone.

Around the same time that mobile phones were shrinking, the personal digital assistant (or PDA) was starting to come into its own. As electronics became cheaper, companies started making electronic organizers that would store addresses and phone numbers, essentially replacing the paper equivalent.

Take European electronics marker, Psion. Its gadgets included extra functions such as a calculator, a world time clock, and basic word processing/spreadsheet functionality. (Whoa!) Eventually they included a full QWERTY-style keyboard for faster text input. In 1992, Apple's stylus-controlled Newton MessagePad was released into a world that wasn't quite ready for tablet computing; despite its advanced capabilities, it ultimately ended up as a punchline on *The Simpsons* TV show.

 ## When Cursive Leads to Cursing

A 1995 episode of *The Simpsons* called "Lisa on Ice," depicted school bully, Dolph, writing on the Newton, which autocorrected the message "Beat up Martin" to "Eat up Martha."

As an aside, the legend goes that Apple engineers were so traumatized by the gag that they slaved over the iPhone keyboard to try to make it perfect.

The 1996 launch of the Palm Pilot was much more successful thanks to its more reliable stylus input, leading to a long line of successors and imitators.

On their own, either the mobile phone or PDA could have laid ample claim to turning humans into cyborgs. Mobile phones enhanced our capability to stay in touch with other people without having to tether to a landline, and PDAs were a handy way to store information in offline memory for quick and easy recovery. But it was when the two finally came together that we found ourselves in a whole new snack bracket.

While pagers had been around in one form or another since the 1950s, Canada's Research in Motion (RIM, later Blackberry) introduced the first two-way pager in 1996, allowing people to respond directly from their hip-slung device. Eventually that led to the BlackBerry device, a full-sized PDA that was always connected to the outside world, enabling mobile email.

Palm responded by adding wireless capabilities to the Palm VII. Shortly afterward, Kyocera release the pdQ, which bundled phone capabilities with the Palm operating system, and though it wasn't a success, it signaled the future of the direction of the phone.

Both Research in Motion and Palm came out with their own phone/PDA hybrids, and the smartphone was born, laying a path leading directly to today's iPhone and Android handsets. With their capability to download apps and connect via Bluetooth to a wide variety of accessories, smartphones are essentially a bridge technology that turns every single one of us into an (improved) cyborg.

Another accessory that's been speeding along the cyborg timeline is health trackers such as the Fitbit, a device that can track the steps you take throughout the day. They use motion sensors embedded inside the device to calculate your progress. By correlating the data gleaned from those sensors with information about your height and weight, the Fitbit app can then calculate your distance walked and the amount of calories you've burned. Newer Fitbit devices contain heart rate sensors, so the device's app can keep a reasonably good record of your health stats throughout each day. It can also sense trends and even compare you against your friends to bolster your motivation to do better. "Hey lazy, Steve is kicking your butt today." Ok maybe not.

But there's an additional incentive, potentially. Your insurance company might give you more favorable rates if they can tap into your Fitbit data to make sure you're taking care of yourself. In 2015, John Hancock became the first insurer to offer policyholders a discount when they use Fitbit wristbands that enable exercise tracking. Imagine your insurance agent calling you to warn you that your rates will be going up if you don't do more sit-ups today.

The Apple Watch also adds health-tracking capabilities when paired up with Apple's HealthKit app. But it's not simply a health-tracking device: The Apple Watch also sends notifications from your phone straight to your wrist, and by using Near Field Communication technology (which enables short distance wireless data transmission), you can pay for items simply by putting your wrist up to a sensor at retail establishments. It can even send messages wirelessly to other Apple Watch wearers.

It's probably fair to say that Google Glass is a device that more closely aligns with the general preconception of what a contemporary cyborg actually is: Worn on the face like a pair of eyeglasses, the voice-controlled device features an eyepiece that can privately impart information to the wearer, and can record a video of whatever the wearer happens to be looking at.

It's probably also fair to say that the public at large isn't quite ready for this kind of cybernetic technology; the device was banned in cinemas, locker rooms in health clubs and bars, and pretty much everywhere else. Why? Google Glass's recording capabilities were considered to be an invasion of privacy and/or a way to subvert copyright laws. Add that to the fact that Google Glass wearers looked like utter dorks, and it was a recipe for failure. And fail it did: Google shelved the Glass project at the beginning of 2015 (see the story of Google Glass at http://superyou .link/brokenglass). That's not to say it's completely dead. In 2015, Google applied for patents on the next generation of its wearable technology, but it might be awhile until it's fully accepted by the public…if it's even made available to them. Right now, the next generation version of Glass looks like it may simply be aimed at increasing productivity in the workplace.

It's not just socially oblivious nerds that want to strap on cameras. GoPro has become ubiquitous in the extreme sports category, with people strapping on a high-def third eye before throwing themselves into death-defying situations, like working as a bike courier. While GoPro has been of special interest to skiers, stunt cyclists, and skydivers, the company has more recently branched out into drone technology—after acquiring Swiss drone company Skybotix—giving people the ability to see at a distance…from hundreds of feet in the air.

Then there's virtual-reality eyewear. It allows users to go places that don't exist. While virtual reality goggles have been available for quite some time now, earlier models tended to have relatively low-resolution displays and slower refresh rates, which could leave those trying to enter into virtual environments headachy or barfy. Now, the Oculus Rift features tech specs that make these virtual environments more tolerable and lifelike.

Combined with a camera, these new virtual reality glasses could functionally put an overlay onto real life, helping wearers to identify what's in front of them and to navigate unfamiliar or dangerous environments. It's worth noting that smartphone apps, such as Layar, have been offering some of this augmented-reality functionality for years, though the smartphone experience is obviously less immersive than a full virtual reality headset.

Can you see the trend? Miniaturized headsets (or soon embedded wetware, maybe?) that is enabled with smart software that overlays helpful guidance through the underwear aisle at Macy's to the discount boxers, and you have the makings of something kind of handy.

While the Apple Watch and Fitbit can alert you that you have a phone call coming in by flashing an alert on your wrist, there are more discreet ways to receive alerts. Kovert's ALTRUIS jewelry line-up includes rings, bracelets, and pendants that will subtly vibrate on predefined events such as an incoming call or a message from a specific person.

The problem with fashion is it's so temporary, so you might want to wait a little bit until you can get a more permanent alert system tattooed to the body part of your choice.

Yep. We went there. Actually it's old hat. In 2012, Nokia applied for a patent for precisely that reason, suggesting that magnetic ink could be compelled to subcutaneously vibrate on incoming calls or messages, when connected to a Bluetooth-enabled phone (we talked a bit about this in Chapter 3 "Beauty Hacks: Becoming Barbie, a Lizard, or Whatever You Want to Be" in the section called Gadget Activated Tattoos). Different vibration patterns could allow you to differentiate between callers just by the different sensations emanating from your tattoo.

While each of these things is high-tech in its own way, you might still be feeling like they're cyborg-lite. Let's take a look at a few people who have really jumped feet-first into the world of the cyborg and are leading the way on how it really could be for everyone in the future.

Super Senses

Most people take their five senses for granted ... at least until something goes wrong with one of them, such as a decline or complete loss of sight or hearing. Similarly, you might start paying closer attention to your senses if you find yourself running up to the limitations of one of them, like being unable to make out the text on a sign that says "Low bridge."

Mr. Cyborg: Steve Mann

If anyone can claim to be the first cyborg, University of Toronto professor Steve Mann can make an excellent case. In many ways, the work that Mann has done over the years has led to modern wearable computing products such as Google Glass. Dubbed "the father of wearable computing," Mann first strapped a computer to his person in 1978. Back then, the technology was heavy and bulky, and the visual component resembled a highly modified hockey helmet with a gigantic eyepiece, spotlight, and antenna. Over the years, the technology has become more compact and advanced, which meant that the technology has called less attention to itself when Mann is in public.

Like Google Glass, Mann hasn't had the smoothest ride, and for some of the same reasons: People tend to be wary when you're wearing a bunch of technology on your head. By 2012, Mann's headgear had shrunk substantially from the helmet-sized assembly he was wearing in the late 70s and early 80s. Even so, his streamlined setup still came with a camera attachment (known as the EyeTap) located directly in front of his eye, and consequently Mann could still be quite conspicuous. Although the technology was surgically connected directly to his skull, employees in a Parisian McDonald's restaurant attempted to forcibly remove it from his face and skull.

Despite Mann's attempts to show them documentation from his doctor explaining the technology, they ultimately ended up ripping up the doctor's note and ejecting Mann from the restaurant. This wasn't even the first time Mann had been assaulted for wearing this technology in public; Mann had also been accosted by a subway security officer a little earlier on his personal cyborg timeline.

Mann continues to move forward with his wearable computing project, and serves as the director of the EyeTap Personal Imaging Lab, which works on ways to provide augmented reality devices in smaller and more user-friendly form factors.

While the original goal of Mann's wearable computing devices was as a "seeing aid" for people with visual deficits, it now acts as augmented reality for anyone who wants it. "Thirty-five years ago I was trying to justify or explain this glass. I think, now, digital eyeglass doesn't require convincing. I think people realize its benefits," he said in an interview posted on YouTube (http://superyou.link/stevetalksatTedx).

Neil Harbisson: The Man Who Can Hear Red

Neil Harbisson was born without the ability to perceive colors—everything appeared in grayscale to him. That started to change when he started wearing a camera that converted color into specific tones, which he could hear through a set of headphones. Eventually, the camera attachment—which Harbisson refers to as the "eyeborg"—was implanted directly into his head, swinging in front of his forehead a little bit like an insect antenna.

 Neil Listens to Color

Watch Neil Harbisson's Ted Talk about hearing colors:
http://superyou.link/neilhearsred

This ability to translate colors into tones has had a few different practical synaesthetic effects for Harbisson. Now, when looking at paintings, he can perceive them on a sonic level, leading him to regard certain paintings with many different subtle shades as sounding more like a horror movie.

But he's also been able to start working in the other direction, rendering the tonal qualities of certain songs or speeches into colors—in fact, he's turned this ability into an art career—and being able to pinpoint the "color" of a phone ringing. He's even joked that he's starting to dress less on how it looks than how it sounds.

While the implant has provided its share of troubles—including a run-in at the passport office where an official didn't want to allow Harbisson to wear the camera for his official photo—it's also had its benefits, including the ability to "see" outside the normal human spectrum. His ability to hear infrared means he immediately knows if there are motion detectors in a room, and his ability to hear ultraviolet means he can tell how safe it is to be outside without having to consult the UV listings from the weather service. The TV weatherman might soon be out of a job.

Jens Naumann: Now You See Me, Now You Don't

When he was younger, Jens Naumann had full use of both of his eyes, but lost his sight in two separate accidents. Though he could sense a small amount of light in his left eye, a series of operations attempting to recover some vision actually made his vision worse. Following a number of years without vision, Naumann hooked up with the Dobelle Institute, which was working on experimental artificial vision technology.

In 2002, Naumann (dubbed "Patient Alpha") was the first of 16 patients to undergo a procedure in which holes were drilled into his skull, wires were connected directly to the brain via implants, and he started to receive a feed from a special camera attachment in front of the right eye. Once he was wired up, he could perceive input via the electrodes in a grid of limited resolution. It was nowhere near the resolution of his original vision, but a damn sight better than the darkness he'd lived in for years. Despite a rocky start—including seizures induced by the electrodes—Naumann was soon able to make out what was going on around him, to the point where he was even able to drive a car around the clinic's parking lot.

Unfortunately, the implant only worked for about eight weeks before deteriorating, and its inventor, William Dobelle, died before being able to fix it or properly document it. Naumann had the jacks in his skull removed in 2010. He readjusted to life without vision, documenting his experiences in *Search for Paradise: A Patient's Account of the Artificial Vision Experiment*. (Read more about Jens at http://superyou.link/meetjens).

Even though the Dobelle vision replacement system ultimately failed, it provided proof that it was possible for people without vision to see again, and many have since benefited from subsequent technologies. In fact, new vision technologies, such as Second Sight's Argus II, are now capable of generating a picture for people who were previously legally blind. Again, the image is crude, but it's a start.

I Wear My sunglasses... for Sight

Second Sight's Argus II uses electrodes placed directly into the retina and wirelessly fed visual information from a camera attached to glasses. See it in action in this video: http://superyou.link/canyouseemenow

Michael Chorost and the 100-Acre Wood

While we've seen a lot of YouTube videos of people having their cochlear implants switched on for the first time, Michael Chorost went through the procedure just after the most recent turn of the century. Born hard of hearing, Chorost was able to use hearing aids to boost his ability to hear until he became "abruptly deaf" in 2001.

However, he had cochlear implants placed directly in his skull, allowing him to hear again. A pair of headpieces were placed on the side of his head, magnetically locking to the implant inside his skull. After creating a radio link to the implant, the headphones sent audio information to the implant allowing Chorost to hear, but not the way people regularly understand hearing.

Because the electrode stimulates portions of the brain rather than moving the bones of the inner ear, Chorost initially found the audio input "incomprehensible." However, he retrained himself to hear all over again with audiotapes of *Winnie the Pooh*. By reading along with the audio version of the book, he was able to teach himself what each of the sounds he was previously familiar with "sounded" like using this direct method of electrode-induced stimulation.

Chorost understood that much of his new experience was made possible by the software, along the way, the software he was using was updated, giving him more access to auditory information.

The entire process led Chorost to switch careers, becoming a technology theorist and science writer. His award-winning book, *Rebuilt: How Becoming Part Computer Made Me More Human*, came out of his experience with his implants, and more recently he's written *World Wide Mind: The Coming Integration of Humanity, Machines and the Internet*. It's safe to say that Chorost has embraced his status as a cyborg. And he has brought Winnie the Pooh into the twenty-first century.

Rob Spence: Life Imitates Art

After Rob Spence lost his right eye in a shotgun accident, he did what any normal person would do. He placed a wireless camera right into his eye socket. Okay, maybe that's not quite what most people would do, but Rob Spence was both a filmmaker and a fan of sci-fi video games full of cyborgs, such as *Deus Ex: Human Revolution*.

From Man, to Machine, to God

Deus Ex: Human Revolution is a cyberpunk-themed first-person action role-playing stealth video game developed by Eidos Montreal. It is set in 2027 at a time when multinational corporations have more power than governments. The main character is Adam Jensen, a security manager of a biotechnology firm. During a terrorist attack on his work headquarters, he is mortally wounded and undergoes radical life-saving surgeries that replace large areas of his body with advanced prostheses and internal organ systems. See a video of game play here: http://www.superyou.link/playgod

The first prototype, also known as the—wait for it—Eyeborg was built right on Rob's kitchen table, but he soon got the company RF-Links involved in custom-building a wireless camera to fit into the prosthetic. While the camera could transmit wirelessly to an external receiver, it was not connected directly to his brain, leaving it as mostly a way to provide a point-of-view shot from the same perspective as regular human eyesight, not to function as an actual replacement organ.

As you'd imagine, these initial prototypes didn't quite match up to the kind of bionic eyes you see in futuristic television shows and videogames. In fact, they looked quite unnatural. In one of Spence's videos, you can see a definite red glow from the eye in lower light, which is certainly attention-getting, but not the sort of thing that would put a Tinder date at ease while sharing a nice Chablis. Ultimately, the goal for Spence's Eyeborg project was to create a model that looks more like a standard eyeball, and to provide the replacement camera with the type of augmented reality functionality typically featured in cyborg-oriented movies and video games. You can watch Rob Spence's documentary on living with a camera in his eye socket at http://superyou.link/robtheeyeborg.

While the Eyeborg isn't connected directly to the brain, it serves as a totem to future models that might be. Researchers have already started work on a 3D-printed eye that comes with built-in wireless, plus zoom and recording capabilities, in a modular format that will connect with the brain yet be quickly upgradable. (No word on whether the French arm of McDonald's will sponsor it.)

Super Strength

How much can you lift? Feel like carrying this case of bowling balls for me? Chances are the answers you just gave were "not much" and "Hell no!" Thankfully there's new cyborg technology that will assist with these tasks, or with health problems that might be preventing you from even lifting *yourself* upright.

Amanda Boxtel: Exoskeleton Pioneer

When soldiers are deployed out into the field, they're often carrying a lot of stuff with them—sometimes 100 pounds or more of gear. So, it's not surprising that the U.S. Army has been exploring exoskeleton options.

As you might guess by pulling apart the word, an exoskeleton is a powered robotic framework a person can wear on the outside of his or her body and clothing, which gives the wearer support, strength, and speed. By distributing loaded weight onto the skeleton, it allows the wearer to carry more than they ordinarily would be able to handle without the need for a lifetime access to a personal chiropractor. It can also help the wearer maintain his or her balance, even while climbing hills or walking great distances.

The powered Human Universal Load Carrier (also known as HULC), developed by Ekso Bionics and licensed by military manufacturer Lockheed Martin, allows the wearer to carry a load of 200 pounds and achieve a sustained speed of 10 miles per hour, while continuing to help the wearer maintain his or her agility. In other words, it's not like strapping on a tank, but rather helium balloons. The company also offers an unpowered model known as FORTIS, which simply provides support by transferring loads to the ground through the metal frame, rather than directly through the wearer's body. We think this would be handy for carrying cat litter, when there's a "buy one get one free" sale on.

While these capabilities obviously have great tactical benefit to soldiers in the field, exoskeleton technology can also return mobility to people who have lost the use of their limbs. In 1992, Amanda Boxtel sustained a spinal cord injury while downhill skiing resulting in paralysis in both her legs. Nearly two decades later, she began to learn to walk again using Berkeley Bionics's eLEGS exoskeleton, and she now has some mobility again, albeit with the use of a pair of arm-crutches for additional support. In addition to regaining mobility, Boxtel notes in one of her *TED Talks* that by standing upright inside the exoskeleton, she also has improved circulation, has better bladder and bowel function, and less pain. (See Amanda Boxtel strap on her eLEGs: http://superyou.link/amandastepsup.)

Ultimately, while the technology was developed for military applications—such as Ripley beating the celestial poop out of the mother of all aliens in *Aliens*—it's clear that civilians will eventually be huge beneficiaries, too. Most of the good stuff—such as GPS and duct tape comes from the military, really. The big hurdle in the short-term is the price. Exoskeleton models cost between $30,000 and $100,000.

They're not something you can just pick up at Target. At least not yet. As the technology develops, however, you know the price will come down. And then wheelchairs will become as antique as the Betamax, 56Flex, and HD-DVD.

People with spinal injuries will be able to function just as well as able-bodied people, of course until we figure out how to repair spines with stems cells and then ... well you get it.

Super Powers

Sometimes, enhancing or replacing one of your natural abilities through cybernetic devices just isn't good enough. Sometimes, in fact, you want to do something no one else can do with their original flesh and blood. Technology to the rescue!

Jerry Jalava: To USB or Not to USB

A lot of people have said they'd like to have data right at their fingertips, but Finnish technophile Jerry Jalava went ahead and did something about it. After losing the end of his left ring finger in a motorbike accident, he was fitted for a prosthetic and got an idea: why not turn the removable prosthetic into a USB drive?

Jalava popped a 2GB flash memory component into the fingertip replacement, and preloaded it with a bootable operating system and several apps. When the USB component was needed for booting or for saving data, he would leave his fingertip behind, connected to the computer. The only problem: 2GB wouldn't always be enough space, and so Jalava immediately started thinking about upgrading to a higher-capacity drive, a wireless component, or an MP3 player ... as always with technology, you want the upgrade—640k, after all, is never enough. (Check out Jalava's USB finger at http://superyou.link/jerryjalava.)

Amal Graafstra: Open Sesame

It all started with a giant ring of keys that Amal Graafstra had to carry around; soon annoyed by office doors that automatically locked behind him, Graafstra started looking for a better way.

Realizing that identification chips use the same general radio-frequency identification (RFID) technology found inside RFID passcards, Graafstra found a biosafe glass-encased RFID tag and in 2005 had his doctor implant it into his hand with a giant syringe. Then, after rigging up his office door with an RFID reader, he was able to automatically unlock his office door with the swipe of a hand, which was useful when he was toting an armload of server equipment. (Anyone who's had to fumble for their keys with an armload of groceries can probably relate.)

After gaining some attention for his innovation, Graafstra sat back while the rest of the world caught up. When he realized there were some grisly do-it-yourself (DIY) body-hacking attempts occurring, he formed the company Dangerous

Things, which is dedicated to providing safe products for people to self-modify (and encouraging people to involve body modification professionals when their conservative MDs are too reluctant to participate).

Unlike people who consider the body a temple (see the next chapter for some of the religious objections to body mods), Graafstra is more inclined to think of the body as a "sport utility vehicle for the brain." Graafstra notes that biohackers are interested in adding other senses such as small fingertip implants that can help the recipient detect magnetism, or implementing carrier pigeon DNA into ourselves to be able to sense directions without having to whip out the iPhone compass app.

Currently, Graafstra is able to use implanted RFID and near field communication (NFC) components to unlock his door, log into his computer, and cue his smartphone to automatically open up contact information, among other things. Given the right gear, the list of things he'll be able to do in the future is pretty unlimited, like to detect and avoid angry French McDonald's employees. Graafstra believes bio-hacking is the future of human evolution—watch his TEDx talk at http://superyou.link/amalgraafstra.

Kevin Warwick: Wife-Fi Connector

Like Graafstra, University of Reading professor Kevin Warwick has also been implanted with RFID to open doors, but he took his research into cybernetics even further. Warwick was so passionate about his work helping to create bionic technology to assist the disabled that he had an implant known as a "BrainGate" placed directly into the nerves in his arm to send data straight to a computer.

Following the procedure, when he closed his fist, a mechanical hand on the desk beside him would also close; when he opened his fist, the fingers of the mechanical hand would retract. While the procedure was controversial at the time, the goal was to analyze the data to determine the specific makeup of each of the signals passing through the BrainGate to better design bionic limbs down the road.

"I was the first human to have that implanted in my nervous system. Which, since then, has been used in about three different people. You know, tetraplegics to control a robot arm, that sort of thing," he said. But Warwick has even bigger end goals. "For me, the next system is clearly brain-to-brain feedback . . . sending signals from one human brain to another."

It's not surprising, as Warwick has already done some of the work, and turned himself into a guinea pig: He and his wife got wirelessly hooked to each other and could communicate in a rudimentary fashion via their implants over the local network.

You can watch Warwick undergo a risky surgery to have the BrainGate inserted into the media nerve in his left hand; then, see it in action at http://superyou.link/kevinwarwick.

(This project was explored in Chapter 5, "The Human Computer: How to Rewire and Turbo-Boost Your Ape Brain.")

Pranav Mistry: Cruise Control

If you remember mechanical computer mice, you know the technology that started Pranav Mistry on his way to being a cyborg: The rollers inside those mice—originally designed to determine which direction the user was moving his hand on top of his desk—were redeployed to measure finger movement as Mistry gestured with his hand. This led him down a long line of experiments with how hand gestures and interactions with other real world objects could be converted to digital information in order to interact with technology.

If you are thinking Tom Cruise in *Minority Report*, then yes, you got it.

This eventually evolved into SixthSense, which uses a camera to track hand movements and projection to display information. Color-coded fingertip markers are tracked, and gestures with the fingers are converted into clicks, zoom, pinch, and other types of input familiar to people working with physical interfaces, all without the need to lay hands on a mouse, screen, or trackpad.

If you're outside, holding your hands up in the gesture of taking a picture will cause your chest-mounted tracking camera to take a photograph. Perhaps, you can use your hands to dial a phone number instead of having to haul out the smartphone and unlock it. Or, by clipping a contact microphone to a standard sheet of paper you could use the vibrations on that paper to track your finger across the surface, effectively turning it into a tablet. And then, when you return to the office, you can pinch the info you have been working with on a sheet of paper and transfer it to your full desktop with a simple gesture. (Or, we wonder, maybe you could raise your middle forefinger to order Le Big Mac on the Champs Elysees.)

Despite the fact this technology ultimately turns the wearer into a cyborg, Mistry notes that its capability to make everyday objects interactive is actually a way to help people reengage with the physical world, rather than forcing them to constantly toil away in front of a machine. (Learn about Mistry's revolutionary tech tools at http://superyou.link/pranavmistry.) A paradox, perhaps, but one that might help keep us human.

A CHIP OFF THE OLD CYBORG

If you've ever used the tap-to-pay option on your credit card, tapped a security card against a sensor, or paid for your transit by placing your transit card against a terminal at the train station, you've made use of RFID technology. RFID uses a radio signal (sent by the terminal) to activate a small chip with identification information embedded in it (found on the card), which then bounces back identifying information. NFC chips are similar.

When you go to the vet to get your pet microchipped, the same technology is used. A small glass-encased passive chip, slightly larger than a grain of rice, is embedded under your cat or dog's skin using a special syringe. Should tiny little Snuggles ever make a break for it and get lost, animal control officers can use a handheld scanner to retrieve the information from the embedded chip. The bottom line: your cat is a cyborg. (Yes, that's right: You don't have to be human to be considered a cyborg. An antipoaching organization has even started installing heart-rate monitors and wireless cameras in rhinoceros's horns to discourage those who'd illegally kill the big beasts for their body parts. Cyborg rhinos: true story!)

In theory, humans could use the same implanted chips under their own skin. Such chips could be helpful, as they can provide identifying information in the case of an emergency, or even medical information (allergy warnings, to give just one example). Or, they could be used as authentication, allowing a chipped user to wave their hand to open a door, unlock their workstation, or start their car.

British teenager Byron Wake recently decided he wanted to get in on that action. After ordering a kit online, the 15-year-old injected an NFC chip into his own hand without his parents' knowledge, becoming the youngest-known cyborg. Currently, Byron can use the programmable chip to unlock his Android phone without requiring a password; in the future, he could well use this implant to unlock doors or pay for a nice cup of tea.

It's easy to see how this type of implant can become useful to more people in the coming years, for just these reasons. With the proliferation of cyber-attacks and data breaches, people are having to change passwords on a regular basis. Add in the fact that "secure" passwords resemble little more than gibberish to most people and you have a big problem, either with security or with users constantly getting locked out of their accounts because they can't remember the gibberish strings they ended up using as passwords on their six billion different regularly used websites. By tying the password to the implant, it could make authentication as simple as the wave of a hand.

Super Body: Wearables for Amputees

If you lost a limb, you might think you would need to become resigned to living without the body part or get by with a clunky artificial limb with next to no functionality. Not so fast, Captain Hook! There's a whole lot of work being done with robotic limbs that replace much of the functionality of the original body part.

Jesse Sullivan: Resume Hugging

Electrician Jesse Sullivan lost both his arms after accidentally grabbing a live wire, resulting in their eventual amputation in 2001. Through the Rehabilitation Institute of Chicago, Sullivan had some of the nerves previously used to control his arms rerouted to spots inside his chest muscles. From there, they were attached via electrodes to a mechanical arm that responded to signals coming from the nerves in his chest. This combination allowed Sullivan to control the arm simply by thinking about moving his arm the way he used to, since the rerouted nerves were now attached to new mechanical parts.

While the initial array of electrodes and connections was fairly unwieldy and unattractive, it certainly allowed Sullivan to conduct some of the tasks of which he was capable before the amputation, including eating, shaving, and vacuuming (see http://superyou.link/jessesullivan). More importantly, it allowed him to play with and hug his grandchildren again. For a man who had no arms at all, that's a pretty big deal. And of course, as technology improves, newer robotic arms will become more natural looking, and the connections to the nerves will become more discrete.

Nigel Ackland: Give this Guy a Hand

In 2006, Nigel Ackland's right arm was crushed in an industrial accident, and after six months of pain and infection, he opted for a voluntary amputation of his arm below the elbow. In 2012, after six years of feeling somewhat like an outcast, he received a call to be the first person to try out the new Bebionic 3 robotic arm prosthetic. With this new hand, each finger can be controlled separately and the motor control is fine enough to crush cans or hold fragile items safely.

With the Bebionic 3, connections to existing nerves aren't necessary; instead, the prosthetic is able to sense muscle contractions at the end of the amputated arm and convert those to the appropriate movement using the servos inside the hand, to a total of 14 different hand positions.

As with Jesse Sullivan, the device allowed Ackland to resume some day-to-day tasks like two-fingered typing and driving a car. While Ackland has been known to rock the plain black robo-hand look, the prosthetic also comes with a skin-toned covering, for those who don't want to call immediate attention to the fact that they're cyborgs. You can hear Ackland's story at http://superyou.link/nigelackland.

DIY Bio-Hacking

The problem with a lot of bionic and cyborg technology is that it requires you to be injured first, and then to have to see a doctor to get replacement parts. That leaves a lot of people out of the equation—unless you're a DIY bio-hacker. Like Amal Graafstra (who we discussed earlier in this chapter), there are others who have taken it upon themselves to improve their capabilities through high-tech means.

Gabriel Licina: Cat's Eyes

If you want to boost your vision, you don't have to pluck your eye out and replace it with a robot eye; instead, why not use an eyedropper to dribble in a liquid known as Ce6 (Chlorin e6), which is typically found in deep-sea creatures? Gabriel Licina, a testing consultant with the group Science for the Masses, had the substance dropped onto his eyeballs, where it soaked down into his retina and increased the sensitivity of his vision to the point where he could see up to 50 meters in the dark.

Don't get any bright ideas, however: the effect of this chlorophyll analogue only lasted a short time, and required Licina to wear black contact lenses to prevent him from receiving too much light during the testing phase: it's not something you want to play around with without precautions. On the other hand, it's ample proof that humans can boost their abilities temporarily without drastic measures, opening up a lot of military and emergency-responder possibilities.

The Grinder Movement: DIY Surgery

Earlier in the chapter we heard about Amal Graafstra and Byron Wake, both who used a syringe to implant themselves with small RFID implants. But these two are by no means the most extreme cases out there. While some doctors might be okay with injecting the RFID implant into a human recipient, for the most part, electronic technology is not rated for implantation into the body, and consequently medical practitioners are hesitant to get involved in these more-advanced modifications, as it could cause them to lose their license.

These do-it-yourself biohackers, known as grinders, are starting to implant themselves with actively powered components that can connect to other equipment with Bluetooth, and possibly (in the future) even interface directly with the brain ala William Gibson's *Neuromancer*.

Tim Cannon had a Bluetooth-enabled temperature-sensing module (known as the Circadia) implanted into his arm as a way to track his health. Despite the unit being much larger than the RFID tags Graafstra and Wake used, Cannon had the entire procedure performed (complete with sutures) without anesthetic by a fellow body-modder. The electronics embedded under Cannon's skin also resulted in a pretty pronounced lump directly under his skin, so there's little subtlety to some of these implants at this stage. But again, as technology advances, miniaturized versions of these electronic components will likely become available in form factors little bigger than the rice-sized RFID tags.

"The medical industry really, truly believes that it is unethical to attempt to supersede your limitations," said Cannon in an interview with *Motherboard*'s Max Hoppenstedt. "I would like to improve a lot of those inefficiencies. I think that's the best course of action for preserving conscious thought in this universe."

The DIY community has to be careful about their modifications, though. Even when it comes to relatively simple procedures like having a neodymium magnetic implant placed into a fingertip (a procedure Cannon also had done previously), it has to be performed without anesthetic or surgical tools—were these things to be used, the person performing the procedure could be charged with performing medicine without a license. Consequently, much of this work has to be done in a gray area, which opens up ethical questions about the procedures, raising concerns that might not be immediately evident to the younger and more enthusiastic modders.

"I think there are a lot of younger people that have more piercings, it is something that is done more now," says Kevin Warwick, speaking of his lab's work with implants. "We've got another student, Matt, who's got all the piercings. He has them all over his ears, his nose, his body. He wonders why we have to bother with all the ethical approval. It was relatively trivial in comparison to what he has already got."

 ## A Different Daily Grind

If you wanted to try these biohacks yourself, you can find out more about these implants at the Grindhouse Wetware site (www.grindhousewetware .com) or through the biohack.me forums. But be warned: This type of modification is almost certainly not sanctioned, can be painful, and unless you already know someone in the movement, you're almost certainly on your own. Proceed with caution, cyborg.

STELIOS ARCADIOU: TRANSHUMANISM PERFORMANCE ARTIST

Now known as Stelarc, performance artist Stelio Arcadiou comes to his work from one perspective: The human body is obsolete. Stelarc's performances often include a cyborg element, with the artist himself strapped into robotic limbs controllable by the human audience. In 1993, he had an installation where a mechanical contraption with a camera was inserted directly into his stomach. In another performance, he was strapped into a six-legged robot that walked around the stage like a spider.

More controversially, Stelarc has started the process of giving himself a third ear, located on his left forearm. The beginning of the process has included the formation of the ear under the skin with the intention of using stem cells to create a more fully formed ear on his arm. Eventually, electronics will be implanted including a microphone that can connect via Wi-fi, allowing people on the Internet to listen in. A speaker will be included in Stelarc's head, allowing him to hear what other people are sending his way—essentially, allowing people to call him without needing a phone to take the call.

Of course, this kind of body-hacking has not been without consequence; Stelarc notes that his romantic partners haven't always been receptive to having a partner with an Internet-enabled third ear. We are not sure why because it could be seen as a potential husband enhancement. Unless of course there was an argument: "You're not listening to me again. Your third ear is turned off, isn't it?!"

The Future of Cyborg?

The cyborg pioneers we visited in this chapter have brought us a long way into the future world of the cyborg, as has the proliferation of wearable technology over the early parts of this century. But it's clear that going further will take a bit of work. Even Norbert Wiener, the father of cybernetics, had his concerns at the dawn of the discipline in 1948: "We have contributed to the initiation of a new science which, as I have said, embraces technical developments with great possibilities for good and evil. We can only hand it over into the world that exists about us, and this is the world of Belsen and Hiroshima."

And, of course, there are many ethical concerns raised along the way. Having an eye or limb replaced when it's lost in an accident is one thing, but what about replacing a perfectly functional body part simply because the replacement part is better? Are we ready for that? Right now, medical practitioners are even unwilling to install nondestructive implants that grinders are eager to use (such as the Circadia), opening up the possibility of back-alley cyborg chop-shops with the inherent risks that come with such nonlicensed activity.

But to a growing number of transhumanists, this is bound to change as the world becomes more familiar with technology. "The distance between the two is so small now. As that blends, where do we begin and where does technology end? It's been getting closer and closer over time," says Amber Case. "I think as people become more moderate in the ways of technology they will start to see it more as a tool to help with creativity."

"When people have more time they will hopefully become more educated and that will have huge ramifications," agrees Zoltan Istvan, who is running for United States President as the leader of the Transhumanist Party.

We'll visit with Istvan again in Chapter 7 "In Hacks We Trust? The Political and Religious Backlash Against the Future" but in the meantime it's worth underlining one point: The fact that a transhumanist is running for President shows there's already been a big movement toward acceptance of these types of enhancements. Can a world where going to get a bionic arm is as easy and acceptable as going for Lasik eye surgery be on the horizon?

We think so.

7

In Hacks We Trust? The Political and Religious Backlash Against the Future

Your baby has a debilitating genetic disorder that runs in your family. She will be born horribly impaired and will likely die in the first weeks of her life. Her doctors have detected it early, in utero. However, there might have been a technology available that could have corrected the issue. But your elected officials—supported by morally opposed religious clerics—took legislative steps that hamstrung the scientific process by making it illegal to pursue the research. It's possible overseas, where there is no preventative legislation, but not at your local hospital.

Robotic technologies are poised to make all cars and trucks driverless. It eliminates once and for all drinking and driving accidents, reducing the number of alcohol-related deaths by more than 10,000 each year in the United States (2013 statistics). It could also eliminate more than $6 billion in speeding ticket revenue for local governments in the United States (data reported by StatisticsBrain.com for 2014). However, the local politicians, police unions and taxi and truck driver unions oppose the legislation that will allow the vehicles. Why? Millions of jobs lost and higher municipal taxes.

Where do you go to church, temple or mosque next week? Who do you vote for in the next election?

What about nanobots in your blood stream cleaning up your arteries? How about lab-grown organs that make transplants easy and major illness survivable? What if these things—as well as living forever thanks to new longevity technologies—were made illegal?

New technologies accelerating us into the future and letting us become the ultimate version of the Super You we've always wanted to be will likely be thwarted not by the limit of physics or chemistry or the nature of the universe, but by the moral majority, religious beliefs and political filibustering.

There's a pretty big chance that it's people—your family, your friends, your neighbors, your political representatives and your spiritual advisors—who want to prevent you from being so damned awesome.

Now let's pause to include a little caveat. If you have detected that your authors are more pro-science than pro-religion, you'd be right—but we also are fair-minded people. We certainly respect the views of others, and enjoy a good friendly debate on these issues. There is certainly room for science and faith to coexist. No one camp has all the answers to the questions we are dealing with today on revolutionizing what it is to be human. We also believe that a little governmental regulation also is somewhat necessary for a society that works. It's a matter of finding the balance, in a respectful way.

History is full of examples of scientific progress being thwarted by the powers-that-be, whether religious, political, societal or a combination of all three.

Once upon a time, the scientist Galileo Galilei spent a lot of time looking into the sky, and through meticulous observation confirmed Copernicus's theory that—contrary to the prevailing wisdom of the time—the Earth actually revolved around the sun, not vice versa. This very much annoyed the Catholic Church, which held that the Earth was the center of the universe and everything else revolved around it. For his troubles, Galileo was put on trial and subsequently confined to his house for the rest of his life. The condemnation forced the famed astronomer and physicist to recant his discoveries as "abjured, cursed and detested." This caused him enormous personal anguish; however, it saved him from being burned at the stake. Galileo died under house arrest at the age of 77.

If you don't remember the big controversy it's probably because it happened well before you were born, way back in 1633. (This disclaimer might not apply to those of you who are vampires or very large trees.)

The Church didn't accept heliocentrism as fact until 1820 and didn't acknowledge having treated Galileo poorly until Pope John Paul II apologized in 1992 and said the scientist was imprudently opposed.

 Heliocentrism versus Geocentrism

In case it's been awhile since you studied astronomy, heliocentrism refers to science that places the sun at the center of our solar system with the Earth—as well as the other planets in our solar system—revolving around the sun. This, of course, flew into the face of conventional "wisdom" of the

day, which was that the Earth was the center of the universe and everything
else revolved around it (geocoentrism).

Still, Galileo got off easy compared to Giordano Bruno, who was persecuted for
such outlandish discoveries that the Earth spins around the sun, that there were
multiple worlds and the fact that stars and planets were not actually fixed in the sky
but moved through the ether. It was scandalous! In the year 1600, Bruno got burned
at the stake for these views (among others with a similar heretical bent). Worse, he
never received a proper posthumous apology from the Catholic Church for treating him like a campfire marshmallow. In the ninth century, some reports claim that
Baghdadi medical expert Razi wrote an extremely large medical text that irritated a
Muslim priest so thoroughly the priest beat him violently on the head—with Razi's
own book—causing Razi to go blind.

You'd think things might have improved by now, but some days it seems that the
only difference is that people are getting beaten over the head with words instead
of the large books that contain them. On the plus side, it's probably easier to avoid
being beaten into blindness.

Prefer an example with a Slavic twist? A whole generation of Russian scientists was
subject to a doctrine that came to be known as Lysenkoism. That name came from
Trofim Lysenko, who rose to a position in the agricultural department of Joseph
Stalin's government. During his time in government, Lysenko worked to suppress
generally accepted information on plant genetics. Lysenko's own dubious (and often
falsified) theories were pushed, counter-evidence was destroyed, and scientists who
didn't toe the party line were persecuted, sent to labor camps or even executed.
This lasted from the 1930s up until 1960.

Over the last century in the United States, persecution of science has been a little
less bloodthirsty. But that's not to say there hasn't been any pushback for theories
that didn't suit the needs of the parties in power.

In a move that would have confounded the GOP of today, Republican President
Richard Nixon founded the Environmental Protection Agency (EPA) in 1970. Since
then, the EPA has remained a target of scorn for business bigwigs, climate change
deniers, and many Republican politicians who often came from one (or both) of the
other two camps. According to NASA scientist James E. Hansen, the White House's
Office of Management and Budget (OMB)—working for Republican President
George H. W. Bush—had Hansen's Senate subcommittee testimony on humanity's influence on climate change altered to include text that, he claims, negated the
entire point of what he was trying to say. As he said to *The New York Times*,
"It distresses me that they put words in my mouth. I should be allowed to say what
is my scientific position; there is no rationale by which OMB should be censoring
scientific opinion. I can understand changing policy, but not science."

Hansen makes an excellent point. Most people are willing to accept that it's possible for intelligent people to disagree in their positions, so long as they are willing to admit to the facts in front of their eyes. Otherwise, you haven't learned the lesson of the clerics who condemned Galileo. And if you're in a position of power, that can hurt everyone.

The Future Is ... Now?

Because you are reading this book, there's a good chance you're interested in ways to use science to make life better for yourself and your offspring, whether that means a life free of disease or the addition of super-snazzy, futuristic implants. Surely, you might think, the examples we presented earlier in this chapter are all locked away in the dusty pages of the science-hating past.

The bad news: People also want to kill your dreams right here in the future. Or, at the very least, they'd like to block them for awhile because they are contrary to personal agendas of people in power or long-held doctrine of powerful interest groups. Need some examples? We happen to have a few.

> "The bad news: People also want to kill your dreams right here in the future."

- Maybe it's the George W. Bush administration's attempts to derail stem cell research.

- Perhaps it's (former) Canadian Prime Minister Stephen Harper's campaign of muzzling scientists whose research disagrees with the official Canadian Conservative Party platform.

- Maybe it's privacy advocates who worry that technology can be used to remove personal freedoms.

- It could be one of the world's religions, working either through its own ministries or through one of the political parties, that adhere to its dogma.

Whichever combination it is, it's clear that when science leads the way to the potential of a Utopian future, some people feel the need to act as roadblocks causing bureaucratic, legal and dogmatic obstacles to your right to a Super You.

Let's take a look at some of the specifics.

 Utopia?

Sir Thomas More coined the term in his book, *Utopia*, which was written in 1516, and chronicled a fictional island in the Atlantic. A utopia is a concept that imagines an idyllic society or world in which life is perfect for all who live in that society. For example, if every person born in the United States truly was equal—regardless of his or her gender, ethnicity, sexuality, or religion, then one could say that our country is utopic in reality and not just in principle. An emphasis on equality in all key manners (government, economic, justice) is the bedrock of the utopic model.

Body Modification

When we talk about body modification, we're not just talking about robotic arms or electronic implants; we actually mean the whole gamut of possible changes that one might want to apply to their own flesh. This runs from commonplace alterations such as cosmetic surgery and tattoos, through to more niche procedures such as purposeful ritual or cosmetic scarring. It even goes to extremes, such as men who have—either for tribal or sexual reasons—split their dangly bits right up the middle. Probably best to avoid Googling this, by the way. (You'll especially want to avoid search terms such as "penile subincision." You've been warned, and your angry mail will be returned with a "told you so" sticky note.)

On the surface, your choice to get a tattoo might seem to be strictly a cosmetic or lifestyle issue, but consider this: Several years ago, cell phone maker Nokia (whose mobile division is now owned by Microsoft) came up with technology that allowed a tattoo to vibrate when a nearby smartphone rang. More recently, Motorola has developed a throat tattoo that can connect via near-field communication with a phone, wearable computer or tablet.

It's easy to imagine this could lead to tiny implants linked to create a complete functioning smart device…all in the shape of, say, the Metallica logo, if you're so inclined. If you were to add a few little scarification-style ridges with antennae embedded so you can connect wirelessly to the Internet, and then you have something that will potentially put you on the watch list of various haughty religious or political organizations.

Curiously, and perhaps thankfully, the major American political parties don't really have much to say about these issues, to be honest. While you can bet your bippy that a lot of individual politicians don't really care for folks with tattoos and self-administered scars, the official party lines don't forbid it.

In fact, in 2014, Republican presidential candidate Rand Paul went so far as to say, "We need a party that looks like America. We need a more diverse party. People with tattoos, and without tattoos. With earrings, and without earrings."

Granted, Rand Paul leans more toward the Republican Party's libertarian side, which tends to believe in the supremacy of individual choice. Still, there are plenty of people out there with tattoos of Ronald Reagan, or of the Republican elephant. And there's even one intellectually challenged individual who had Mitt Romney's

2012 presidential campaign logo permanently inked onto his temple. Try explaining that one to your McDonald's crew chief.

The Democratic side is no brighter. There are Democrats who have inked their arms with the Democratic donkey, Obama's O logo, and one guy who permanently inked Jimmy Carter's face onto his butt cheek. (Maybe that guy was not a supporter, come to think of it.)

The major political parties don't really have a stake in this. The scriptures (for whichever religion you choose) are generally a bit less tolerant, and there's probably a good reason for that. Body-modification techniques such as scarring and tattoos have a long history in tribal societies, as part of rituals and to mark achievements and milestones. Because these societies have been (in many cases improperly) regarded as uncivilized, it's always been in the interest of the new, modern, super-shiny religions to avoid such barbaric rituals. The specifics, however, vary by religious affiliation.

Judaism

When it comes to religious restrictions, a lot of things date back to a little text called Leviticus, which makes up one of the five books of the Torah. This book, to put it mildly, features a whole lot of serious *don'ts*, said to have passed directly from God to Moses. Apparently, a Judaic God prohibits messing around with your body. Take Leviticus 19:28, where God proclaims "Ye shall not make any cuttings in your flesh for the dead, nor print any marks upon you."

On the other hand, it's worth mentioning that this is just after He's just dropped this one on our man Moses in Leviticus 19:27: "Ye shalt not round the corners of your heads, neither shalt thou mar the corners of thy beard." In other words, don't cut your hair or trim your beard. Barbers be damned!

While that no-trim proclamation tends not to be strictly adhered to (apart from the Orthodox Jews), tattooing and scarification still tend to be looked down upon. Indeed, there has been much talk about the inability of Jews with tattoos (now there's a band name for ya) to be interred in a Jewish cemetery, including a famous comedy bit in which Lenny Bruce riffs on how he came home from the Navy and his mother screamed about his arm tattoo worrying that he would never be able to be buried with his own kind. "OK. Maybe I will be buried in a Jewish cemetery. They can bury my arm in a Catholic cemetery," he riffed.

Like so many things, however, that particular nugget is urban legend rather than an official position of Judaism.

Of course, there's always room for disagreement, and naturally there's some divergence here, in this case around the interpretation of Shulchan Aruch (Yoreh De'ah 180:2), "If it [the tattoo] was done in the flesh of another, the one to whom it was done is blameless."

Some have interpreted this to mean that tattoos are only forbidden when self-administered, and that sitting in a chair while someone else wields the tattoo gun is generally quite godly. But Orthodox Jews think that that's just taking advantage of a ruling that's meant to provide forgiveness for those who have had tattoos forcibly applied, such as survivors of the Holocaust. In other words, it would be like saying it wasn't your fault you got drunk, as long as someone *else* poured your favorite beer into your mouth.

On the other hand, male circumcision is not only acceptable by those of the Jewish faith, but it's considered de rigeur, not only for cleanliness but to symbolize God's covenant with mankind (Genesis 18:14). As part of the Abrahamic tradition, this exception has also become prominent in both Christianity and Islam, though it generally remains optional—and is elective by parents of boy children. The babies, of course, don't get a vote.

Christianity

In addition to its prominent placement in the Torah, Leviticus is a part of the Christian Old Testament. While many of the teachings found in it have been somewhat deprecated and displaced by those found in the New Testament—in fact, it doesn't take much Google searching to uncover tattoos with a Christian flavor— Leviticus still casts a somewhat long shadow, especially for those who are more fundamentalist.

 Messiah in Ink

You can find some pretty impressive tattoos depicting Jesus Christ at http://superyou.link/jesustattoos

It's when we get into later books of the Old Testament—ones that fall outside of the scope of the Torah—that we get a few hints as to *why* some of these restrictions are in place. In 1.Kings 18:28, for example, we learn one rationale for the prohibition on cutting is to distinguish the individual from the follower of a forbidden religious group, such as the prophets of Baal: "And they cried aloud, and cut themselves after their manner with knives and lancets, till the blood gushed out upon them." In other words, we're not savages and pagans like *those bozos*.

On the issue of tattoos, however, there's slightly more confusion elsewhere in the Bible. For example, in Isaiah 49:16 we read "Behold, I have graven thee upon the palms of my hands," which could literally be interpreted as having tattooed God's name straight onto the hands. Or it could be a metaphor. Biblical interpretation is a slippery business.

The New Testament is a bit less rigid when it comes to issues of body modification. One of the oft-quoted sections of the Bible comes in 1. Corinthians 6:19–20: "What? Know ye not that your body is the temple of the Holy Ghost which is in you, which ye have of God, and ye are not your own? For ye are bought with a price: Therefore, glorify God in your body and in your spirit, which are God's."

In other words, rather than tattoos and scars being considered mortal sins, humankind is asked to consider the reasons for making these modifications. Getting a tattoo because you want it? That's not gonna buy you a ticket to Hell by itself. Getting a tattoo because you're trying to annoy your parents? Maybe it's time to don some asbestos underwear because the Hell Express is on the way and you've got a first class ticket (thanks to the Fifth Commandment, which requires children to honor their parents).

Islam

In Islam, body modification is also considered an affront to the will of the creator, as outlined in 4:119–121 of The Qur'an, where Satan is detailed as claiming: "I will mislead them and incite vain desires in them; I will command them to tamper with God's creation," following up with "Whoever chooses Satan as a Patron instead of God is utterly ruined," and "Such people will have Hell for their home and will find no escape from it."

But not so fast. While this might seem pretty straightforward, there is some debate as to whether tattoos are halal (permissible) or haram (forbidden), and some of the debate comes down to … water. While hardliners might not bend on tattoos (being a modification of God's creation), others think it's allowed because tattoo ink is subdermal (ink is injected under the skin) and thus there is nothing coming between water and a person's skin, meaning proper cleansing is possible. Ergo, the tattoo is not a problem. It would appear that Shia Muslims generally seem to lean more toward allowing tattoos, while Sunni Muslims seem to lean more toward forbidding them.

As for circumcision, while Islam is part of the Abrahamic tradition, it is not considered mandatory. And while female circumcision tends to be more associated with Islam than other religions, it's also not considered mandatory.

Hinduism

After this long list of negative reactions from the world's religions, you're probably expecting the Hindus to frown on tattoos and body mods, right? Well, nope.

In fact, there's quite a long history of Hinduism not only allowing tattoos, but in some cases actively encouraging them. While the red dot many Hindu women place on their forehead (known as the bindi) is often applied using a temporary powder, other women have had it tattooed on permanently. Hindu women have also been known to tattoo other dots around the eyes both for cosmetic reasons and as a method to "ward off evil."

It's not just women, however. Both men and women have long gotten "Aum" tattoos on their arms, again to ward off evil. And piercings, such as nose rings, are commonplace.

In short, Rand Paul's statements should draw Hindus to the Republican Party.

As for body modification, India leads the way in skin whitening, with the country's residents purchasing a stunning 258 tons of bleaching cream in 2012 alone. But this is, perhaps, unsurprising in a country haunted by a caste system where lighter-colored skin suggests power.

ORGANS ARE OKAY

Oddly enough, despite the many restrictions on adjusting parts of a person's outward appearance via ink or blade, the major religions are pretty tolerant on organ donation and transplants, even considering it virtuous to donate an organ if it can save a life. The main opposition to organ transplantation comes from those that follow the doctrine of Jehovah's Witness, but it's not even a ban. Their main concern is the transfer of blood from one person to another; so long as the organs are drained of blood before transplantation into the recipient, it should be considered acceptable.

THE BIGGEST ENEMY OF TATTOOS: YOUR OWN TECHNOLOGY?

The Apple Watch launched with great fanfare in the spring of 2015, but a few of the Apple faithful were left frowning at their wrists. Why? Because they were also fans of inking the skin on their wrist.

One of the key features of the new Apple Watch is the capability to track the user's health statistics. The rear surface of the watch, which rests against the wearer's wrist, has LED lights that shine onto the skin. Sensors also track blood flow through the area. I think you see where this is going.

People with a true predilection for inking themselves are rarely satisfied with a butterfly tattooed discreetly on their bums; in fact, many go so far as to ink their arms from wrist to shoulder, a technique known as "sleeves." And that's where the problem arises: If the ink along the wrist is too dark, it can prevent the sensor on the Apple Watch from reading the blood flow below it, which makes several of the health-tracking features on the watch useless.

It's worth noting that this isn't a problem exclusive to the Apple watch. Users of other health trackers, such as the Fitbit Charge HR, the Microsoft Band and the Scosche Rhythm+, have noted similar problems making the trackers work on inked skin.

Ultimately, heavily inked tech fans might have to wait for implanted fitness trackers until they can fully take part in the new fitness craze.

Genetic Engineering

The idea of inheritable traits first came to public attention in the 1800s with Charles Darwin postulating his theory of evolution in the 1859 book *On the Origin of Species.* Around the same time, an Augustinian friar named Gregor Mendel was conducting experiments on pea plants, noting how their traits changed when hybridized.

Ultimately, while Mendel's research was largely ignored when released in 1865, his conceptualization of "discrete inheritable traits" became the basis of genetic study in the early 1900s. When you consider how much religious handwringing about genetics has arisen since then, Mendel's calling as a friar seems somewhat ironic.

Although it took until 1953 for James Watson and Francis Crick to discover the structure of DNA—which contains the specific information necessary to determine traits of an individual—the idea of manipulating those specific traits entered the public consciousness well before then. In fact, in the 1931 novel *Brave New World*, Aldous Huxley described a world where all reproduction took place outside the human body and babies were engineered to produce humans with specific traits and capabilities.

While *Brave New World's* vision of engineering of the characteristics of all babies was perceived as a dystopian vision of the future, people have seemed more receptive to the idea of engineering their babies on a case-by-case basis: It's a much different situation if you're determining your own baby's future than if the state is doing it for everyone, after all.

The arrival of CRISPR-Cas9 technology, which allows the editing of DNA—and the potential for potential parents to "edit" the DNA of their unborn children—is causing ripples of discontent. The concerns are not just about potentially breeding a generation of genetic monsters. There are concerns that only people who can afford the technology will have access to it.

The potential upside is massive. It offers the promise of erasing devastating hereditary diseases such as Parkinson's or Alzheimer's from the family tree. If only the wealthy can afford the procedures, the fear is it will create a two-tiered society of the healthy rich and the sickly poor. When you add in the possibility of enhancing traits such as intelligence or height, you further run the risk of disadvantaging the offspring of those without financial means.

There are many books, video games, and movies that talk about the problems with genetic engineering causing inadvertent stratification of society, including Margaret Atwood's *Oryx and Crake*, Kim Stanley Robinson's *2312*, the BioShock video game series, and the 1997 movie *Gattaca*.

Critics of genetic engineering claim there's no need to edit DNA because screening during the in vitro fertilization (IVF) process can accomplish many of the same results for the gene pool, and at much lower costs.

In an article for *MIT Technology Review*, Sangamo Biosciences CEO Edward Lanphier argued, "People say, well, we don't want children born with this, or born with that—but it's a completely false argument and a slippery slope toward much more unacceptable uses."

Zoltan Istvan, the 2016 United States presidential candidate from the Transhumanist Party, disagrees: "Critics—many of them fundamentally religious—worry that genetic engineering will create a race of non-human beings who resemble monsters. Their fears are overblown and tied more to Hollywood horror movies than actual science. The far greater likelihood is that genetic engineering will create a populace free of diseases and ailments that have plagued humanity for tens of thousands of years. In fact, genetic engineering could change the very nature of healthcare."

Nonetheless, in an editorial in *Nature* magazine, Lanphier has joined colleagues in calling for restrictions on editing the human germ-line, noting that there's a difference between genetically engineering a noninheritable change for therapeutic purposes and genetically engineering a baby with changes that will be passed down to *its* future offspring as well.

"In our view, genome editing in human embryos using current technologies could have unpredictable effects on future generations," Lanphier said. "This makes it dangerous and ethically unacceptable. Such research could be exploited for non-therapeutic modifications. We are concerned that a public outcry about such an ethical breach could hinder a promising area of therapeutic development, namely making genetic changes that cannot be inherited."

It's worth considering that the success rate of genetic engineering is still low—MIT researcher Guipong Feng has pegged genetic engineering tool CRISPR's (clustered regularly-interspaced short palindromic repeats) success rate in deleting/disabling a gene in a zygote (the single-cell union of sperm and egg) at about 40 percent.

We won't have to worry about making these decisions about our offspring in the very near term, but given Kurzweil's assertion that technology improvements are ever accelerating, chances are this will be an issue in the next decade, and certainly, we assert, by 2030.

That's assuming, of course, it isn't banned by legislation. As we write this, 15 of 22 European countries have already banned such human genetic modification, and while there's been no decision in the United States yet, but that could change. Blocking genetic engineering through legislation would come with a hefty social and financial cost: A rise in medical tourism to other parts of the world.

"People are going to go overseas if they don't allow it here," stresses Istvan. "It's just like we go overseas to have a kidney replaced because it's so much cheaper, we are going to go overseas to have these babies done this way. And you can already see people talking about it. There is money going to be put into it."

THE VATICAN'S STANCE

In Chapter 2, "Baby Science," we heard from Dr. Jeffrey Steinberg, who discovered a way to isolate the gene that determines eye color. That attracted the interest of a lot of potential customers who wanted to start tinkering with their offspring's eye color. But it also attracted the attention of the Vatican, as well as a number of people on both the left and right, who had strong opinions over this.

In 2008, the Vatican issued a bioethics document called "Instruction Dignitas Personae on Certain Bioethical Questions." It outlined Roman Catholic teaching on the latest procedures concerning human reproduction and condemned most forms of artificial fertilization and genetic engineering. It also urged Catholics to oppose them.

Cloning

There's a rationale for genetic engineering. It can eliminate hereditary disease through DNA editing. However, cloning is much more problematic at both a conceptual and ethical level.

In *Brave New World*, Huxley imagined a cloning-like method he called Bokanovsky's Process, which took a single fertilized egg and caused it to divide multiple times into dozens of identical embryos. In 1996, however, the birth of Dolly the sheep proved the process could start with DNA from another living being, not just a traditional egg/sperm pairing.

While the capability to clone genetic material opens up the possibility of creating gene-based therapies to halt or eliminate debilitating disease, it also opens up the fear of the unethical use of such genetic material by unscrupulous researchers or governments. That's a theme explored extensively in the television show *Orphan Black*. The show has also examined the ethics of cloning, unforeseen health problems, religious persecution, and issues of personal identity.

In the United States, politicians on both sides of the aisle share concerns about the future of cloning. According to the 2012 Republican Party platform, "We urge a ban on human cloning and on the creation of or experimentation on human embryos."

In a 2009 address, President Barack Obama claimed, "... we will ensure that our government never opens the door to the use of cloning for human reproduction. It is dangerous, profoundly wrong, and has no place in our society, or any society."

The United States House of Representatives has faced legislation aimed at banning human cloning at least four times, but at this point, no legislation has cleared both the House and Senate. Most of these bills have been voted down or, as the saying goes, died in committee.

That said, while there have been no bills passed at the federal level to forbid cloning, 15 states have passed laws on the issue, with some banning reproductive cloning outright, and two forbidding the use of public funds for cloning research. Just under half of these states also prohibit cloning for therapeutic reasons.

As you'd expect, the take on cloning from the major world religions is mixed:

- **Christianity**—The Roman Catholics are firmly against it. "Halting the human cloning project is a moral duty which must also be translated into cultural, social and legislative terms," notes a 1997 report from the Vatican. "In human cloning the necessary condition for any society begins to collapse: that of treating man always and everywhere as an end, as a value, and never as a mere means or simple object." Protestants, on the other hand, have been known to encourage the use of cloning for therapeutic purposes. In the 2008 Lutheran document *Genetics and Faith: Power, Choice and Responsibility*, the church's take was to "encourage individuals, corporations, and institutions to set public policy that will [...] encourage stem cell research and, if necessary, therapeutic cloning."

- **Hinduism**—In Hinduism, there are unofficial guidelines prohibiting the cloning of humans, however there are no laws on the books as of yet prohibiting the practice in India. In fact, India has been at the forefront of research in this area, with one group having cloned a bison and another group of researchers working at resurrecting the Asiatic cheetah, an extinct species.

- **Islam**—When it comes to Islam, it's less clear. According to the Qu'ran, (Al-Zariat 51:49), all things must come in pairs, and because reproductive cloning is dependent simply on a single gender, it is not acceptable. On the other hand, cloning of single body parts for therapeutic purposes would arguably be considered acceptable.

- **Judaism**—Judaism also considers cloning to be acceptable under the right circumstances as well. In the article "Cloning People and Jewish Law: A Preliminary Analysis," Rabbi Michael J. Broyde says, "… it would appear that Jewish law accepts that having children through cloning is perhaps a mitzvah in a number of circumstances and is morally neutral in a number of other circumstances. Clones, of course, are full human, and are to be treated with the full dignity of any human being. Clones are not robots, slaves, or semi-humans, and any attempt to classify them as such must be vigorously combated."

Stem Cell Research and Genetic Therapy

There's been a lot of angst over the use of stem cells in scientific research over the last two decades. Initially, these wonder cells were harvested from human embryos. That landed squarely in the same moral gray area that plagues the abortion debate: Are embryos alive? Or are they only *potential* life? What are the moral implications of creating embryos strictly for the purposes of research if we can't even determine definitively if they qualify as life? There's a lot at stake here. Some think that embryonic stem cells have the potential to unlock a cure for diabetes, for example. However, some people just can't accept their use given their source.

After the discovery of human embryonic stem cells in 1998, a debate started about what types of research the government should be allowed to fund. The Bill Clinton presidential administration ultimately decided that the best compromise was to only allow federal funding for research to be carried out using embryos discarded after in vitro fertilization treatments, but not for research on embryos created specifically for research purposes. But implementation got delayed until Clinton left office, leaving the issue to fall in the lap of the incoming administration of George W. Bush.

In July of 2001, Pope John Paul II weighed in on the matter: "Another area in which political and moral choices have the gravest consequences for the future of civilization concerns the most fundamental of human rights, the right to life itself. Experience is already showing how a tragic coarsening of consciences accompanies the assault on innocent human life in the womb, leading to accommodation and acquiescence in the face of other related evils such as euthanasia, infanticide and, most recently, proposals for the creation for research purposes of human embryos, destined to destruction in the process."

Former President George W. Bush authorized funding for stem cell research, but only for stem cell lines that were already in existence, and not for any future potential sources.

While he originally claimed there were more than 60 stem cell lines available for research after this restriction was put into place, many of the lines being counted were actually considered to be dead ends, and unviable for research.

There was also a lack of diversity in the genetic material found in the existing embryos: All originated from parents who were seeking IVF procedures. That meant the embryos came from stock that had potential fertility issues, among other limitations.

While the Bush restriction effectively cut off the possibility of creation of embryos specifically for stem cell research, it meant that researchers weren't allowed to avail themselves of embryos generated by in vitro fertilization procedures that were still being created—and still being discarded—effectively wasting any potential they might have had when no longer needed for IVF.

It also had the side effect of driving stem cell research abroad to countries without these restrictions in place.

"This was absolutely a tragedy for science," said Zoltan Istvan, leader of the Transhumanist party. "Stem cells have proven and continue to prove everyday that they offer huge hope, and could be one of the biggest and most important medical projects in the 21st century for human beings. And to have it stopped in America, and to have some of our best scientists leave America because they weren't able to get the funding was, like I said, a tragedy for transhumanism."

Between 2004 and 2007, Congress passed several bills designed to expand the parameters for federal funding of stem cell research beyond the 2001 restrictions, but Bush vetoed all of them. This effectively limited researchers to use of the same limited run of stem cell lines being used since the restrictions were imposed, which researchers continued to argue was not varied enough.

When Barack Obama took office, he used Executive Order 13505: *Removing Barriers to Responsible Scientific Research Involving Human Stem Cells*, to lift the restrictions on federal funding of new stem cell lines, noting, "When government fails to make these investments, opportunities are missed." That opened research up to an additional variety in stem cell lines, but again, only for embryos that were otherwise going to be discarded anyhow. On the other hand, the order continued to reinforce the ban on funding of any research using embryos created strictly for research.

The good news for stem cell research is that the United States Supreme Court has been reluctant to reopen the possibility of further restrictions on the use of embryonic stem cells. In 2013, opponents felt there was no need to continue using embryonic cells thanks to the discovery of (adult-derived) induced pluripotent stem cells (iPS).

Current research makes it possible to turn any of your own cells back to a cell that can be used anywhere. When this technology become possible, a cell from, say, the liver, could be altered and used as a brain cell by turning it back into a pluripotent stem cell. (To refresh your memory on this reference Chapter 2, "Baby Science: How to Conceive a Tennis Star and Other Procreative Miracles," and Chapter 5, "The Human Computer: How to Rewire and Turbo-Boost Your Ape Brain," where we talk about this process in detail.)

Today, it is possible to take any cell from a particular area of the body and convert it into a stem cell that can only be used from the part of the body where it was sourced.

The Supreme Court refused to hear a case that could have led to further restrictions of government funding of embryonic stem cell research. "We couldn't be happier that this frivolous, but at the same time potentially devastating distraction is behind us," noted Doug Melton, Harvard Stem Cell Institute's codirector, "and we can once again focus all our attention on advancing all forms of stem cell science, including research using embryonic stem cells—which are the gold standard against which we measure other types."

Even techniques that don't utilize embryonic stem cells, such as nuclear transfer, have opened up a whole new can of worms, as the technique requires donated eggs, which again leads to worries about human cloning.

Political Views on Stem Cell Research

So that's a whistle-stop tour of the recent history of opposition to stem cell research, but where does each of the political groups stand now?

Republican Views

When the 2012 presidential election rolled around, the Republican Party had this to say about it in the party's official party platform:

"We call for a ban on the use of body parts from aborted fetuses for research. We support and applaud adult stem cell research to develop lifesaving therapies, and we oppose the killing of embryos for their stem cells. We oppose federal funding of embryonic stem cell research [...] We call for expanded support for the stem-cell research that now offers the greatest hope for many afflictions—with adult stem cells, umbilical cord blood and cells reprogrammed into pluripotent stem cells—without the destruction of embryonic human life."

It's basically a revised stance held by George W. Bush, except for complete opposition to any federal funding and a seeming unwillingness to even make use of the embryonic stem cells already in the research pool. But at least there's recognition that stem cell research has great promise in finding cures and/or therapeutic treatments for conditions such as Parkinson's disease. Perhaps there's a small victory there.

As this book goes to press, the 2016 election season is in full swing. Republican frontrunner Donald Trump isn't quite sure what to make of the issue, claiming, "I'm studying it very closely [...] It's an issue, don't forget, that as a businessman I've never been involved in." His closest competitor for the nomination, Ted Cruz, strongly opposes the use of embryonic stem cells in research.

It'll be interesting to see what ends up being in the 2016 Republican Party platform.

Democrat Views

As for the Democratic Party, its 2012 platform didn't outline any new restrictions or expansions on the party's position on stem cell research. However, it reiterated its already extant stance: "... the President issued an executive order repealing the restrictions on embryonic stem cell research and signed into law the Christopher and Dana Reeves Paralysis Act, the first piece of comprehensive legislation aimed at improving the lives of Americans living with paralysis."

Under the Democrats, decisions about funding of stem cell research and the addition of new stem cell lines has now fallen to the National Institutes of Health (NIH). The organization works under the restrictions enacted in President Obama's 2009 signing ceremony.

For the 2016 nomination race, there are a few small differences between the two frontrunners. Hillary Clinton has supported stem cell research, even promising to lift the ban on embryonic stem cell research. And while Democratic leadership contender Bernie Sanders has voiced general support for stem cell research (including embryonic), he's also opposed the use of cloning, even for therapeutic reasons.

Religious Views on Stem Cell Research

You might suspect that stem cell research would get a tougher handling by religious leaders. But your suspicions would be unfounded. (Of course, there are hardliners in all these religious groups who consider the embryo to be potential life, and consequently embryonic stem cells to be contrary to God's will. However, the views presented here represent the current thinking from the leadership of each dogma.)

Christian Viewpoints

The Catholic Church has come out firmly in favor of stem cell research, with a few caveats, of course. As with Pope John Paul II, the Church hasn't changed its stance on research using embryonic stem cells, but in recent years it has been receptive to research using adult-derived stem cells.

During a 2006 address, Pope Benedict XVI had this to say:

"Progress becomes true progress only if it serves the human person and if the human person grows: not only in terms of his or her technical power, but also in his or her moral awareness […] In this light, somatic stem-cell research also deserves approval and encouragement when it felicitously combines scientific knowledge, the most advanced technology in the biological field and ethics that postulate respect for the human being at every stage of his or her existence."

"The fact that you at this Congress have expressed your commitment and hope to achieve new therapeutic results from the use of cells of the adult body without recourse to the suppression of newly conceived human beings, and the fact that your work is being rewarded by results, are confirmation of the validity of the Church's constant invitation to full respect for the human being from conception. The good of human beings should not only be sought in universally valid goals, but also in the methods used to achieve them."

Pope Benedict later went on to write the introduction to a book on stem cell research, *The Healing Cell: How the Greatest Revolution in Medical History Is Changing Your Life,* coauthored by Dr. Robin Smith, Max Gomez and Monsignor Tomasz Trafny.

Among the Protestants, there seems to be a level of acceptance of these new technologies, at least from one camp.

In the 2008 document "Genetics and Faith: Power, Choice and Responsibility," published by the Evangelical Lutheran Church in America, the church said: "The truth is, though, that for better or worse, our society, indeed the human race, is going to be living in the age of genetics for good. We must take responsibility for this new living space because it's not going away."

Speaking about genetic manipulation and stem cell research more specifically, the document refers back to a 2004 ELCA Social Policy Resolution: "The human capacity for genetic manipulation should be understood, in principle, as one of God's gifts in the created order to be pursued for the good of all. As with any such gift, it must be used responsibly and tested for its contribution to justice and stewardship."

Islamic Views

Similarly, Islam appears to be quite open to stem cell research with many of the same caveats. The basic thought is that embryos created outside the womb are not considered human beings until they have grown inside the womb.

On the other hand, it's still suggested that only embryos that would otherwise be discarded following IVF treatment be used. One bit of guidance that has been used as justification for allowing stem cell research comes from The Qur'an 5:32: "... if any saves a life it is as if he saves the lives of all mankind."

Jewish Views

The Jewish position stakes out much of the same territory. Rabbi Levi Yitzschak Halperin writes in Ma'aseh Chosev, Vol. 3 2:6:

"As long as it has not been implanted in the womb and it is still a frozen fertilized egg, it does not have the status of an embryo at all and there is no prohibition to destroy it...it is preferable not to destroy the pre-embryo unless it will otherwise not be implanted in the woman who gave the eggs (either because there are many fertilized eggs, or because one of the parties refuses to go on with the procedure—the husband or wife—or for any other reason). Certainly it should not be implanted into another woman. ... The best and worthiest solution is to use it for life-saving purposes, such as for the treatment of people that suffered trauma to their nervous system, etc."

Hindu Views

Hindus are struggling with the issue. In Hinduism, the belief is that life begins at conception, which means that embryonic stem cells are problematic; on the other hand, the faith is not completely against abortion (it's acceptable in situations where the mother's life is in danger, for example).

"Many Hindus see the soul—the true Self (or atman)—as the spiritual and imperishable component of human personality," wrote Pankaj Mishra in the *New York Times*. "After death destroys the body, the soul soon finds a new temporal home. Thus, for Hindus as much as for Catholics, life begins at conception."

It's worth noting that this core belief hasn't prevented the Indian biotech industry from rocketing ahead with genetic research—especially benefiting from Bush's funding restrictions in the United States. It's worth saying, though, that India is not strictly a Hindu state; nearly 15 percent of the population identifies as Muslim. And of course, business rarely saddles up closely to every tenet of a dominant religion when there is money to be made. In the boardroom, beliefs can often be selectively ignored if they get in the way of profit.

> "... business rarely saddles up closely to every tenet of a dominant religion when there is money to be made."

Emerging Technologies

Let's talk a bit about the good stuff such as nanotechnology, implants, cyborgs, and bio-automation. Of all the sections of this chapter, these emerging technologies have had the least vocal pushback from the usual religious and political sources. Why? In some cases it's because there's no particular reason to forbid the technologies. However, we suspect there's another reason for the lack of reaction: These technologies are either so new that there hasn't been enough time for these groups to develop a reaction (negative or otherwise), or they've seemed so far out that people don't think they will actually become a reality anytime soon, or if ever.

It would be like asking a senior cleric at a church, mosque, or temple to comment on the *Star Trek* communicator and its impending impact on society in the 1960s (when the original show aired). The device was science fiction then and yet a half-century later, members of most congregations on the planet have a smartphone in their pocket. And the device does far more than the *Star Trek* communicator ever could and has had a further-reaching impact on culture and society than anyone could have ever imagined.

Istvan believes the pushback and the impending new technologies will soon come as the technologies find their way into our every day life. "I would not be surprised if over the next five to 10 years you see a huge shift of these religious people implementing laws that absolutely make a lot of this stuff illegal," he says. "It's because it's the very first start of people saying 'well, lets put a moratorium on something when we don't understand it or it seems unethical'. But, what's unethical is very much open to question. Obviously, making babies smarter or healthier or completely eradicating disease, these are great things, but a lot of Christians don't look at it that way."

In the meantime, we're starting to see early reaction—and some opposition—coming from society and technology critics. While the technology industry often roars ahead with new technology simply because it *can* be done, some are starting to ask whether simply being able to do something is enough justification to do so, especially when there can be unintended consequences.

Writer Nicholas Carr has explored this topic for a number of years now. In his 2008 *Atlantic Monthly* article "Is Google Making us Stupid?" (and further in his book *The Shallows*), he argued that our recent habit of always turning to online sources for answers to our questions has impaired our ability to retain information in the long-term, or to engage with deeper, longer-form arguments at all.

More recently, in *The Glass Cage*, Carr worries that automation has been having much the same affect our ability to think and troubleshoot real world problems. "Automation complacency takes hold when a computer lulls us into a false sense of security. We become so confident the machine will work flawlessly, handling any challenge that might arise, that we allow our attention to drift."

While the end result of that lack of focus could be as simple as trusting your phone's autocorrect feature to the point of sending nonsensical (yet formally correct) messages, the stakes could be much higher as all these technologies converge.

Imagine, for example, following turn-by-turn directions beamed straight into your eye by something such as (the now defunct) Google Glass or more likely some future optical implant. Now imagine finding yourself walking straight into a location you shouldn't actually be in, owing to faulty GPS location information, out-dated map data, or map data that's been maliciously hacked. Who hasn't already been advised to make an illegal U-turn by a GPS device?

The consequences of these seemingly innocent technology errors could range from simple disorientation to getting robbed—or worse killed. This gets worse if the error happens while you are at the helm of an airliner or space shuttle filled with hundreds of passengers.

In *The Glass Cage*, Carr relates the story of the ocean liner *Royal Majesty* which ran aground after the cable to the GPS unit came loose; despite many visual indications that the ship was off-course, the crew continued to trust the technology was correct.

Critics of many of these technologies designed to augment our capabilities, then, are concerned that improper implementation might ultimately have the opposite effect.

More problematic, at least in the short term, is the concern that these new robotics and automation technologies will threaten the average person's ability to have a job and earn a living. While self-driving cars seem to offer many positive benefits, such as the elimination of fatalities caused by impairment or inexperience, extending the concept to self-driving transport trucks means a loss of jobs and income, at least until the truck drivers can retrain for another occupation.

But what if all jobs become automated to the point there's less need for a human workforce? What if manual labor is replaced by robots and automation? What if a robot makes your pizza and delivers it by drone? Wait, that would be cool.

Istvan says part of his party's platform is to support a universal basic income, to ensure that everyone has their basic needs taken care of, even as automation and technology takes over the actual labor.

"You are going to see a huge outcry and this will probably be one that will cause the most civil unrest," he says. "Universal welfare is a basic system but everyone needs something if they don't want to work because the rich keep getting richer and the poor keep getting poorer but maybe we can make it so no revolution can occur and give everyone a—what I call—a luxury communism. I'm not supportive of this type of communism but quite frankly it's just better, robust and it will work."

Yes that's a current presidential candidate advocating a new kind of communism. Fear not, however, because this is just semantics. No need for neo-McCarthyism here.

This redesign of America would require a complete restructuring of how society currently works, which is the sticking point.

Istvan said he wouldn't be surprised if some politicians adopted an anti-robotic stance, at least if those robots start replacing human jobs. That said, you can't win Wall Street with that platform. "And if you can't win Wall Street you can't win the election," he explained.

"It's really a matter of making sure that economies run smoothly. We can't have something like in 2007 where all the banks collapse. If all the banks collapse, America collapses," muses Istvan. "If that happens we are going to go back to the real world, which means you might not be able to get tomatoes at your local Albertsons. You may not be able to get fresh water, even in a first world country. We get this sort of dystopian society and that could stop transhumanism in its tracks."

This sentiment is echoed by David Wood in his book *Transpolitica*: "... there are many uncertainties that influence technology—both how it is developed, and how it is deployed. Technology does not determine its own outcome. Instead, the allocation of resources to technological development is strongly impacted by the operation of markets, incentives, subsidies, regulations, and public expectations. In turn, all these factors are impacted by politics (either in commission or in omission)."

In short, even if the technology is there to lengthen life or enhance human capabilities, it will require bringing all parties onboard to be sure that funding and research continues on the technology itself, and that financing continues to be available so people can continue to live at a certain standard, even if robotic automation technology obviates the need for humans in the workforce.

When you consider the history of political and religious reaction to the other body modification and technological innovations detailed through this chapter, it seems rather unlikely that the path forward will be obstacle-free. Ironically, science is not in the way of people becoming super.

People are.

8

Hyper Longevity: How to Make Death Obsolete

If you came to this chapter expecting to read the magic 10-step certified Super You process that will help you live a very, very long time, then here it is; although, notice that it's only three steps.

1. Don't get sick.

2. Avoid accidents.

3. Wait.

Easier said than done, right? Don't get sick? That's not a step. But you need to invest effort into avoiding it at all costs because unless something bad happens, such as a Whole Foods truck taking you out as you cross the road to buy a Twinkie, then you can pretty much be sure some nasty disease will end your life at some point. Don't get sick. We'll show you what we know about not getting sick in more detail later in this chapter. We'll then show you how technology (and its accelerating improvement) is going to help you stay healthy. Or cure what ails you.

Step 2 is less controllable. Still, here is the advice, avoid the following: Falling down, guns, cars, poison, suffocation, and water (drowning). Avoid people because they statistically kill the most people, by accident or on purpose. People also kill themselves. It's hard to avoid yourself. But be vigilant with your mental health.

 Depressed? Please Read This

If you are depressed and you want to get undepressed, author Kay Walker has built a self-help website to aid people who suffer from severe depression. She uses the latest research and techniques to help people through this devastating and sometimes fatal condition. For more information, visit http://Depression.Zone.

If you are successful with Steps 1 and 2, and most people are because even though people do die of accidents and disease, the average human life span worldwide is 71 (based on 2013 World Health Organization (WHO) data). By the time you read this, it will be pushing toward 75 and in the next decade on its way to 80. Australians live to 83. Canadians live on average longer than Americans. Their life span is on average 82.5 years. The Japanese are the longevity champs at 84.6 years. Average American life expectancy in 2014 was a rather sad 79.59.

Step 3 is "wait." This is deceivingly simple. But we really mean it: Let time go by and stay alive as best you can. Waiting is important because as time goes by, the acceleration of technological improvement will bring new therapies to stave off and eventually mitigate death. We will talk more about this in detail a bit later.

When these handy steps help you live a very long time, you can send us a nice thank you card when you turn 100 or 200 … you'll see.

Table 8.1 shows the top ten things that kill people in the United States.

Table 8.1   Leading Causes of Death in the United States

Rank	Cause
1	Heart disease
2	Cancer
3	Lower respiratory (lung) disease
4	Stroke
5	Unintentional injuries (accidents)
6	Alzheimer's
7	Diabetes
8	Kidney disease
9	Flu and pneumonia
10	Suicide

Source: National Vital Statistics System, United States, 2010.

The History of Aging

Here's the best news of all: Life expectancy rates—the median age of death for most people in a given population—has been consistently increasing since early man dropped out of the trees and moved into subdivisions.

Technology helps increase life span so it is no surprise that the trend in technology is similar to the graph of life expectancy rates.

Let's start as close to the beginning as possible. Research efforts to plot the growth of human life span in early human development have been somewhat daunting.

The passage of time has erased remains of early humans from the prehistoric era. With access to only the fragments of skeletal remains from archaeological digs, scientists are limited by the resources at their disposal to conduct their research.

Until recently, this impeded scientists' ability to uncover the average life expectancy rates of the earliest humans. However, in 2004, anthropology professors Rachel

> "Life expectancy rates have been consistently increasing since early man dropped out of the trees and moved into subdivisions."

Caspari of Central Michigan University and Sang-Hee Lee of the University of California-Riverside established a revolutionary method of fossil analysis. This allowed them to track the first significant life span shift in the history of mankind.

They use what is called the OY ratio (old to young ratio), which uses bone analysis to measure relative age—instead of the exact age of a person—at the time of death. Using this method, they were able to group bone fragments for the fossilized remains of 768 humans in four regions of the world into categories of "young" and "old."

When the researchers measured the proportion of young to old in each population, they discovered life span rates increased marginally across most of the time periods except one. Humans living 30,000 years ago had an OY ratio five times greater than earlier populations. For the first time, three generations of the same family coexisted, and humans lived long enough to become grandparents. Naturally, this discovery is called the "evolution of grandparents." And the 5 P.M. early bird dinner special was not long behind. (This last bit is speculation, don't write that in your thesis.)

This sudden increase in life span appears to be related to knowledge transferred from earlier generations. Over time, the oldest people transferred the tools they had for survival to the youngest people. Then the youngest people used existing knowledge to improve upon those tools. Eventually, family members could survive long enough so three generations were living at once.

Caspari and Lee identified the impact of knowledge transference on longevity in another study they published in 2006. They reviewed the OY ratios of Upper Paleolithic Europeans to understand whether life span increases in the population were a result of their biology or their culture. They found that life span increased when modern humans arrived in Europe, bringing new knowledge with them.

The earliest information available that verifies the exact ages of humans at death comes from epitaphs of those who died during the Roman Empire. These show an average life span for Romans was 20 to 35 years old. Infectious diseases or infected wounds from accidents or conflicts were the major causes of death. Child mortality was high as well. However, if Romans survived birth, didn't contract a deadly disease, or get skewered with a spear, they could live into their 60s and 70s.

Major killers included cholera, tuberculosis, and smallpox. Plagues such as the bubonic plague in the fourteenth century—also known as the Black Death in Europe—wiped out as many as one-third of Europe's entire population.

Between 1500 and 1800, life expectancy rates rose to between 30 and 40 years. By the 1800s, life expectancy rates had doubled thanks to the Industrial Revolution. Major innovations in manufacturing sparked improved health care, sanitation, access to clean water, and better nutrition. Inventions in transportation, such as the steam engine, also increased the dissemination of knowledge across continents.

Life expectancy rates have improved gradually in the last couple of hundred years or so between 1800 and 2012. There was a small dip from 1918 to 1919, when the influenza outbreak (disease) and World War I (other people) killed large numbers of the population before age could get them. As we said, it's illness that greatly shortens most people's longevity. However, technological innovation in science and medicine is the great tool against illness. For those of you that were around in the 1970s or 1980s, you'll recall (or if you don't, ask someone who does) how people related to cancer and AIDS in those decades. These diseases were once pretty much death sentences if you became ill with them. After a diagnosis, you cleaned up your affairs, told the people around you that you love them, and then sooner or later you succumbed to the disease. There was little medical science could do for you, except perhaps help you to suffer less at the end.

In the nineteenth and early twentieth centuries, there were a series of major health innovations that helped prevent and manage acute and chronic disease. Among them were antibiotics, vaccines, pharmaceuticals, and various medical instruments that advanced treatment capabilities.

At the turn of the twenty-first century, the average global life span was 75. That number is still improving, as has been the long-term trend. As we said earlier, the average American life expectancy in 2014 was 79.59.

And guess what's going to happen to the trend as nanotechnology, stem cell research, robotics, and genetics all continue to progress? Humans will live longer, especially those in developed nations with access to wealth, and of course the technology to spend that wealth on.

The Methuselah Award Goes to ...

If you are going to live a very long disease-free life, then it's probably helpful to understand who has done the best job at it. And it would be logical to copy that person's habits, even if that logic is flawed.

Here's a little story. Once upon a time there lived three very different people who lived on three different continents and led three very different lives. They all were named in the Guinness World Records as record holders for longevity.

- **Jeanne Calment**—Let's start with Jeanne Calment, a French woman who lived to 122. Upon her death in 1997, she was referred by the newspaper *Le Monde* this way: "Elle était un peu notre grand-mère à tous," which means, "she was a little bit grandmother to us all." Calment holds the Guinness World Record as the oldest person ever to live. She was born in the south of France and spent her entire life there. She witnessed the building of the Eiffel Tower and met Dutch artist Vincent Van Gogh. Each week, she purportedly ate 2.2 pounds of chocolate paired with a daily glass of port wine. The rest of her diet was rich in olive oil, which she also slathered on her skin in an effort to fight wrinkles.

- **Jiroemon Kimura**—While Calment ate chocolate and drank port, Jiroemon Kimura, from Japan, was restricting his diet. Kimura is the oldest man that ever lived. He died at age 116. Kimura believed in eating small food portions every couple hours. During his life, he witnessed the reign of four emperors and saw 61 Japanese prime ministers hold office.

- **Sarah Knauss**—Then there was American Sarah Knauss. She lived three years more than Kimura, dying at the age of 119 in Pennsylvania. The Ford Model T was introduced while she was growing up. She also lived at the time when the *Titanic* sank in 1912. She was a homemaker and her hobbies included needlepoint and watching golf. Her favorite snacks were milk chocolate truffles, cashews, and potato chips.

So what do all three have in common? They were supercentenarians—people who live to the age of 110 or more.

But besides that, they seemingly have few lifestyle commonalities. For scientists that have studied these long livers, the conclusion is mostly a collective shrug. Nothing about these supercentenarians outwardly suggests any set of strategies that can be copied to produce a longer life in another person.

That said, there has been some significant research in longevity that has produced some interesting results, and this work does suggest actions anyone can take to extend their natural life span.

What We Know About Super Agers

According to the New England Centenarian Study, as of 2014, one person for every 5 million people on the planet will live to 110 or more.

People who live to the age of 100 are called centenarians and their ranks are much more common than they used to be. In 2014, the incidence of centenarians was one for every 5,000 people and that number is steadily growing.

In 1999, data compiled by the United States Census Bureau shows that during the 1990s, the number of American centenarians nearly doubled: 37,000 at the beginning of the decade and 70,000 by the end.

Then there are the predictions.

A 2010 edition of *TIME* magazine reported that by 2050 more than 800,000 Americans will live into their second century of life. Remember, at the start of this chapter, we pointed to a long life strategy called: "Step 3." It was "wait." If you can hang on for another 30 years or so, you might be in luck. Compare 800,000 centenarians to data from 2010 when there were only 80,000 centenarians in America. That's a forecasted tenfold improvement.

In 2013, *National Geographic* launched a longevity issue with a baby on the cover and a headline that read "This baby will live to be 120." If that proves to be true, there will be at least 4 million supercentenarians in the United States by 2133.

A more aggressive prediction comes from controversial gerontologist Aubrey de Grey, a world-renowned longevity expert. He is also the chief science officer of the Mountain View, California-based SENS Foundation, a research-focused outreach organization that educates policymakers and the public about how humans can live longer through the "damage-repair" approach to treating age-related disease.

"We are looking at a divide and conquer approach. Dissecting the problem of aging, accumulating (the) damage of old age into sub-problems and addressing those sub-problems individually," de Grey said. "The problem with aging is that so many things go wrong that we can't control and fix them all."

De Grey predicts by 2030 there will be 3 million centenarians worldwide. Based on accelerating technological improvements, he is likely to be right in his forecast, although not everyone agrees with him. A lot of that longevity progress will depend on breakthroughs in heart disease, cancer, and other life-ending diseases. Stem cell and genetic therapies, organ regeneration, and nanotechnologies will drive the trend.

Longevity Research Is Still Young

Humans only began living to the age of 100 in the twentieth century, so the longevity research field is a relatively youthful one. Most of the work has been done in the last two decades. Aubrey de Grey said, "It's only been about the last 10 or 15 years that we have been able to start talk about actual theory of concrete plans for delivering medicine that postpones aging."

The majority of the knowledge about longevity today has been obtained by studying super-agers like Calment, Kimura, and Knauss (see "The Methuselah Award Goes to…" earlier in this chapter). The intention is to find out what these people do that the rest of us don't do that makes them live exceptionally long lives. Researchers study their daily habits to isolate common lifestyle choices.

What they have discovered is that gobbling chocolate and slurping port, taking olive oil baths, or eating potato chips while watching golf isn't the fountain of youth. (But it sounds fun though, doesn't it?) What we do know is that what you eat and how you live your life is certainly a major factor. However, your genetics play a big role in how long you live as well.

Research from the world's largest centenarian study, the New England Centenarian Study, shows the ability for an individual to live a long time is 20 percent to 30 percent attributable to their genetic makeup. The remaining 70 percent to 80 percent relates to what you do regularly to stay healthy.

One of the core fields of study in longevity science is genetics. If scientists can master genetic engineering and bring everyday genetic therapies to the masses—especially to those that have short genetic fuses and have a family history of short livers—then we are starting to unlock the puzzle on a key factor that can extend longevity.

The Kay Walkers of the world don't have to worry much on this genetic front because she has two grandmothers that are still alive at 87 and 90. That suggests Kay has good longevity genes.

The Andy Walkers have it a little tougher. His paternal grandfather died young, in his 20s (as a result of war), and his maternal grandfather is unknown. That said, Andy's father and eldest paternal uncle are mostly healthy in their mid 70s.

In Sean Carruthers' case, it's a wild card situation. His father died of a heart attack/stroke at 60, after fighting hypertension most of his life. His mom on the other hand is a hardy 71 and her mom is 95. Sean's other three grandparents made it to their late 70s or 80s.

Like us, if you want an indicator of your own longevity genetics, look at your grandparents and parents and uncles/aunts and you'll get a sense of what programming is likely nestled in your genes.

One study shows your life span can somewhat correlate to your parents's life span. It has a minimal impact though, and is not a guaranteed indicator of your longevity. It should be factored in with your lifestyle. Disease in your immediate family might be more of an indicator that your longevity fuse is shorter thanks to your genes. But again, no one factor will predict your life span. Twin studies, however, suggest genetics only account for approximately 20 percent to 30 percent of your predictable life span.

You can play against a bad hand dealt to you with two strategies:

- Wait for genetic therapies to adjust for your genetic deficiencies, and

- Adjust your lifestyle to buy time until longevity-extending technologies arrive.

Lifestyle Secrets: Live Long and Prosper

If genes are 20 percent to 30 percent of the longevity equation, then what accounts for the other 70 percent to 80 percent?

Simple. How you live your everyday life.

Your lifestyle significantly affects how long you live. There are some very specific choices you can make that will give you a better shot at a long life than if you are more cavalier about it. It's supposed to be quite simple: eat nutritious food and engage in regular physical activity. These are the basic health rules most medical professionals advise today (and most of us ignore). To a certain extent these ring true from the research that has been done, though drinking diet "anything" and getting your exercise from video games, doesn't appear to be the solution to a hyper-long life. Jane Fonda aerobics and fruit smoothies aren't either, at least not according to the longevity doctors.

Most of the common sense rules (that are backed up by longevity science) we list here are generally intuitive, although you might find a few surprising:

- Eat less of everything

- Eat nutritiously and consume less animal protein and more beans

- Avoid obesity

- Live with purpose

- Live in a supportive community

- Stay married

- Drink wine in moderation

- Stay active

- Manage your stress

- Don't smoke or abuse drugs, including alcohol

- Buy lots of books by Kay Walker, Sean Carruthers, and Andy Walker

Although this is by no means a complete and definitive list, it is a general snapshot of what scientists know about lifestyle choices that extend longevity. All except buying our books. That one will just keep you entertained, and it helps us pay for our fleet of yachts.

Longevity scientists have gleaned a lot from their research in the last couple of decades, which extensively includes the study of centenarians. Following is a look at some of the more significant research done on Super Agers.

Centenarian Studies

Longevity research that has studied centenarians and supercentenarians has led to some major findings that support our list of tips for longevity (see the previous section). In the next section, we've summarized some of the most prominent research that has been done to date.

The Blue Zones

It could all start with where you live. In 2004, one of the most extensive longevity studies was conducted by American explorer Dan Buettner in partnership with *National Geographic* and a team of the world's top longevity experts. The premise of the study was to map out areas of the world where the highest concentration of people are living abnormally long and healthy lives, and then, to explore those regions to isolate potential longevity factors. The project identified the following five regions, which Buettner calls the Blue Zones:

- **Icaria, Greece**—This Greek island is home to approximately 8,500 people that enjoy an isolated culture built on tradition. They value family and spend time with the people in the community. They drink wine together, play games, and socialize daily.

- **Loma Linda, California**—A city in Southern California with a big slice of the population that are members of a Protestant sect known as the Seventh-Day Adventists. The religion focuses on principles of healthy living, exercise, vegetarianism, and avoidance of tobacco, alcohol, and illegal drug use. The focus on the California community comes from the concentration of Seventh-Day Adventists, however the lifestyle the religion promotes extends the Blue Zone merits outside that geographic area to include any Seventh-Day Adventist congregation.

- **Sardinia, Italy**—Inhabitants of the highlands of this Mediterranean island hunt, fish, and cultivate most of their own food. While they live a very healthy life, their superior longevity could be related to a genetic marker, M26, that might be a factor linked to long life spans.

- **Okinawa, Japan**—Okinawa has long been a focus of longevity research. It's where the world's longest-living women live. However, men live longer there, too. The Okinawans value community. They work hard with a sense of purpose and eat healthy diets. They exemplify the Power 9, which are longevity-promoting lifestyle choices that Buettner recommends (see the following section entitled "Power 9: The Nine Lifestyle Choices that Promote Longer Life Spans").

- **Nicoya, Costa Rica**—An 80-mile peninsula south of the Nicaraguan border. People who live there have access to water rich in calcium and magnesium. It strengthens their bones and protects them from heart disease.

 Why Blue?

When Dan Buettner was asked during a *TED Talk* why he called his famous longevity study the "Blue Zones," he answered it this way: "I wish I had a wonderfully exotic and scientific answer for that but the reality is, was when we were honing in on it in Sardinia, our team of demographers were drawing concentric circles with blue ink on a map and we just started referring to the area inside the smallest circle as the Blue Zone. And the name stuck."

Power 9: The Nine Lifestyle Choices that Promote Longer Life Spans

People in the Blue Zone regions are ten times more likely than the average American to reach the age of 100. When they did a cross-comparison of the five zones, Buettner and his team identified nine reasons for higher rates in life span. Buettner calls them the Power 9. The following nine factors are suggested as essential for a longer than average life:

1. **Move naturally**—A super-ager fitness regime involves daily physical activity that occurs spontaneously. The exercise is incorporated into activities that have to get done.

   Sardinians are a sheep-herding culture. Many of them get their physical activity from their work. Many Okinawans work in agriculture. Their day involves gardening and tending to fields.

2. **Live purposefully**—The French refer to it as "raison d'etre." The Nicoyans call it "plan de vida." In English, it's the "reason for being." People who live long believe there is a reason for their life. They believe they are meant to achieve something. Waking up with this mindset encourages them to take action. It gives them a reason to want to live.

> "People who live long believe there is a reason for their life. They believe they are meant to achieve something."

World-renowned Austrian psychiatrist Viktor Frankl would probably agree. He was a Holocaust survivor who studied the mental health of fellow prisoners in the concentration camps where he was imprisoned during World War II. In 1959, he published a book that established a revolutionary new therapy called logotherapy. It was built on the fundamental principle that people with a life's purpose deal better with difficult life circumstances. He found that religious Holocaust prisoners were more apt to live through difficult times than commit suicide. He asserted it was because they believed they were there for a reason.

3. **De-stress regularly**—The world's longest living people have routines that help them de-stress regularly. This makes sense if you consider what the American Medical Association (AMA) says about stress. It's the number one proxy killer in America and is linked to 60 percent of all human illness and major disease. Stress produces chronic inflammation in the body, which can lead to chronic illness and early death.

   To de-stress, Dr. Mihaly Csikszentmihalyi, Distinguished Professor of Psychology and Management at Claremont Graduate University in California, suggests that people should engage in activities that encourage what he calls "flow." Experiencing flow involves engaging in an activity that encourages a total involvement with life in the moment.

   Here's an example of flow. When a musician learns to play a song, the musician must concentrate on the notes he or she plays. However, once the song is learned, he or she can play without thinking. In a sense, the musician becomes one with the music. This experience of becoming completely entrenched in an activity so thoughts are concentrated on that very activity is "flow."

4. **Don't over-eat**—Blue Zoners stop eating when they are 80 percent full. The Okinawan people have a name for it: "hara hachi bu." Roughly, translated it means "eat until you are eight parts (out of ten) full." Resist the temptation to yell out "HARA HACHI BU!!!!" at the pimply McDonald's cashier the next time you are asked to super-size it.

5. **Eat more beans**—Inhabitants of the Blue Zones eat a lot of beans and plant-based foods. Buettner says that Super Agers only eat meat about five times per month and it is usually pork. And no, that's not a free pass to an all-bacon diet.

   Buettner's diet suggestion is consistent with a 2014 study by Dr. Valter Longo, Professor of Gerontology at the University of Southern Carolina. He suggests that eating too much protein before age 65 can lead to an early death.

   The research study analyzed the diets of 6,831 middle-aged and older adults who responded to a national dietary survey. The findings indicated that people who ate high levels of protein before age 65 died sooner than those who didn't.

   Longo suggests that an unhealthy growth hormone, 1GF-1, is activated in the body when a person eats large amounts of protein.

   However, adults older than 65 with little body fat are encouraged to eat protein to gain weight during a time when their body needs excess fat.

6. **Drink wine**—We didn't make this one up. And yes, we agree, it's a good idea. People living in the Blue Zones provide solid evidence that moderate wine drinking increases life span. It turns out moderate drinkers outlive nondrinkers, especially when they enjoy their drinks with friends.

7. **A sense of belonging**—Attendance at religious gatherings are falling but there might be a reason to start going to church, synagogue, or a mosque again. People who attend religious gatherings (no matter what their denomination) at least four times per month, are more likely to live longer than noncongregants. In fact, Buettner says they might live up to 14 years longer.

   A 2013 study by the Pew Research Center reported that only 37 percent of Americans attend church on a weekly basis. In 2014, 16 percent of the population said that they had no religious affiliation.

   In an interview about his documentary "The Happy Movie," director Roko Belic said that during a visit to Okinawa, he learned the Okinawans see all members of their community as part of their family. They have a strong sense of belonging. He believes this is "one of the cultural traits that relates to why Okinawans are so happy."

   It suggested this sense of incredible community has members with a strong sense of belonging always feel supported and able to take on tragedy easily with the support of others.

8. **Put loved ones first**—Centenarians believe in building a close-knit community. They marry, live close to grandparents and relatives, and bear children. Remember that the next time you plan a convenient business trip on the same week your spouse's parents are coming to town. Drinks with the in-laws equals a longer life.

9. **Get lucky**—Becoming a super ager is ten percent luck. People born into a society that cares about health usually live longer. Their environment teaches them how to be healthy. Good news if you live in Monaco, Macau, or Japan. According to the CIA, those countries are top three for life expectancy rates.

In 2009, Buettner partnered with the American Association of Retired Persons (AARP) and the United Health Foundation in a pilot project to test the Power 9 on a community: the city of Albert Lea, Minnesota.

The project involved the enrollment of 20 percent of the city's members, 50 percent of the top 20 employers, and 25 percent of the city's restaurants, schools, and grocery stores. The Power 9 principles were incorporated into day-to-day life. The focus was on building sustainable health initiatives and making it easy for members of the community to make healthy choices.

After just 1 year, participants added an estimated 2.9 years to their life span and health care claims for city workers dropped 49 percent. Efforts to implement the Power 9 into communities continue. There's more about this at BlueZones.com.

 BadAss Centenarians versus the Early-Grave Sprout Eaters

You have heard of healthy people who die young and really old people who eat fried food and chain smoke. Dan Buettner, in his *TED Talk*, had an explanation for these genetic outliers: "To clarify ... there's a small percentage of people who could eat bean sprouts and walk every single day and be completely engaged and die of cancer at 50. And there is another end of the spectrum of people who can smoke two packs of cigarettes, drink a fifth of whiskey and live to a 100. They (give) centenarians a bad name. Most of us, 80% of us, are in the middle."

The New England Centenarian Study

The New England Centenarian Study is the oldest, largest, and most prestigious centenarian study in the world. The study was established by the Boston University School of Medicine in 1995.

The premise of the original study was to understand the incidence of Alzheimer's in centenarians living in eight towns in the Boston area. However, initial findings showed that centenarians with Alzheimer's are rare. So, the study's main focus shifted to isolating potential lifestyle and genetic factors that allow a person to live to the age of 100 or more.

When the study began in 1995, the prevalence of centenarians in industrialized regions was 1 per 10,000 people. In the United States specifically, that number has changed to 1 in 6,000, making centenarians the fastest growing segment of the population.

The original group studied consisted of 46 people. Today, the study group has grown to 1,600 centenarians, 500 children of centenarians aged 70 to 80, and 300 younger people. Participants no longer include only people in the Boston area. The expanded group lives around the world. It also includes 107 supercentenarians.

The most current findings from the New England Centenarian Study suggest that there are three lifestyle factors that contribute to longevity.

Centenarians have the following things in common:

1. They are not obese.

2. They don't smoke.

3. They've learned to deal with stress better than most people.

The study also uncovered some genetic findings:

1. Fifty percent of centenarians have more Super Agers in their families.

2. Centenarians have extroverted personalities (results based on personality tests).

3. Centenarian women have had babies after the age of 35 years old.

This study shows that longevity is 70 percent to 80 percent lifestyle and 20 percent to 30 percent genetics. These statistics were gathered from studying twins separated at birth. They compared twins who were brought up in different environments to see how it affected their life spans.

The study also suggests that it is physically possible for all Americans to live to the age of 88 or 89. Researchers discovered this biological capacity by closely examining the Seventh Day Adventists in Loma Linda, California. Members of this group have an average life expectancy rate of 88 (see the Blue Zones section earlier in this chapter if you missed it).

In January 2014, the study published the results of their latest genetic findings. To date, they have isolated a total of 281 genetic markers. These markers are

61 percent accurate in predicting who will live to 100 years old, 73 percent accurate in predicting who will live to 102 years old or older, and 85 percent accurate in predicting who will live to 105 years old or older.

These same genetic markers play roles in many of the genes involved in old age diseases such as Alzheimer's, diabetes, heart disease, cancer, and high blood pressure. They also play a role in the biological mechanisms that create aging in the body.

WHY MEN WHO MARRY YOUNGER WOMEN LIVE LONGER

If you are a man and you want to live longer the answer, it seems, is simple. A study produced by the Max Planck Institute for Demographic Research (Rostock, Germany) claims that men who marry younger women live longer.

Researchers looked at the deaths of the entire population of Denmark between 1990 and 2005. The study showed that Danish men who married women who were 15 to 17 years younger than them lived longer than men with wives who were closer to their age.

There are three hypotheses for the findings:

Natural selection—It is possible that younger women choose healthier, better-maintained older men as their husbands. These men naturally take care of their health so they typically live longer.

Financial wealth—Men with considerably younger wives are often rich. Because of their financial means, these men enjoy a more comfortable lifestyle. Factors such as the ability to afford a healthy lifestyle are possible contributors for a longer life span.

Emotional support—A younger woman will care for a man better than spouses who are older, so he lives longer. A younger spouse might also have a beneficial psychological effect on the older partner. (And, we assert, his libido, too.)

This proves that Groucho Marx was right: "A man's only as old as the woman he feels."

The Longevity Genes Project

The Longevity Genes Project was launched in 1998 by the Albert Einstein College of Medicine (AECM) at Yeshiva University in New York City. The project, led by Dr. Nir Barzilai, Director of AECM's age-related diseases. The project has looked at more than 600 centenarians and their offspring. All the people in the study are

Ashkenazi Jews, a group originally from Eastern Europe, who have been identified as being able to live extraordinarily long and healthy lives. Barzilai also needed a population with the same genetic background to better isolate specific genes.

In 2013, CBS News interviewed Irving Kahn, one of Barzilai's study participants. At the time, Kahn was 107 years old. His brother was 105. His sister lived to 109. And, his son was 69. (Kahn died in 2015 at the age of 109.)

When asked what he saw when he walked past Central Park in New York on his way to school as a young boy, Kahn told CBS News he would see "things you would never see (today) … cows, sheep on the lawn."

The study has revealed that lifestyle has nothing to do with the family's exceptional longevity. (Kahn said his favorite food is "A rare hamburger … and a good cheese.")

Phase 1 of the study produced significant gene-related findings. The analysis of the blood samples of the Ashkenazi centenarians and their children showed high amounts of good cholesterol called HDL. The average amount found in their blood was 80 mg/dL to 250 mg/dL. The normal range for HDL is 35 mg/dL to 65 mg/dL for men and 35 mg/dL to 80 mg/dL for women.

They also discovered three genes that centenarians have that protect their bodies against age-related diseases such as diabetes, cardiovascular disease, and inflammation.

The research produced by the study is going into the development of drugs that will help people deal with age-related diseases. Barzilai says the future lies in developing drugs that could give everyone the same advantage that Super Agers have.

Strategies for a Longer Life

There are other strategies that can affect how long you live. An obvious behavior is watching your calorie intake. Another strategy that has been touted all over the Internet is drinking red wine—in moderation, of course.

Calorie Restriction

One actionable strategy that appears to extend human life is perhaps not that appealing for cultures that like to eat a lot. It's certainly not compatible with the Super-Size Me America we live in today. The technique is to eat 30 percent less than you might normally eat. Calorie restriction (CR) as a strategy to live longer has been around for more than 500 years.

Luigi Cornaro, a fifteenth century Venetian nobleman, is the first person who increased his life span by eating fewer calories. He ate 350 calories and drank

414 milliliters (about half a bottle) of wine per day, and he lived to 102. Before he died, he published a book about calorie-restrictive eating called *Discourses on the Temperate Life*. There was not much evidence to back his theory until the twentieth century when studies with mice popularized the idea of eating less to live longer.

In 1934, Clive McCay and Mary Crowell, two nutritionists from Cornell University, published a breakthrough study that found mice fed a calorie-restrictive diet almost doubled the typical life span for their species. The same findings were reproduced in similar studies in the 1980s by two notable longevity researchers, Richard Weindruch and Roy Walford.

Walford used the results of the study to formulate a calorie-restrictive diet that he sold to humans (because mice can't read). He called it "The 120 Year Diet." And he outlined the process in a book by the same name, which he published in 1986.

A calorie-restrictive diet is simple. It suggests eating about 30 percent less calories than nutritionally advised. That would mean if the recommended daily minimum for an adult male is 2,500, Walford's diet would have him eat 1,800 to 2,000 calories per day.

In 2000, Walford's follow-up book did better than his first as it contained tangible evidence related to humans that backed his theory. He was involved in a research study, called Biosphere 2, that required eight bioscientists, Walford included, to eat calorie-restrictive diets.

From 1991 to 1993, the participants lived in a three-acre, self-contained greenhouse in the Arizona desert. The intention was to test the survivability of a small group of people in a man-made colony for a long period of time. The group was forced to live on only what they could grow, and naturally, their diets were limited to approximately 1,500 calories a day.

When they left the biosphere, lab tests showed dramatic health improvements. Their glucose and insulin levels were down, their body fat was reduced, and the process of cell loss had slowed. This gave Walford credibility.

Unfortunately, he died in 2004 at the age of 79. (And, let's face it, it's hard to be a credible longevity doctor if you don't live a long time.) In all fairness he had suffered from the brain disease ALS (amyotrophic lateral sclerosis, also known as Lou Gehrig's disease) and it's likely that this caused his premature death. A later ALS research study that used mice found that calorie-restrictive eating might provoke an early death in ALS patients. The irony.

Many people still eat calorie-restrictive diets today. This way of eating involves calculating calories before consuming food. Calorie-restrictive eaters also avoid using extreme heat cooking methods. No frying, grilling, roasting, barbecuing, or smoking food. (We can't imagine there are many CR diet proponents in the

Southeast.) Heating food introduces harmful compounds into the food, called advanced glycation end products (AGEs). This occurs when carbohydrates and proteins combine without any enzymes. When glucose in carbohydrates combines with protein, cells go stiff. It's been suggested this process leads to cellular damage and premature aging. Eating raw and unprocessed foods helps calorie-restrictive eaters avoid AGEs.

TheCalorist.com is a site dedicated to teaching people today how to live the CR way. In 2012, its founder, Joe Cordell, was featured on *The Oprah Winfrey Show* when he was 51. He was described by medical professionals at the time to have the body of a 20-year-old athlete.

Cordell told Oprah the whole idea of calorie restrictive eating is "about getting the most nutrients per calorie." For breakfast he'd eat the peel of an apple, for its nutrients, mixed with berries and walnuts that he weighed. Lunch was a giant family-sized bowl of salad, and dinner was something similar.

 A Day in the Life of a Calorie Restriction Eater

Breakfast—A glass of freshly squeezed juice with a bowl of homemade cereal topped with a banana.

Mid-morning snack—A protein shake and a handful of dried fruit

Lunch—Steamed vegetables

Dinner—A vegetarian dish, such as steamed fish with homemade tomato sauce

Cordell's site shares references to positive research studies and quotes from qualified experts that reinforce the benefits of calorie-restrictive eating. However, recent findings conclude that it is doubtful that calorie-restrictive diets are an effective life-extending strategy.

In 2012, *TIME* magazine published an article with this headline: "Want to Live longer? Don't Try Calorie Restriction." The article referenced two calorie-restriction studies with conflicting results done on the rhesus macaque, a small monkey.

The Wisconsin National Primate Research Centre (WNPRC) in Madison published the results of a 20-year study in 2009. It reported that monkeys fed calorie-restrictive diets had lower rates of diabetes, cancer, heart disease and brain disease. It also found that a lower number of monkeys on calorie-restrictive diets died from nonage-related causes. It found 13 percent of the CR diet monkeys died from age-related causes compared to 37 percent of monkeys in a control group that were not fed a restricted diet.

This all sounded pretty good, until the National Institute of Aging (NIA) published its conflicting results three years later. That research involved a 25-year study that followed a very similar format to the WNPRC. However, the NIA concluded that genetics and dietary composition matters more than calorie restriction for prolonging life.

When both studies were examined more closely, the qualifying difference was the quality of food both groups of monkeys were fed. The WNPRC monkeys were fed an unhealthy diet, high in sucrose (table sugar). And their calorie-restricted monkeys ate less of the bad food.

The NIA fed their monkeys food composed of fish oils and antioxidants. Monkeys in the NIA control group were also fed fixed amounts of food versus the WNPRC monkeys in the control group who ate what they wanted when they wanted.

NIA's research produced no connection between calorie restriction and health. They suggest that genetics could be more important than diet.

The debate continued and in 2014 a further examination produced different results. Because the WNPRC monkeys were allowed to eat all the food they wanted, they naturally had a higher body weight than the NIA monkeys who lived longer. Monkeys that ate what they wanted had a three-fold higher risk of death. This means that calorie restriction can still make a difference in increasing life span.

 A heaping bowl of lactalbumin, please

The difference in the two conflicting CR research studies was as follows: The NIA diet contained protein and fat from natural ingredients including wheat, corn, soybean, and fish. The WNPRC study used a purified diet with a single protein source (lactalbumin) and fat derived mainly from corn oil. The WNPRC food also contained much more sugar.

The NIA and WNPRC might collaborate in a further study in an effort to fully understand why such different results were generated and to get closer to accurate, measurable results. Until then, calorie restriction might be a strategy to consider. The lesson: Eat less, but ensure it is a high-quality diet. So put that chocolate bar down and go eat half a kale sandwich.

Red Wine and Resveratrol

If you are unwilling to restrict your diet (or are dubious that it will work), you might want to simply drink more wine. Well, kind of. You'd have to drink a lot of red wine to reap its longevity benefits. Hooray!

"If you are unwilling to restrict your diet (or are dubious that it will work), you might want to simply drink more wine."

The health benefits of red wine come from a compound called resveratrol, which is found in grape skin. It is as you might guess a key component of red wine. Sorry white wine and liquor drinkers, your beverages don't contain resveratrol. But take heart, resveratrol is also found in peanuts (and peanut butter), dark chocolate, and blueberries, as well as, you guessed it, red skinned grapes.

Red wine contains at most 12.59 milligrams of resveratrol per liter, so to get 500 milligrams daily, you'd need to drink almost 40 liters daily (that's about 53 bottles).

It was first isolated by Japanese researcher Michio Takaoka in 1939. Newer contemporary research confirms the theory that resveratrol can promote longer cell life by stimulating the cellular proteins known as sirtuins.

Dr. David Sinclair of Harvard Medical School originally discovered resveratrol's effect on sirtuins in 2003. (Sirtuins are proteins that regulate biological processes linked to aging.) He and his team recently discovered that resveratrol appears to help increase the activity of mitochondria, which produces energy within cells, which extend the cell's lives.

In scientific circles, resveratrol is what's known as a synthetic Sirtuin-activating compound (STAC). This means that it can be removed from its originating source (grapes, peanuts, or berries) and made into pills. When it's ingested, resveratrol activates a specific gene linked to longevity called SIRT1.

In a 2013 *TED Talk* for *TEDMED*, Sinclair recalled the initial study. "I thought the mice would die. Resveratrol wasn't known to be safe in those days. But, what happened was really surprising. The mice fed resveratrol stayed healthy and had the physiology of a lean mouse."

Sinclair discovered that mice put on regular fatty diets and fed resveratrol were as healthy as mice that were fed a lean diet.

A follow up study, conducted in Switzerland, fed resveratrol to mice who became unusually healthy. The resveratrol-fed mice could run faster and for a longer time than a control group and became a breed of high achievers.

> "Resveratrol could be a modern day fountain of youth."

Some suggest resveratrol could be a modern day fountain of youth. In 2008, pharmaceutical company GlaxoSmith paid $720 million to acquire Sirtris, a research company co-owned by Sinclair. The company intended to use his research to develop new drugs that act on sirtuins. In 2013, initial drug trials were successful so they moved the entire operation to Philadelphia. Research continues today.

Sinclair is still studying the sirtuin genes. When we asked him what he does to stay young, he told us, "I take resveratrol."

We Want to Live Forever, Too

After interviewing David Sinclair, Super You authors Kay Walker and Andy Walker started taking daily resveratrol supplements in capsule form every day. They take a daily dose in pill form of 1,000 milligrams. Just to be sure, Kay also supplements her resveratrol intake with plenty of red wine.

Current Bodies of Research in Longevity

Longevity is a complicated field of scientific study because there is no single cause of aging. Sinclair said, "Our body functions better when we are young and better when we exercise and diet. Conversely, when we get old and more sedentary these genetic pathways are turned down and their ability to protect the body diminishes."

He went on to explain there are four major areas of genetics currently being researched linked to longevity. They are:

1. Sirtuin genes

2. The mTor pathway

3. The AMPK pathway

4. Insulin signaling

The following is a brief discussion of each of the four major areas of genetics, as well as the latest research pointing to a link in extending longevity. None of the genetic research has been applied to humans yet. All are still at an early stage of testing, which is why you'll notice the studies only involve laboratory mice.

Sirtuin Studies

Sirtuins are proteins that regulate biological processes linked to aging. There are seven sirtuin genes and they're found in different parts of the cells in the human body. Their functions include regulation of cell death as well as repair, insulin secretion, metabolic processes, and gene expression. Sinclair called sirtuins "protectors of cell health."

The term Sir2 genes is used interchangeably with sirtuin. The name "Sir2" comes from the initial discovery of this group of genes. In the mid-1990s, the Sir2 gene was discovered by Dr. Leonard P. Guarente, a biology professor at MIT. His team (which included Dr. Sinclair, a graduate student at the time) was studying yeast cells to isolate potential longevity genes.

Guarente's group split a group of the yeast cells up and removed the Sir2 gene from one of the groups. Both groups were fed a calorie-restricted diet. This stressed the cells and triggered the Sir2 gene. That gene expression halted the production of waste material in the cell, which allowed the cell to work more efficiently and for longer. The yeast cells lived longer than they should have.

Mammals, such as humans, don't have Sir2 or (SIR2), but they do have SIRT1, which works in the same way, protecting cells by suppressing specific genes that when activated produce a malfunction in the cell. It's suggested that this error could lead to Alzheimer's, diabetes, and other genetic conditions.

SIRT1 is the sirtuin we know most about at this point. And as you read earlier, resveratrol targets SIRT1. It is the one substance we know that can activate it. Although, calorie restriction might also activate SIRT1.

The most recent findings with the SIRT1 gene provided breakthrough insights into how it is linked to aging. The research garnered Sinclair a spot on *Time* magazine's list of the top 100 most influential people of the year in 2014.

Sinclair's SIRT1 Breakthrough

For a long time, it was assumed that SIRT1 protected the function of the mitochondria. The mitochondria, as mentioned in Chapter 2, "Baby Science: How to Conceive a Tennis Star and Other Procreative Miracles," is the cell's energy turbine. When they aren't functioning optimally the body's motor functions slow down. This is why seniors move more slowly than 20-year-olds.

Why mitochondria break down over time is still unknown. If you think of the human body as a car, the process of natural wear and tear overtime is quite similar. Parts wear out. Humans happen to have the ability to regenerate themselves to a point (some cells die and are replaced) until aging catches up to them.

However, Sinclair's team discovered that the SIRT1 breakdown does not directly cause mitochondrial breakdown. The process involves a chain of chemical events. SIRT1's role is intermediary.

They discovered that SIRT1 is affected by a chemical called nicotinamide adenine dinucleotide (NAD) that determines whether SIRT1 functions at normal levels.

 You Don't Know NAD

NAD is short for nicotinamide adenine dinucleotide. Because your authors are simple minded, we will simply refer to it as NAD henceforth. There will, however, be a test later in this book, and you will be expected to know what NAD means, and be able to spell it correctly.

When SIRT1 is at normal levels it protects the cell from harmful intruders, like a chemical called hypoxia inducible transcription factor (HIF) that destroy the mitochondria. When HIF gets into the cell it causes disruption—much like a drunk at a party.

Think of it like this, SIRT1 is like a bouncer at the door of the nightclub. The SIRT1 bouncer keeps HIF-1 out of the club and protects the mitochondria inside. Think of NAD like a fitness coach in that the fitness coach (NAD) is the determining factor whether SIRT1 will be in shape enough to keep the HIF-1 out of the club.

Normal levels of NAD = Normal levels of SIRT1

When NAD malfunctions, SIRT1 can't do its job to keep HIF-1 out and the intruder attacks the mitochondria, or at least barfs on his new shirt.

Sinclair's team made this connection when they removed the SIRT1 gene from a group of mice, expecting the mice would show signs of aging and mitochondrial dysfunction. However, the researchers found the mitochondrial proteins remained at normal levels. When the research team investigated this further they discovered NAD. Research efforts are now focused on creating NAD-producing compounds that might one day help slow the aging process.

During preliminary trials, a NAD-producing compound was given to a group of mice for one week. When the research team examined the test rodents, their bodies had reverted back to a younger state. In human years, Sinclair said, "This would be like a 60-year-old converting to a 20-year-old in specific areas of the body."

SIRT3 Breakthrough

In 2013, another sirtuin breakthrough was made, this time by a team at the University of California at Berkeley. The research dealt with the SIRT3 gene. It was led by Danica Chen, UC Berkeley Assistant Professor of Nutritional Science and Toxicology.

At the time of the announcement, Chen said, "We already know that sirtuins regulate aging, but our study is the first one demonstrating that sirtuins can reverse aging-associated degeneration." Chen's lab reversed the aging process in a group of mice by manipulating the SIRT3 gene in a two-part process. The SIRT3 gene produces protein that helps blood cells cope with damage that automatically occurs when a cell produces energy.

SIRT6: Helpful, But Not

A team at the University of Chicago is actively researching the SIRT6 gene. Their most current research studies have provided insights into the role of its anti-aging properties and exciting prospects on treating cancer.

One study with transgenic mice showed that by overexpressing the SIRT6 gene in male mice, the mice lived 15 percent longer than normal.

In 2014, a second study with SIRT6 pointed to the gene as a cancer-causing agent. A research team at the university found a higher level of SIRT6 protein in sun-damaged skin cells compared to healthy skin.

To better understand the connection, they removed the SIRT6 from the skin cells of cancer-infected mice. Tumor production in those mice decreased.

Amazing mTOR

Another focus of research in longevity science is the mTOR protein, or mechanistic target of rapamycin. Say that ten times fast!

The protein regulates cell growth in mammals. At the most basic level, the mTOR receives information from a number of growth-related biological processes. Then it makes a decision whether to start or stop the body's growth response.

If this process malfunctions then diseases such as diabetes, obesity, depression, and various cancers might develop. Longevity research suggests that aging processes occur via similar cellular malfunctions. And those malfunctions occur when specific mTOR chemical pathways in the body become hyperactive.

Scientists first learned about mTOR while studying a molecule called rapamycin, back in the 1970s. It is a bacterial component that was first discovered in soil samples on Easter Island (a Polynesian Island in the southeastern Pacific Ocean. It's the site of all those huge stone heads.)

 A Molecule-Sized Easter Egg

The indigenous name for Easter Island is Rapa Nui, hence "rapamycin."

The molecule has since been engineered into an FDA-approved drug. Rapamycin is an immunosuppressant, a drug that reduces activity in the immune system. It's been used to treat cancer and prevent transplant rejection. It affects the mTOR protein specifically by binding to it and stop it from performing normal cell-regulation processes. Kind of like how an octopus on your face would stop you from driving.

Methuselah Mice and New MTOR Research

In 2013, Dr. Toren Finkel, Chief of Molecular Medicine from the National Heart, Lung, and Blood Institute, made a breakthrough discovery with the mTOR pathway. Finkel's team bioengineered a group of mice that lived 20 percent longer than normal. Finkel's mice lived to 28 to 31 months-old. Their average mice counterparts only lived 22 to 26 months.

The difference between Finkel's "Methuselah Mice" and normal mice was a scaled-back mTOR gene. Finkel's mice had a mTOR gene that had been cranked back to a quarter of its normal operating level using rapamycin.

Finkel's mice had better balance and memory and also had improved organ function. However, they also had a greater loss of bone mass and more infection. It took them longer to age, but they weren't healthy. This is due to an aging process that is not uniform. Aging occurs at various speeds in different parts of the body. When the study was published in 2013, Finkel told the *Scientist* magazine that he's interested in learning how mTOR reduction affects aging cells. "Perhaps cells get rid of cellular garbage at a faster rate. That's our best guess. But it's a complete guess," he said.

Insulin Signaling (Long Live the Worms)

When a species of worms lived to double their normal life span it piqued the interest of Tom Johnson and David Friedman, two scientists at the University of California at Irvine, who made a breakthrough discovery in 1988. The team identified two gene mutations (daf-2 and age-1) that contributed to the worm's longevity.

More importantly, later research connected these genes to a pathway in both worms and humans called the insulin signaling pathway. Scientists are actively studying this pathway as a potential major player in longevity in humans.

The insulin signaling pathway is triggered by the hormone insulin when it's released from the pancreas in the normal process of sugar metabolization that occurs in a healthy human. The release of insulin causes a number of chemical processes to be sent into motion that affect the body's metabolism.

Insulin regulates the body's ability to absorb sugar (glucose) that enters the bloodstream through food. During the process of digestion, insulin facilitates the body's ability to absorb sugar. It allows sugar from the blood to be distributed to the skeletal muscles and fat tissue. That sugar is used for energy.

Restricting food intake by means such as calorie restriction affected the insulin signaling pathway of the worms that Johnson and Friedman were studying. When food was abundant the worms developed normally. When food was unavailable or the worms were overcrowded, they entered a long-lived larval state called dauer in which the aging process slowed down. Worms with this muted gene are also resistant to oxidative stress, along with environmental and bacterial pathogens.

The Curious Case of Laron's Syndrome

An unusual group of people with a rare disease called Laron's syndrome also connect the insulin signaling pathway to longevity. The disease impairs an important growth hormone affecting their ability to grow to normal adult height. Individuals with Laron's syndrome are easily identified by their characteristic dwarfish appearance. However, their short stature is a bit of a trade-off if you consider the mutation they all have blocks life-threatening diseases such as diabetes and cancer.

Dr. Jaime Guevara-Aguirre, an Ecuadorean physician, has been studying Laron's syndrome patients since the mid 1980s. He found the incidence of Laron's syndrome was most common in remote villages in southern Ecuador, which were populated by descendants of conversos (Sephardic Jews from Spain and Portugal). In 1994, he learned that people with Laron's syndrome were curiously immune to cancer and diabetes.

Dr. Valter D. Longo, of the University of Southern California, discovered that IGF-1, a hormone that makes normal children grow, was not present in Laron's syndrome patients. He also learned that the process of IGF-1 creation in the body is directly associated to the GH receptor gene. To test his theory, he gave doses of IGF-1 to prepubescent Laron's patients and found they grew to almost normal height.

Longo took a look at a laboratory roundworm that like humans, has what is known as the IGF-1 pathway. He found that knocking out a receptor known as DAF-2 (which basically makes IGF-1 work) allowed the worms to live longer.

And, like the DAF-2-less worms, Laron's patients lack a normally functioning IGF-1 pathway. This is another possible reason for why they live longer.

In 2011, Longo and Guevara-Aguirre put their heads together. They took a genetically compounded serum from Laron's patients and added it to human cells in a petri dish.

There were two key findings:

- The cells were protected from genetic damage when the serum was added.

- Any cells that were already damaged were destroyed. This mechanism is used by the body to stop the spread of cancerous cells.

When IGF-1 was added to the petri dish both the effects noted above were reversed. The scientists suggest that lowering the level of IGF-1 could be beneficial. A drug that does this could prolong life span.

AMPK, the Cellular Housekeeper

Let's talk a bit about the adenosine monophosphate-activated protein kinase pathway. But, er, best we call it AMPK because otherwise this book is going to be much longer than it needs to be.

The AMPK pathway, similar to the mTOR pathway, is a regulator. It's a master switch that handles metabolic change and protects cells by blocking unhealthy intruders.

Humans have the AMPK gene but in most people it's turned off. However, it can be activated with drugs, diet, and exercise. When it gets turned on, it regulates the release of a molecule that allows cells to transfer energy. Simply put, it's really good at cellular housekeeping and helps cells survive stress.

The good news is there is a drug on the market called Metformin that turns the AMPK pathway on. Metformin was invented in 1922 and has been in use since the 1950s to treat type 2 diabetes. The drug got some flak for a long list of side effects. But multiple studies proved it could decrease the incidence of cancer, heart disease, and even better, it can reverse aging.

 Andy's Off-Label Experiment

Inspired by news of Metformin's magical properties, author Andy started taking it daily for its off-label longevity-enhancing properties. He got his doctor to approve its use and discovered with a prescription in hand the drug was easily accessible from his local pharmacy for free. It's true that it can have some unpleasant side effects, including diarrhea and occasional dizziness, but you can eliminate them by managing doses and the time when you take the pill. We offer a free step-by-step guide to buyers of this book at http://ReadSuperYou.com/metformin

 Long Live the Fruit Flies

In 2014, UCLA announced a successful study using AMPK to increase the life span of fruit flies. A team of UCLA biologists took a group of 10,000 fruit flies and fed them Metformin to activate AMPK. The flies with the AMPK gene lived eight weeks. Flies in the control group lived the usual six weeks. The AMPK flies were also healthier for a longer time.

Living Forever: The Research of Dr. Aubrey de Grey

One of the people who is all for the therapies being developed in the previous pages by scientists such as David Sinclair is longevity researcher Aubrey de Grey. He is a theoretician of gerontology and chief officer of the SENS Foundation which was founded in 2009. It actively studies how to preventing age-related physical and cognitive decline.

"The work that David is doing and the approach we are doing at the SENS Foundation should be pursued full-tilt," he said.

The two approaches are complementary. On one hand the "simple" therapies will keep you alive long enough to benefit from the more-complicated approaches studied by the SENS Foundation.

The two work hand in hand. "David (Sinclair) is looking at magic bullets, single interventions that have global impacts on the whole of aging. The SENS Foundation is looking at a divide and conquer approach to solving the problem of aging," said de Grey.

The aging process is a series of degradations, so solving one issue won't solve the entire problem. de Grey said, "The SENS approach is the sweet spot between stopping damage that occurs in the first place versus fixing the diseases and disabilities that result from old age."

"There are seven problems targeted by the SENS Foundation research," said de Grey. Here are the causes of aging and their proposed solutions:

1. **Cell atrophy**—Cell death that occurs naturally in the heart and the brain as the body ages. Suggested solution: Stem cells can be added to the body to replenish parts of the system.

2. **Unwanted cells**—The body contains unwanted cells, like fat cells that "poison" the body overtime. Suggested solution: A procedure called "suicide gene therapy" could be tried. Suicide gene therapy involves the injection of a viral or bacterial gene into the body causing unwanted cells to destroy themselves.

3. **Protein cross links**—The loss of elasticity that occurs when protein links—components that hold together cells—are overproduced. Suggested solution: Drugs that counteract this process could be created.

4. **Internal cellular garbage**—Cells are in a constant state of action, breaking down proteins and molecules that do not serve the body. Molecules that can't be digested become "junk" inside the cells. Suggested solution: Enzymes that break down this "junk" could be developed.

5. **External cellular garbage**—"Junk" accumulates on the outside of cells. Suggested solution: Development of enzymes to break down this "junk."

6. **Mitochondrial damage**—Damage over time to the power supplier in cells slows energy production. Suggested solution: Gene therapies could repair and prevent this damage.

7. **Chromosomal mutations**—Mutations to DNA components cause diseases like cancer. Suggested solution: The regular replenishment of cells via stem cell therapy could help.

 ## The Secret to Longevity

So what does Aubrey de Grey do to extend his own life? "I don't accept invitations to speak at seminars in dangerous countries," he said. "I am well built, I eat, drink what I like and nothing happens. Every few years I have a check up and I always end up doing really well in terms of my biological age: that is, younger that I am. I am lucky that way. The only real generalization is to pay attention to your body, to just understand everyone is different, and what you need to do is work with what your body is telling you it needs."

De Grey proposes that in 25 to 30 years we will have the technology to solve aging. "However," he cautioned, "one big unknown in terms of time prediction is not just science, but feedback from the public." The process to approve new therapies involves the development of the technology and then the social, legal, or political policy to allow the technology to be used.

The second issue that could slow progress is the lack of funding. De Grey said, "The world's best people working on these projects are hot to trot but they are working much more slowly than they could due to limitations on funds. As far as I am concerned, if we had the money, we would be done. When I say 'the money' I am talking a ridiculously small amount of money. At the moment the key work that needs to be prioritized the most is the work at the earliest stage of development. That is, at the stage of cell culture and experiments with mice as opposed to clinical trials. That means, much cheaper."

"Right now SENS has a budget of about $4 to $5 million per year," said De Grey, adding that there is another $10 million being spent by other similar organizations studying the same things.

"That is unbelievably tiny. We probably only have to multiply that by ten and we would pretty much be able to do all we wanted to do in the short term," he said.

"In the longer term, 8–10 years from now, we will want to be doing more clinical trials, which will be more expensive. But, that kind of doesn't bother me because it seems to me that the achievement of specific results in the lab with mice to motivate clinical trials from a scientific perspective was going to be ample to also motivate the funding of those trials," he concluded.

Once this is demonstrated, he believes that acquiring the funding will be trivial because the work and results will garner the interest and enthusiasm of the public and policy makers.

Incidentally, de Grey believes he will live forever, but just in case he doesn't, as a precaution, he has personally invested in a specific type of cryonic preservation known as neuropreservation. During this procedure the head is removed from a deceased person and frozen to protect the face and brain. When the specimen is revived years later it is attached to a new body.

Cryonics: Freeze Me When I Die So I Can Live Forever

In the 1992 film, *Forever Young*, Mel Gibson plays a pilot who dies in 1939, is cryonically frozen by his best friend, and wakes up in 1992. And it's not just a Hollywood invention—cryonics is real. People can choose to become life-sized popsicles when they die with the promise of being brought back to life when mankind figures out how to revive them and nano-repair their disease or damaged bodies and then—ZAP!—bring them back to life.

Actually, we shouldn't use the word "frozen." The correct term is vitrified, which is to be cooled to a glassine state at a temperature of −140°C (−220°F). To set it all up, you join a cryonics non-profit (there are four in the world) and pay an annual membership while you are alive. You provide a cash allocation that can be funded by a life insurance policy, before you die, which goes to the cryopreservation company.

On your death, the cryo company sends in a team to cool you, package you, and ship you to a facility where you are cryopreserved in liquid nitrogen until the technology is available to bring you back in 100 or so years. (Or maybe sooner.)

We are going to oversimplify this here, but it helps to understand the process by thinking of the food in your freezer. Consider a bunch of peppers that are on the verge of the compost bin. Chop the veggies up, throw them in a freezer bag, and weeks later they will still make a pretty good pizza topping.

Unlike freezing vegetables, the cryopreservation process is used to freeze a recently deceased person into a glassine state. The key here is to first remove all water from their body and then freeze them in liquid nitrogen. The problem with freezing someone in water is that ice crystals pierce the cell walls. A defrosted frozen pepper is never as vibrant and crunchy as its prefrozen self.

This is why the cryonics process is carefully done slowly and without water to preserve the structure of your cells, and actually most importantly the structure of your brain cells. If you die and are preserved, it means the structure of all that is you is still there. The cryonics process halts physical decay and preserves you as you were at the moment of legal death.

Robert Ettinger founded the Cryonics Institute in 1976 in Clinton Township, Michigan. They house 150 patients in cryostasis. A monthly membership fee (usually financed from an annuity) pays to house them in cryostasis.

The other well-known cryonics facility is Alcor in Scottsdale, Arizona. As of July 2015, it has another 150 patients in-house who have been cryopreserved. Also, there are facilities in Russia and Switzerland that have a handful of patients. Another facility is being built in Australia and should be opened by 2017.

There are an estimated 300 people in cryostasis in facilities across the United States. Not all have their whole bodies vitrified. Some have opted for preservation of just their heads. This is called *neuropreservation*. The theory is all that is you is encoded in your brain tissue, and in the future you will be able to grow a new body to replace the one that isn't frozen.

One of the most notable patients at Alcor is baseball great Ted Williams, whose head was cryopreserved by his children.

 ## The Costs of Cryo

Costs to be cryopreserved upon your death vary wildly. However, there are three components: 1) An annual membership to the non-profit for between $150 and $700. 2) An allocation to the company of $28,000 for the Cryonics Institute or $200,000 for Alcor. 3) Cost to transport your body to the facility, although Alcor includes transportation in its $200,000 fee. Costs to cryopreserve just your head is $80,000, and is only available at Alcor.

 ## Cryo-Pets, Anyone?

For those of you that haven't been able to get over the loss of a furry family member, there's good news. You can opt to freeze Spot or Fluffy for a future reunion, as well. In the various facilities around the world there are more than 50 pets that have been cryopreserved by their owners.

 ## Get Frosty

Do you find all this cryonics info super interesting? Learn more details in this great blog post: http://superyou.link/getvitrified and http://waitbutwhy.com/2016/03/cryonics.html

Reports of Your Death Are Greatly Exaggerated

So what does Ray Kurzweil say about all this? In a 2013 interview with the *Wall Street Journal*, he said his baby boomer status puts him in the last generation that might have to worry about natural death.

"I'm right on the cusp," he told the newspaper. "I think some of us will make it through." The "us" are the baby-boomer generation that includes Kurzweil. He was born in 1948, making him 65 at the time of the interview. Practical immortality, he believes, might be possible for boomers that can hang on for another 15 years to about age 80.

"By then," the article continues, "Mr. Kurzweil expects medical technology to be adding a year of life expectancy each year. At that point we will be able to outrun our own deaths."

When we spoke to Kurzweil, he pointed to two of his books, which he wrote with Dr. Terry Grossman: *Transcend: Nine Steps to Living Well Forever* (2010) and *Fantastic Voyage: Live Long Enough to Live Forever* (2005).

"We talk about three bridges to radical life extension," he told us.

Transcend spells it out like this:

- **Bridge One**—Bridge One represents the strategies you can use right now to slow down, and in many cases stop, the processes that lead to disease and aging.

- **Bridge Two**—Bridge Two is access to the coming biotechnology revolution. By 2030, "we will have means," say the authors, "to perfect our own biology by fully reprogramming its information processes." Kurzweil said recent developments (in 2014) in technology had launched us into an early Bridge Two phase.

- **Bridge Three**—Bridge Three is the endgame in the longevity strategy, and by end, we don't mean death. Bridge Two adds decades to our lives so that we can blow past traditional life expectancy to access the coming nanotechnology revolution "where we can go beyond the limitations of biology and live indefinitely," says Ray and Terry in their book. That's Bridge Three. And that's when the fun begins.

The rest of the book is a guide to how to set yourself up to live forever and take advantage of Bridge One and Bridge Two technologies.

Extendgame, Not the Endgame

Now you might get the sense that the avant garde thinking among the people we studied and spoke to for this chapter is that death is generally curable. And aging is not only stoppable but reversible. You'd be right.

However, the cynics out there don't believe any of it. Some say it will be hundreds of years before any of this is feasible. Some even say it will never be possible and that Mother Nature will find a way to do in the oldest of us with some virus or other biological process that clears the way for future generations.

If you are onboard with the cynics, go book your casket this weekend and pick out your funeral flowers. If you're not that cynical, here's what there is to do: Stay alive for 15 more years. American life spans on average, as we said earlier, are just shy of 80 years. And that is increasing every year.

> "If you are onboard with the cynics, go book your casket this weekend and pick out your funeral flowers."

If you are a baby boomer or younger that should be relatively simple to do. Then by 2030, there should be technologies around to hyper-extend your life long enough for you to be around when nanomedicine and nano-engineering technologies are commercialized such that we can fix all this old person business and become young again.

We'll give the last word here to Kurzweil, who said in his 2006 book *The Singularity is Near*, the following: "We have the means right now to live long enough to live forever. Existing knowledge can be aggressively applied to dramatically slow down aging processes so we can still be in vital health when the more radical life extending therapies from biotechnology and nanotechnology become available. But most baby boomers won't make it because they are unaware of the accelerating aging process in their bodies and the opportunity to intervene."

9

Human 2.0:
The Future Is You

It might come in 2030. It might come in 2045. Or, somewhere in between.

The Singularity, as defined by Ray Kurzweil, will happen (see Chapter 1, "The Emergence of (You) the Human Machine"). He asserts there will come a time where "man will become a hybrid of flesh and machine, and ultimately, the non-biological portion will dominate." Many of the world's top technology experts agree.

Certainly none of the experts we interviewed for this book dispute the fact that one day strong artificial intelligence (AI) will happen. That is to say, a machine with the same and inevitably better intelligence than a human.

Some experts have made similar strange and perhaps "radical" predictions and watched them realized. Take Kevin Warwick. "Fifteen years ago I was saying we are going to have little aircrafts that fly around autonomously and folks were saying 'Oh, don't be silly.' Now, we have had drones for so many years you wouldn't even imagine that we never had them at some time."

So, *Super You* author Andy Walker's belief that he will live forever might not be as far-out an idea. Live forever? That's ludicrous! It's what coauthor Kay Walker originally thought prior to the writing of this book. A few interviews later, she had completely abandoned what she had previously believed. Now she's trying to decide if she'll go back to school at age 100 and get that Ph.D. she "missed the boat on" getting.

Sean's still not sure if he'll be able to live forever—or even if he wants to—but he admits it would certainly make it easier to catch up on his Netflix queue and backlog of books.

Kurzweil's assertions are what have given him clout as a technology futurist. It's why your *Super You* authors have written this entire book around his now famous theory.

Kurzweil's books such as *The Age of Intelligent Machines* (1990) and *The Age of Spiritual Machines: When Computers Exceed Human Intelligence* (1999) made predictions about today's technology back in the 1990s. The majority of them were right.

It's the nature of man to be curious, to build new tools for optimizing his world. It's how man got to where he is.

> "It's the nature of man to be curious, to build new tools for optimizing his world. It's how man got to where he is."

It's as Kurzweil said: "We create these things. They are part of human civilization. They are part of humanity. Man couldn't reach the food on the highest branch thousands of years ago, so he invented tools to expand his physical reach. Now, those tools have been adapted to allow us to build skyscrapers. Man's also created mental tools so he can access all of human knowledge with a few keystrokes from his smartphone."

So, we believe the Singularity will happen. The data supports it. It's the next evolution of man, already in progress. And even though when the time comes, as Transhumanist Party leader Zoltan Istvan predicts, "There will be clashes in the street," the truth is that once man creates a technology that makes his life superior and the consensus agrees, that technology will be adopted.

For instance, the day a robot can operate surgically on you with a lower rate of error than a human surgeon, there will be little reason to choose a human surgeon, other than in perhaps a supervisory role. That's especially true when it comes to critical surgeries such as heart replacement.

Of course the technology laggards—those people who are slow to change—will resist it—because some hate and even fear technology. These are the same people who still prefer a human teller to an ATM machine for everyday transactions.

But largely, it is Hollywood that has planted the seed that technology change is not good. The reality is that, as Kurzweil puts it, "Technology is part of who we are. It's not us versus machines."

Bring it on. Peace, love, and nanobots!

Still, there are some high-profile naysayers. If you ask Tesla CEO Elon Musk, physicist Stephen Hawking, or software billionaire Bill Gates—and a handful of other smart, but perhaps less optimistic prognosticators—the Singularity and all its trappings might be our undoing. They think machines are going to rise up and take us all out, then stack our bodies like cordwood and use us as door stops.

Is that valid?

All this is mostly conjecture, with some Kurzweilian data points supporting it and some emotional fear mongering that is against it. So let's consider both the dystopian and utopian views to help understand what you need to know in the next 30 years to 2045 and beyond.

And at the end of sections throughout the chapter, we have added some of our thoughts, musings, and "wish lists" about the future.

 Dystopia

A "dystopia" is the opposite of "utopia," which you might remember meaning, "an imagined place of perfection." That means a "dystopia" is an imagined place of despair and ugliness.

Look Like Who You Want to Be and Be Who You Want to Be

The future looks good, quite literally. In the next few decades, cosmetic and plastic surgery procedures are going to be more accessible than ever before. So, if you've been wavering on whether to spend the $12,000 in your savings account on a bifurcated tongue and eyebrow implants, consider waiting five years. Future technologies will be safer and more cost-effective, as they'll need to be to keep up with consumer demand.

The cosmetic industry has been on an upward trajectory since 2000 when the American Society of Plastic Surgeons (ASPS) began tracking the number of cosmetic and reconstructive plastic surgery procedures in the United States. In 2010, the total number of cosmetic procedures totaled: 13,117,063. By 2014, the number grew to 15,622,866. That's 2,505,803 more people in only four years.

In 2014, Dr. Patrick J. Byrne, Director of the Division of Facial Plastic and Reconstructive Surgery from John's Hopkins Medicine told *The Wall Street Journal* that "30 years from now, there will be much more demand (to look good) as our society not only ages, but also does so with greater vitality than ever before."

A growing field of plastics means the continued development of more advanced technologies that cater to the masses. Current research efforts are attempting to make future cosmetic and plastic surgery procedures less invasive than they are today. This means moving away from "slice and dice" tactics. In the next five to ten years, it's likely that surgeons will be using their patient's stem cells to grow or print (on 3D bioprinters) organic tissues and organs in the lab.

Dr. Robert Murphy, former President of the ASPS, said, "One of the hot topics that is near term is stem cell therapy." Stem cell technology will dominate the cosmetic and plastic surgery field. They will allow surgeons the ability "to grow cells that morph into a number of different things."

The list ranges from "cardiac tissue for people with heart damage" to cosmetic solutions that treat "damaged tissue from radiation therapy in people who have had cancer," or also to help people maintain "a youthful appearance, by filling both volume and improving the quality of the overlying skin."

Murphy said in the next three to five years, the plastic surgery industry is "talking about being able to extract stem cells, process them in a laboratory, and reimplant them."

While stem cells are already being used in institutions across the United States and Europe, Murphy said what's required is more "scientific rigor." The technology requires perfection, so that it "becomes a technique that has defined applications and accepted risks, benefits, and outcomes." This, he asserts, could happen in the next two to five years.

It's likely that stem cell technology will reinvent the entire plastic surgery industry, causing current surgical techniques to become obsolete. Murphy said, for example, "Implant augmentation of breasts (will become) passé because we could use tissue instead of using a foreign body."

The nature of these procedures would require a surgeon to harvest stem cells from a patient's body, process them in the lab, and reimplant them. Even more exciting is the prospect of "marrying genetics with stem cells in vivo," he said. This allows for the "genetic manipulation of a stem cell in the body without extracting it." He said this could happen in the next ten years.

Using stem cells from a patient's body to create new tissues and organs has massive advantages. The patient's immune system has a better chance of accepting the material. It's likely these procedures would be done faster and easier with no need to cut open the body. A new firmer bum on your lunch break, perhaps?

And for those not interested in surgical procedures, other "looking good" technologies—from makeup, to diet shakes, to lotions will soon be smarter, too. Many will have dual functions. Nails with radio-frequency identification (RFID) tags and conductive polish used to unlock doors? Tattoos made with magnetic ink that buzzes when someone texts? Clothing that reads vital signs? All possible.

Many will become more accessible to everyone. In 2015, Tina Alster, MD, Clinical Professor of Dermatology at Georgetown University Medical Center in Washington, D.C. told Oprah.com, "In the not too distant future, lasers and intense pulsed light machines that remove pigment, soften lines, and treat acne will be sitting on bathroom counters next to our toothbrushes."

In other words, Super You will be a canvas that you can change at will with easily-accessible, and low-cost technology. Just make sure you don't opt for the bargain bin version. If you have opted for the Etch-A-Sketch budget package versus the premium *Venus de Milo* package from your cosmetic surgeon, you might be out at a dance club, shake your buns a bit too hard ... and erase them.

Kay says: "I can grow bigger breasts using my own stem cells? Sign me up!"

Sean says: "Can I get lasers shooting out of my eyes? Or do I have to wait until we talk about cyborgs?"

Andy says: "Everyone can have webbed toes like me!"

Customize Your Children So They Live a Life Free of Disease

In the future all babies might get their start in a petri dish. Transhumanist Zoltan Istvan approves and suggests: "Who doesn't want the best traits in their children? Who doesn't want to increase their child's intelligence level? Who doesn't want their kids to avoid disease?"

Quick clarification: That's not to say sex won't happen. Though it will likely be more about the recreational component of intercourse than procreation. And, of course, it might include a robot as a partner.

Engineered babies are already happening, as you learned in Chapter 2, "Baby Science: How to Conceive a Tennis Star and Other Procreative Miracles." Test tube babies have been conceived out of the womb since the 1980s.

Today, for $18,000, a couple can go see Dr. Jeffrey Steinberg to guarantee the sex of their baby. Other companies, such as GenePeeks, use technology to genetically match two hypothetical parents on a computer to ensure the baby they create will be disease free.

Though, some would argue, guaranteeing the genetic viability of the world's future generation is a more ethical use of the technology than manipulating individual traits, such as hair texture or eye color.

The Center for Bioethics & Human Dignity (CBHD) of Trinity International University sees genetic manipulations for health purposes as somewhat reasonable,

"... the big concern is if genetic manipulation is allowed to produce a disease-free baby, then that opens up the opportunity for trait manipulation, too."

though the onus for procedures such as preimplantation genetic diagnosis (PGD) still lies with the parents. And ultimately, the big concern is if genetic manipulation is allowed to produce a disease-free baby, then that opens up the opportunity for trait manipulation, too.

An article on the Circumpolar Health Bibliographic Database's (CHBD) website written in 2004 by surgical pathologist Samuel Hensley argues that designer baby technologies have implications that could alter the future health dynamic of "child-parent relationship." Hensley writes: "The oppressive weight of parents' expectations—resting in this case on what they believe to be undeniable biological facts—might impinge upon the child's freedom to make his own way in the world."

Okay, so in the future, kids might resent their parents for giving them athletic abilities when all they want to do is top the charts with their death metal band Nanobot Annihilator. And parents might resent they made their kid extra tall so he'd be a basketball player because Dad didn't make the NBA. Is that so different from now? Most kids hate their parents for something they did or said. Nothing new there.

Peter Lawler, a Dana Professor of Government and former Chair of the Department of Government and International Studies at Berry College who served on President Bush's Council on Bioethics from 2004 to 2009, speaks to this concern. In a 2010 article published on BigThink.com he wrote, "I think that enhanced people will become, in some ways, more miserable than people ever have."

Istvan mentions Lawler as one of a group of futurists that have written blogs about designer babies that argue, "If you edit the genome, future generations could have new diseases pop up." But Istvan suggests this is a linear view of the future. Thinking exponentially, he suggests, "Future generations—we're talking in 70 to 80 years out—humans won't be human beings anymore. We may have made new types of DNA. We may have synthetic DNA."

(This is a classic example of arguing about future technologies from the perspective of current technology.)

Ultimately, the future of designer baby technology is up to policy makers. Recall Steinberg's story from Chapter 2 about the Vatican suggesting he shelve his plans to offer baby trait biohacking. But even further, new policies are enacted when large groups of people argue for their personal rights.

To that end, Istvan argues: "It comes back again to this idea of morphological freedom. We want the choices to do these things." From this standpoint, denying the choice to have the technology to enhance the next generation for what some would argue "frivolous" trait selection such as height or hair color, could step on issues of personal freedom. So ultimately, it's up to what the people want.

If the era of designer babies ever comes, Larry Arnhat, the Presidential Research Professor of Political Science at Northern Illinois University in DeKalb, suggests there might be implications the world has not yet considered. On his personal blog he wrote: "Manipulation of human nature to enhance desirable traits while avoiding undesirable side effects will be very difficult if not impossible."

But once again, it can be argued that Arnhat is considering a linear view of the future.

"Eventually, the whole designer baby argument will be old news," Istvan says. "The age is going to come and go so quickly. This idea that we can change hair color is going to be here for three years. Then, we are going ask ourselves in 10 years, or maybe 20 years, whether we want to have hair that day because, we will be able to grow it in two minutes. Or, we are going to have some type of helmet that gives us virtual reality all day long."

And it's a possibility that when we get to the era of designer babies, we will already have the technology to handle any unexpected issues.

Kay says: "I don't need designer baby tech. My offspring is perfection. Though I would like to ask the universe for a mini-Kay the second time around because putting my son in frilly dresses will give him a complex."

Sean says: "Can I get bright yellow skin, four fingers per hand, and only two hairs on the top of the head? I'm gonna call him Homer."

Andy says: "I think we will need to regulate gender selection somehow. Anything other than 50 percent male and 50 percent female is going to mess with high school dances and Tinder."

Live Your Life Disease Free

It's one thing to want to live forever, but it's probably a safe bet you wouldn't want to spend that eternity battling various forms of crippling disease or conditions that will reduce your ability to play golf in your extended dotage. It's an implicit part of the deal: If you want to stay alive, you want to do it in as fine a state of health as you can possibly manage.

As research into the human genome continues and scientists find ways to make mapping it less and expensive, it's almost certain that we're going to unlock treatments for afflictions that are bound to ravage us as we grow older—whether it's figuring out a way to turn off cancer cells or reverse or prevent neurological conditions such as Parkinson's disease or Alzheimer's. Or for that matter, to replace failing organs with 3D-printed or artificial versions, or to ameliorate chronic conditions such as hypertension, diabetes, asthma, or chronic obstructive pulmonary disease (COPD).

In some cases, we'll apply treatment to people who already have these conditions as adults, but almost certainly there will be a push to use clustered regularly-interspaced short palindromic repeats (CRISPR) technology (or its successors) during the prepregnancy planning stages, by editing out potentially problematic genomic sequences in the parents' genomes … even before the sperm hits the egg.

While recognizing we're not quite at that point yet, Dr. Bertalan Meskó thinks we're going to be there soon … perhaps even sooner than you think. "Based on my experience with genomics, if in ten years' time I don't get treatment based only on my bioparameters, health parameters and my genomic profile, I will be quite surprised and actually angry, to be honest with you."

As ever, the ethics of this are going to be a big question and for more than one reason. As we touched on earlier, one major concern about adjusting our offspring's DNA is economic: If only the rich can afford to give their babies a disease-free future, we would further exacerbate the health divide that has already stratified much of the world.

While countries with single-payer health care programs might have an advantage in this regard, it's worth noting that not all procedures are covered under these systems: Even in Canada, which has been looked upon favorably by single-payer advocates in the United States, coverage of procedures varies from province-to-province for some lifesaving procedures and drugs. If you can't even agree on treatment coverage in cities that might be only 20 miles apart, what hope is there for trying to convince a wide array of private insurers to get onboard?

Istvan believes that much of the solution lies in changing people's minds across the board. "The question is: How do we move forward as a culture to allow these things in general in a broad sense? This is why I've made most of my (presidential) campaign," notes Istvan, "on generally changing cultural viewpoints. The more we talk about the benefits of transhumanism as a whole, the more people just say 'ah, screw it, let's just go with it and if we don't like it, we won't have a robotic heart, or a designer baby'."

Naturally, the same would go for prevention at the genomic level; people are almost unanimous that they'd like to live disease-free, but get squeamish when you start talking about tinkering with DNA to make that happen. But opening up the discussion might change minds.

As we noted earlier in this chapter, others are worried that tinkering with DNA might have unforeseen consequences that might not reveal themselves until it's too late to undo it, and that rewiring humans at this level is problematic, especially if those genetic changes are then passed on to future offspring.

It's harder to argue with this type of assertion; sure, it's partially driven by fear of the unknown. On the other hand, there are plenty of times when science got it completely wrong, such as proclaiming DDT and thalidomide safe, or nuclear-age claims that exposure to radiation was actually *good* for your health. It's easy enough to take this argument into tinfoil hat territory, but the doubts from more reasonable people might be enough to throw a wrench into immediate social acceptance.

Whether we come to a consensus on rewriting DNA before conception, we're going to see a lot of advances on how to fix people up after they're born. In fact, the process will become even easier as more research can be done using adult stem cells, which will remove many potential religious and political objections from the mix.

Meskó thinks this is an inevitability. "I believe that—this is a very strange prediction of mine—but by the time I have my first child I will not have to freeze the cord tissue because it won't be required, because in my kid's adult life we will be able to use the adult skin cells and convert it into any kind of tissues."

That will certainly unlock a lot of potential treatments that could render some of our current medical procedures outdated. "It means we should forget about organ transplantation in five to 10 years' time because we will have at least the tissues which we will be able to print out as biomaterials quite fast," predicts Meskó. "I'm optimistic that this would be available in ten years. Right now the technology is there, we just have to fine tune it and find a solution."

Or, of course, we can inject ourselves with nanites designed to keep an eye on what's going on inside our body and take corrective action, such as releasing oxygen, making minor repairs to blood vessels, or terminating unwanted cells (cancer, viruses) or foreign materials (toxins, plaque, or beer).

Dr. Robert Murphy, President of ASPS, thinks we are close to such a reality. "We have been talking about—in terms of nanotechnology and stem cells—injecting cells to go through the body to go into the heart and form new heart muscle," he said. But he notes that it doesn't stop there. "To be able to manipulate certain cells to go into the body and clean out plaque in blood vessels that is essentially inner space so we are actually on the cusp of doing some of that. How we can purify and perfect, that is more forward thinking than conjecture. We are on the cusp."

We're already tracking our heartbeat from our Apple Watches and Fitbits, so people are definitely interested in living longer, healthier lives. Now it's just a matter of convincing people to take it to the next level, whether it's by injecting nanites into our bloodstream to monitor our health from the inside, printing replacement parts as they wear out, or rewiring us from the ground up before we're even born.

Kay says: "I like the idea of health being easier. If I can have a device inserted into my body that monitors my blood glucose levels and manages my metabolism, I wouldn't have to kill myself at the gym as much and as often."

Sean says: "If it keeps my hypertension in check, sure, I'll take the nanites in the blood!"

Andy says: "I think about the cardiovascular disease that runs in my family all the time, especially since I am approaching my 50th birthday. So this gives me peace of mind. Bring on the nanites!"

Be Superhuman

We've said it before and we'll say it again: The cyborg revolution is in full swing and you're almost certainly already a part of it. The thing is, if you're carrying around a smartphone, you're only mildly involved. If you have a robotic leg or hand, you're certainly much further along. Either way, are you ready to really leap into the fray and join the Borg?

For those who are trying to regain some lost functionality, the answer is almost certainly yes, which is why artificial limbs, pacemakers, and hearing aids have become relatively commonplace. These bits of technology are nonthreatening because at a fundamental level, people intuitively understand them and what they're meant to do. Very few people would begrudge an amputee a robotic prosthetic if it allowed them to become mobile again. People get what mobility means to them and how it would feel if it was taken away.

Biohacks that add capabilities you didn't previously have are a different story.

The grinders and body-hackers we looked at in Chapter 6, "Franken-You: A Better Life Through Cyborg Technology," can be a bit off-putting to people in much the same way that you'd be alarmed if the Terminator walked into your workplace and demanded your clothes and the key to your Harley: It's unnatural, and people tend to react to the strange with suspicion or fear.

Plus, it doesn't help that electronic implants might make you look dorky... hell, all it takes for that is Google Glass. But we ignore these enhancements at our peril.

Remember Kevin Warwick? He knows a thing or two about how implants can change your basic capabilities. First, he got an RFID implant that could open doors and control lights, and then he installed the BrainGate implant, which allowed him to receive signals directly from his wife over WiFi.

While he admits that some of these things are strange, he says that you can get used to their new capabilities rather quickly: "Oh yes, if you link into your brain or the peripheral nervous system you can start to give yourself a new sensation. It takes a while, but meaning, a few weeks, before you can understand all types of different things. So we can extend people's sense, we can have new forms of communication, there is a lot more we can do with the brain."

Warwick sees a future where people can interact with their world simply by thinking about it, without all that pesky, talking, typing, or button-pushing. "Thought communication in some basic form will be enormous, just as the telephone has been enormous. Though, the telephone will only be a tenth of what thought communication will be like."

But when asked, Warwick is unsure how quickly this revolution will come about. "In terms of a commercial product, I don't know. I think the first experiments will have happened [by 2024], but how long until it's a commercial success, I don't know."

You can bet there are a lot of people who would love to start controlling things with their mind, such as changing the channel on the television without having to lift that onerous remote. Or thinking about a nice cold grown-up beverage and having your spouse (or bartenderbot) deliver it right to your comfy chair.

On the other hand, if you put an implant into your brain allowing you to connect directly to the network, how long will it be before you have to install an antivirus program into your wetware? The last thing you want is to be trying to pick someone up at the nightclub and have some pimply dingleberry hack directly into your brainpan. Your pickup lines are terrible enough already without being biohacked into blurting out random Madonna lyrics.

It's no surprise that Ray Kurzweil is more positive about the possibilities of melding the biological with the technological and what it's going to offer humanity.

"We won't be constrained by physical parameters. We'll become a hybrid of biological and nonbiological thinking," he predicts. "The most important part is we will expand our neocortex again. And if you remember what happened the last time, two-million years ago, it allowed us to take a qualitative leap. And that will allow us to take another qualitative leap."

Kay says: "When you can't beat 'em, join 'em."

Sean says: "Ooh, ooh, lasers! Shooting out of my eyes! Plus an implant that allows me to stream music directly into my brain 24 hours a day."

Andy says: "♪ Like a virgin, touched for the very first time. ♪ ♪ ♪"

Live Forever, If You Choose

In his book, *Stumbling on Happiness*, author and Harvard professor Dan Gilbert writes: "Each of us is trapped in a place, a time and a circumstance and our attempt to use our mind to transcend those boundaries are more often than not ineffective." He's learned this from decades of research in a field of psychological study called affective forecasting. In other words, he studies man's ability to accurately predict his own future.

From a plethora of research, he's learned that man uses his present emotional state and his past experiences to consider his future. Not surprisingly, this makes man a very unreliable fortuneteller.

This inherent biological shortcoming is one potential reason most people find it hard to believe that we will ever have the choice to live forever. It's not possible now, so how could it be in future? Birth and death is as much a reality as gravity is. Right?

Well, there are experts, one being biomedical gerontologist Aubrey de Grey from the SENS Foundation, that would argue death is only unnatural now because technology hasn't figured out how to solve it. But perhaps one day it could make the list of impossible conundrums that technology solved alongside flying (airplanes), long-distance communication (telephones), and sending information without wires (radio, cell phones) to name a few.

So perhaps living forever is not such a strange prediction. Though, in terms of the research required to get there, de Grey told *London Real* (listen in at http://superyou.link/maverick), "Irrationality is slowing things down and costing lives … 100,000 people die from age-related issues each day." This lack of forward thinking has a direct impact on funding.

If de Grey could wave his magic wand and get what he needs so that SENS could develop a cure for aging, he said what they need is a "$50 million budget rather than just a $5 million budget."

The scientific thinking has been done. The theories are in place. But years of research to affirm and develop technologies, and the money to do so, is in the way.

When it comes to timelines, de Grey always makes sure to reaffirm the speculative nature of these predictions. He told *London Real* in 2014 that he believes SENS has a "50/50 chance of getting these technologies into a decisive level of comprehension in 20 to 30 years." Though, he admits there's still a "10% chance of 100 years if we encounter something along the way we didn't expect." And ultimately, he said, "The only way we'll get there is if the 5 to 10 years constitutes good proof of concepts and the funding improves."

Funding and technological innovations aside, a third hurdle is social policy. Legal issues relating to living forever would need to be examined. New laws would have to be created to regulate the new issues that arise.

One for example, is population control. de Grey is often asked about this concern to which his response is "the carrying capacity of the planet is not fixed," so he believes the "prospect is very overblown." And perhaps, there would be new laws that would govern who could have babies to accommodate the new change.

Lastly, policy only changes when large groups of people demand they do. For that, minds will need to change. People will need to want to live forever. They'll need to understand the benefits.

The shift in thinking will likely come in time. It takes decades to shift world conversations. Though, it has happened. At this point, it's left up to speculation what will happen in future. In the next four decades, it's unlikely you'll need to be concerned.

But, if you're like Andy Walker, Kurzweil, and de Grey, who want the ability to choose to live forever, here's your game plan:

> **Step 1**—Stay healthy so you can live as long as possible and make it to the day technology ends death.

> **Step 2**—Get a "living will" now that ensures you'll be cryonically frozen in the event of an accident or death.

> **Step 3**—Be an advocate, and encourage others to learn and help forward longevity research efforts.

And, in the event you're cryonically preserved, don't be upset if when you wake up you're living with more robots than people, drinking Soylent at mealtime, and a flight suit is your main mode of transportation.

Kay says: "I don't believe this will happen in my lifetime. But, I'll prepare my son."

Sean says: "Even if it was possible to live forever, I'm not sure it's all that great an idea. It's already hard enough to find a seat on the subway. Also, just because someone can live hundreds of years doesn't mean their attitudes will shift with the times. It could get ugly."

Andy says: "Death is curable. Why would you accept the false inevitability of becoming plant fertilizer? I don't get it."

Robots Replacing Jobs—Less Work and More Fun

The opening to this chapter mentions that the Singularity will inevitably happen. In truth, there's no 100 percent guarantee. No expert, even as knowledgeable, meticulous, and mathematical as Kurzweil can accurately predict the future in full. (Although he's done a very good job in his first couple of books.) But, unless technology ceases to progress, or some severe environmental catastrophe takes out the entire world within the next 30 years, strong AI is likely going happen.

In that case, it's time to get comfortable with the idea of bumping into C3P0-V2 at the company water cooler. Or, sending your kid to school in a driverless car. Or, spending an intimate evening with SugarBabe3000 whom you recently purchased from your local Stag shop (no judgment here!).

Many people fear this reality. Or they don't give it much thought and tend to think: AI happens in the movies, not in real life. But consider what you used the last time you needed to find some information quickly on the Web. A Google search bot perhaps (which is an information processing robot)? Or the last time you asked your car's GPS module for directions. Also a robot. Or pull cash out of an ATM machine—robot too.

More robots, and smarter ones, means less need for humans to exert their own powers on tasks that range from the mundane (taking out the household trash or filing taxes) to eventually more specialized procedures (performing a kidney transplant operation). Istvan predicts that, "Five percent of jobs will be gone in the next five years." Which he asserts is "sooner than most people realize."

Robots taking jobs is nothing new. Consider how some of your favorite foods are processed and packaged. In most cases, an automated machine is performing a job a person used to do. The baker that used to bake the bread by hand has been replaced by giant mixing machines and conveyor belts that whip a few food ingredients into a loaf of Wonder bread you pick up at your local grocery chain.

This progression didn't happen overnight. Which is why it's likely robots running the corporate world will progress in stages:

> **Stage 1**—Humans will continue to use bots to better perform in the workplace.
>
> **Stage 2**—Robots will replace jobs that do not require an aptitude for conscious thinking.
>
> **Stage 3**—AI robots will replace information-based jobs that are subject to human error, where a bot would be more effective.
>
> **Stage 4**—Robots will replace most (if not all) humans at work.

Stage 1: The Bots You Know

We're already well into Stage 1. Robots have been used in the workplace for decades to increase efficiency and productivity. This is, once again, a good example of the role of robots in factory work.

The invention of the Internet and mobile devices has also enabled workers in many industries to access information using a robot assistant. A warehouse manager, for example, on the ground floor can get a custom quote on a new forklift purchase using a Google search bot from his mobile phone. Or, consider a chef that needs to look up a recipe for hollandaise sauce. He could do so, using a tablet from his kitchen counter.

In the medical community, Meskó said: "IBM's Watson has been tested across the U.S. in the decision making process … Watson can make a few suggestions to a doctor about course of treatment." Meskó said he knows doctors that have used this technology and they "love this method because they feel more certain (because), for humans, it's impossible to keep in mind all the millions of papers."

Stage 2: The Bots Are Coming

Robotic applications in the workplace will only continue to help workers perform across all industries. In Stage 2, robots will replace jobs that do not require an aptitude for conscious thinking and this is also already happening, but at a smaller level. Your household iRobot Roomba, for example, has yet to replace the need for your cleaning lady. Amazon Prime Air drones have yet to deliver toilet paper to your house. But, that's not to say that robots designed to replace these types of jobs aren't currently in development.

In 2015, Amazon CEO Jeff Bezos told CNN that one day robots delivering your mail will be "as common as seeing a mail truck." To that end, in July of 2015, an Australian startup called Flirtey received clearance from the Federal Aviation Administration (FAA) to test the first delivery drone in America.

In their 2013 paper, "*The Future of Employment: How Susceptible are Jobs to Computerisation?*," economists Carl Benedikt Frey and Michael Osborne write that jobs most vulnerable to robots are those that "mainly consist of tasks following well-defined procedures that can easily be performed by sophisticated algorithms." Popular ones that made their list are: telemarketer, retail sales associate, accountant, technical writer, and insurance adjuster.

Istvan agrees about writers: "The way stories are written today, there is no reason a person is going to be able to write a quicker and better story than a bot. Half of today's stories are rehashed … they take sentences from other places and reorganize them so it's not plagiarism. A computer can do that."

This is good news to your *Super You* authors who would all like some help from a bot if they're asked to write *Super You v.2*.

Kevin Warwick asserts that within the next ten years, "If you are communicating with something (without seeing it) you won't know if it's a machine or a human." The job telemarketer from Frey and Osborne's report would likely be on Warwick's list.

Stage 3: Smarter, Strategic Bots

Replacing jobs that require a higher-level aptitude of physical or mental intelligence (Stage 3 on the preceding list) will only happen when computers match or exceed human capability. Jobs in this group include occupations that make use of traits that are unique to humans, such as creative thinking, manual dexterity, and emotional intelligence.

Frey and Osborne cite jobs such as: dentist, nutritionist, athletic trainer, elementary school teacher, and mental health therapist, as having a less than 1 percent chance of being replaced by a machine. They also suggest a firefighter as a job that has a lower chance of becoming automated than a pilot. Though a pilot requires more technical ability, a firefighter has physical ability combined with strategic thinking, so it's harder to be replaced by a robot.

Stage 3 will only happen when strong AI arrives. When that happens, it's been suggested a robot will have a superior capability than those of humans. A machine will be unaffected by emotion (unless it's been programmed with an "emotions" feature), will have a higher capability to access information, and its motor skills would be faster and more precise.

Stage 4: Here Come the Lawyer Bots, et al

Zoltan Istvan predicts in the future there will be "robots delivering babies as opposed to people delivering babies." It's likely a robot would be able to achieve a level of accuracy that exceeds what man can do. Istvan argues "robots are just simply better. They make less mistakes. They don't have fights with their spouses the night before (surgery)."

He suggests a "doctor standing behind a machine in case of a weird emergency … could be here in 10 years." In 20 years, he suggests, a robot alone might perform operations on its own.

Istvan cited another power position that could easily be replaced by strong AI: "Attorneys are one of the easiest jobs to replace by AI because you are just talking laws." But that's when Istvan believes there will be some resistance at the level of government.

He said: "40 percent of the U.S. Congress is controlled by attorneys … and they are absolutely going to make it so that their jobs don't get replaced."

But people will only fight for their jobs (including attorneys) until they learn, firsthand, the value in being replaced by bots. Ultimately, technology is made by humans for the purpose of making life better and easier. Robots are no different.

Consider that in the last decade technology has brought more fun, more play, more joy into the workplace, and ultimately to life.

In a 2015 interview, cyborg anthropologist Amber Case told us, "The division between work and play is more blended together. You go home from work and you look at your television through Netflix, and then you have your iPhone and your laptop. Sometimes we answer our work mail in the evening. Sometimes we enter Facebook replies at work. So all these things that we had stable fences between are no longer there."

A 2014 study from the Telework Research Network reported that 30 million Americans work from home at least one day a week. The number is expected to grow by 63 percent in the next five years. And get this, 54 percent of the 30 million of the work-from-home respondents reported that they are happier than ever before.

Technology has provided the workplace with many improvements and the benefits of robots will be similar. Some experts argue it will give rise to a new and improved world. Istvan suggests the day self-driving cars replace human drivers will immediately prevent 100,000 deaths a year by ending drunk driving. Not to mention, passengers can spend their car ride playing cards, talking with loved ones, or catching up on a good book.

It's possible when humans put robots to work the world will be a lot more pleasurable. Istvan said: "People could potentially become perpetual students or whatever it is that they want to do."

For this to happen, the government would need to create new laws and policies to ensure the unemployed would be supported. It's why Istvan says the "Transhumanist Party supports a universal income." Which would mean income would be disbursed by the government to ensure citizens can afford to live and enjoy life while robots work.

Your *Super You* authors can't dispute a world where robots work and people are paid to learn, travel, or drink wine on a patio with their friends—do whatever it is you want to do—is a world we'd like to see happen.

By the way, for those Type A's who are freaking out right now: Steady your gin and tonic for a sec. That doesn't mean to say you couldn't work if you wanted to. Or that you wouldn't be able to earn supplemental income over your allocation of universal income. You could choose to work, if that floats your catamaran.

Kay says: "A world with universal income sounds blissful. I'd spend my days learning stuff. Student forever!"

Sean says: "There are definitely a few things I'd like to get robots to work on around my house."

Andy says: "Robotic lawyers...what will they think of next?"

Restructured Society, Economy and Political System

All the great ideas in the world aren't going to get very far if humans don't get out of the way of progress. And, unfortunately, they have shown an incredible ability to, yep, get in the way of progress.

Or rather, we should say that certain parts of the world and elements of society have shown that particular nasty trait. Although it's true that significant resistance comes from many of the religious leaders all across the spectrum, it's our political leaders that have the power to prevent us from adopting some of the technologies that will lead to a transhumanist future. Although they might think they're doing the right thing for their constituents (or congregants), the resistance to technology will spill into unintended sectors. Such as the national economy. Or the ability to retain intellectual talent inside the country.

The bottom line is this: If people want a procedure that's available in other parts of the world and not their own country, they will book flights to get the procedure done offshore. Of course, that takes money out of the country, too.

According to Zoltan Istvan, it might not even be simply a matter of unavailability of the procedures; sometimes it's also a matter of a local jurisdiction making it so expensive to perform a new procedure that it might as well be unavailable. "It's just like we go overseas to have a kidney transplant because it's so much cheaper, we are going to go overseas to have the (designer) babies done this way. And you can already see people talking about it."

In the next decade or two, technology will explode in many super ways, but perhaps only for those that can afford it initially. Technology drops in price as it is adopted and as new successive generations of technology replace it. But the bleeding edge is always going to be expensive.

Amber Case explained: "As long as you have enough money and you are in the right social class you will have the privilege of the better interfaces as they come out. If you do not have enough money you will have to use the same phone for three or four years. Interfaces will turn against you."

Istvan cautioned slow governmental process might also hinder adoption of new technologies. "When you look at how quickly drugs get passed in China, and how quickly medical endeavors get passed in China versus America, it's something like 70 percent quicker. The FDA has an eight-year limit to pass a drug for the public. Whereas, in China it's like 18 months or something."

This government inaction will mean a lot more research and implementation of these new transhumanist technologies will happen in countries other than the United States, says Istvan, and that will cause a drag on the economy. "People are going to go overseas if they don't allow it here. Which again this is part of this great brain drain out of America, which is very sad because if we want to keep up in the future—and we are already getting our butts kicked if you look at the timeline between how quickly Russia has been growing, or China has been growing. It's quite possible that in 10 years China and India will have more GDP than America and will become more influential."

Countries such as India and China have already demonstrated a great willingness to invest in these technological areas, and have been spending money in areas of research that are still dead-ended in North America.

It doesn't really matter whether the cause of the roadblocks is religious in nature or a question of ethical reservations; the fields are moving forward in some parts of the world but not in others, which means a drain of talent and money from one part of the world to the other. Whether it's even an area we'd ultimately want to be involved in from an ethical perspective is almost beside the point.

This disparity in tolerance levels for these new avenues of research has already led to some stark contrasts in how different countries are progressing in these new technologies, with much of this research moving to China. Basically, the more we resist these new technologies, the more another area of the world will grab the ball and run with it.

That's not even getting into the disparity that happens within a country's own borders and the income inequality between different groups of people living in the same area. Not having to wait in the Apple Store line to get your new Apple gadget the minute it hits the market might not be an issue to some. But consider the inability to purchase a genetic test that would have you learn if you are going to die from cancer if you don't do anything about it. This will expand the already giant gap between social classes in the United States.

In a 2011 report, which later became a video that went viral (see the video: http://superyou.link/whereisthewealth), Michael Norton, a business professor at Harvard University, and Dan Ariely, a behavioral economist at Duke University reports that the top 20 percent of American households own more than 84 percent of the world's wealth. This economic disparity has been growing steadily since the 1970s. It's likely to continue to grow as the wealthy become super human and naturally have an advantage to continue to succeed, where others might not.

Meskó agrees: "People who have the financial background will be able to be super humans, with perfect eyesight, digital hearing, better strength, and more intelligence. They will have access to these quality features just because they can afford them."

He said this is a cause for concern: "It's going to cause amazing changes in how we think about ourselves and the human race. Differences between two individuals have never been greater than before as they will be when this super era happens."

With that stark disparity between people's opportunities to access the technologies, it might motivate people to start throwing up roadblocks, especially when coupled with possible moral objections. "I would not be surprised if over the next five to 10 years you see a huge shift of these religious people implementing laws that absolutely make a lot of this stuff illegal," says Istvan. "Within five to 10 years, we're going to be seeing clashes when people start saying 'we don't want genetic engineering, we don't want this type of radical technology so let's just make it illegal.' And of course it's never going to become illegal in other places, and then China is going to get the upper hand on us just like they did with genetic engineering."

This becomes more problematic in a world where more work is in danger of becoming automated, such as telemarketing, package delivery, and airline piloting. Although that's the natural evolution of the workplace through the years—though no one is really lamenting the loss of telephone company switchboard operators or lack of work for people that deliver big hunks of ice for your icebox—the sheer amount of jobs that can be automated with smart algorithms will leave a lot of people without jobs, and without skills to get jobs that are still beyond the range of automation.

"As soon as the robots come to take jobs the entire society is going to have to be restructured in a different way," stressed Istvan. "It's been people's jobs and careers that have kept the fabric of society together. Pretty soon we are just going to be human beings that don't know what to do with ourselves." This is one of the reasons that Istvan's Transhumanist party supports the concept of a universal basic income.

"I think for sure you are going to see a huge outcry and this will probably be one that will cause the most civil unrest is this idea that people are losing jobs," says Istvan. "But I think very quickly candidates are going to have to address these issues or increase welfare. I mean universal welfare is a basic system, but everyone needs something if they don't want to work because the rich keep getting richer and the poor keep getting poorer. But maybe we can make it so that no revolution can occur and give everyone—what I call—a luxury communism."

Did he just say the "C" word?

Yes, he did, but don't summon your grandfather's McCarthyism here.

Luxury communism is going to be a great big social pressure valve. It will give those that don't want to work the ability not to. Those that want to retire, an early option. And those that do, new challenges to provide a contribution as a volunteer or possibly in infrastructure and innovation design. The dreamers can dream. The artists can create. The writers can ruminate. And everyone else? Istvan says education will be free. So you'll be able to go back to school. Or raise your children. Or … well I am sure you will figure something out.

Now some people will say where will the money come from? Remember, cost of production will diminish drastically without hourly labor costs and output will massively increase. In a robotic run AI economy there will be revenue streams from the robot productivity. There will be massive gains in production and the old school notion of human labor-driven capitalism will become obsolete.

That said, the bottom line here is that the United States, inclusive of its trading partners, will need to step in with progressive solutions.

Universal basic income will need to be built for the people by the people. And there will need to be a plan to keep technology jobs in the country. Otherwise the power base for these new technologies might well shift to other countries, leaving America and the West in the dust instead of joining the rest of the world in a bright, new transhumanist future.

Kay says: "The greatest factor here is going to be what the majority of people think. When a technology has undeniable advantages, it will be fought for by the masses. This will put pressure on policymakers to accept it. Governments that take too long to approve new technologies will face losing power to competing countries that are progressing at a faster rate. I am very concerned about the United States from this perspective. Countries like China are much more forward thinking when it comes to technology."

Sean says: "I think this is the biggest thing standing in the way of transhumanism. Some people and organizations have too much invested in the way things are, and restructuring the world goes against their interests. They're going to put up a hell of a fight."

Andy says: "The dystopians will be all over this, predicting a grim and bleak future with war and famine ravaging the nation. And they would be right if a technologically linear future was ahead of us. But you can't base the future on present technologies that are only incrementally improving. As always, technology will provide the solutions to all that will ail us. It will be the resistors that will suffer the most. They might fancy themselves as postmodern Kim Davises, but time marches on and so will technology, and they will not be the revolutionaries that they think they are but will become footnotes on the wrong side of history."

A Final Word

Two and a half years ago we made a bold and confident prediction. That this book would be written in five months and you would have it in your hands by the end of 2013. We were very wrong.

Writing books is hard. And writing books about the future is even harder. And books such as this one are often wrong and sometimes right. Since then, we (Andy and Kay) moved our company from Canada to the United States, and we connected with Sean to fill out key sections of this book and get the final chapters complete. Along the way we lost a dog and a cat, and gained a baby.

Still, we were also right about a lot of things in two short years. We assumed self-driven cars were a decade away, and yet self-driving trucks have been authorized to navigate Nevada highways. Personal drones were expensive toys and by December 2013, Amazon CEO Jeff Bezos had announced Amazon's intention to fly your purchases by Amazon Prime Air drone to your home in 30 to 60 minutes. As we complete this book in the spring of 2016, commercial domestic delivery drones are an almost certainty in the next year or two. And how long before they are taking our children's lunch to school?

We could go on with hundreds more developments that have occurred in under three years, but it would beleaguer the point. Technology in the next half century will blossom. Life for everyone will change … mostly for the better. You see what's happening today is exciting and as predictable and measureable as we have tried to make it in this book, on its unpredictability is certain as it drives along the exponential curve.

Ray Kurzweil told us: "Exponential growth is surprising and seductive and people use their linear intuition even if they're used to thinking exponentially."

> "The only thing to be sure about the future is that it will be fantastic."

His critics and those who dismiss this book will have missed the point. We are living in the most extraordinary of times. And life as we know it is getting better and better every day. Those who refuse that are living their present into a linear future. And that's their biggest miscalculation.

With that, we will leave you with a word from author and futurist Arthur C. Clark who said in a 1964 BBC documentary: "The only thing to be sure about the future is that it will be fantastic."

It was. And it will be, Arthur. It will be.

Index

Numbers

2FL (2-fucosyllactose), 30–31
3D printing
    for babies, 51–52
    limb replacement, 219–220
    in medicine, 149–152
3D ultrasound, 31
4D ultrasound, 31
23andMe, 63
32-bit processors, 192
64-bit processors, 192
The 120 Year Diet
    (Walford), 279

A

abdominoplasty, 109
Abilify, 140
ABOi organs, 50
abstinence, 36
acetaminophen, 184
Ackland, Nigel, 235–236
adenosine monophosphate-
    activated protein kinase
    (AMPK) pathway research
    studies, 289

adipose stem cells, 122–123
Adolphs, Ralph, 199
Agassi, Andre, 94
*The Age of Intelligent
    Machines* (Kurzweil), 298
*The Age of Spiritual Machines:
    When Computers Exceed
    Human Intelligence*
    (Kurzweil), 298
aging. *See* longevity
AI (artificial insemination),
    44–45
AI (artificial intelligence),
    203–205
    future of, 310
        *Stage 1 (current
            technology), 311*
        *Stage 2 (repetitive job
            replacement), 311–312*
        *Stage 3
            (information-based
            job replacement), 312*
        *Stage 4 (total job
            replacement), 312–314*
    relationship with humans,
        205–206
    robot consciousness,
        208–209

technology singularity,
        209–210
albinism, 58
Alcor, 293
Allen, Paul, 208, 210
alopecia, 93
ALS (amyotrophic lateral
    sclerosis), 279
Alster, Tina, 300
amblyopia, 188
amniocentesis, 55–56, 63
AMPK (adenosine
    monophosphate-activated
    protein kinase) pathway
    research studies, 289
amyotrophic lateral sclerosis
    (ALS), 279
Andersen, Charlotte
    Hilton, 105
anesthesia, history of, 127
anti-aging products. *See*
    cosmeceuticals
antibiotics, 128
antioxidants, 82
antivirals, 128
appearance. *See* beauty
Apple Watch, 223, 249–250
Arcadiou, Stelios, 238

"Are Humans Still Evolving?"
  (Stock), 13
Ariely, Dan, 315
aripiprazole, 140
Aristo, 210
Aristotle, 42, 166
Arnhat, Larry, 303
arterial perfusion, 33
artificial heart
  continuous-flow heart, 126,
    131–132
  as cyborg technology,
    220–221
  history of, 130–131
artificial insemination (AI),
  44–45
artificial intelligence (AI),
  203–205
  future of, 310
    Stage 1 (current
      technology), 311
    Stage 2 (repetitive job
      replacement), 311–312
    Stage 3
      (information-based
      job replacement), 312
    Stage 4 (total job
      replacement),
      312–314
  relationship with humans,
    205–206
  robot consciousness,
    208–209
  technology singularity,
    209–210
artificial limbs. See limb
  replacement
artificial reproduction
  technologies, 42–47
  artificial insemination (AI),
    44–45
  DNA sequencing, 45–47
  in vitro fertilization (IVF),
    42–44
Asprey, Dave, 186–187
assistants, robots as,
  158–159
Atala, Anthony, 67, 150–151
Attenborough, David, 11
Austin, Steve, 216

B

babies
  birth control technologies,
    33–41
    history of, 33–34
    male contraception,
      38–39
    remote control birth
      control, 37–38
    types of, 34–36
  conception
    artificial reproduction
      technologies, 42–47
    erectile dysfunction (ED)
      treatments, 39–41
    lab-grown vaginas,
      67–68
    male pregnancy, 66–67
    pregnancy tests, 36–37
  designer babies
    future of, 301–303
    gender selection, 54–58
    genetic engineering,
      58–65, 250–252
    Morphthing.com,
      52–54
  feeding choices, 27–31
  life-saving technologies,
    47–52
    3D printing, 51–52
    cord blood, 47–48
    genome sequencing, 48
    low-cost incubation,
      50–51
    nanotechnology, 49–50
  ultrasound technology,
    31–33
Bach-y-Rita, Paul, 170
baldness. See hair, loss and
  restoration
Barbie doll look-alike, 78
Bardwell, Vivian, 118
barrier methods (birth
  control), 34–35
Barzilai, Nir, 278
BBB (blood-brain barrier), 184
Beatie, Nancy, 66
Beatie, Thomas Trace, 66

beauty
  body image, 71–73
  body-shaping tools, 90
  cosmetic surgery
    abdominoplasty, 109
    blepharoplasty, 109
    breast augmentation,
      110–117
    breast reduction, 110
    cosmetic dentistry, 110
    dermabrasion, 109
    double eyelid surgery,
      119
    extreme changes,
      75–80
    foot surgery, 120
    future of, 299–301
    history of, 106–107
    iris surgery, 119–120
    limb lengthening, 120
    liposuction, 108
    palm alteration, 119
    plastic surgery versus,
      107–108
    popularity of, 120
    rhinoplasty, 108
    rhytidectomy, 109
    sex reassignment,
      117–118
    stem cells in, 121–123,
      299–300
    trends in, 60, 73
    vaginoplasty, 110
  cosmetics, 81
    Botox, 83–84
    cosmeceuticals, 81–83
    cosmetic stickers, 86
    electro cosmetics, 85–86
    future of, 300
    for men, 84–85
  hair, 91
    loss and restoration,
      93–97
    removing, 91–93
  tattoos, 86–89
    common symbols, 87
    gadget-activated tattoos,
      89
    invisible tattoos, 88
    LED tattoos, 89

political viewpoints, 245–246
religious viewpoints, 246–249
removable ink, 88
white tattoos, 89
weight loss, 97
non-surgical techniques, 103–106
Soylent, 98–101
surgery for, 101–103
Beauty Pays (Hamermesh), 72
Beckham, David, 191
Belic, Roko, 274
Benedict XVI (pope), 258
Bezos, Jeff, 311, 318
Bina48, 205
bio-hacking. See cyborg; do-it-yourself bio-hacking
bionic
  cyborg versus, 213–215
  limb replacement, 219–220
The Bionic Woman (television series), 214
Biostamp, 86
birth control technologies, 33–41
  history of, 33–34
  male contraception, 38–39
  remote control birth control, 37–38
  types of, 34–36
blepharoplasty, 109
blindness. See vision replacement technology
blood
  clotting, 152–153
  synthetic, 151–152
blood sugar, 104
  glucose-sensing contact lens, 153–154
blood tests in prenatal testing, 62
blood typing in organ donation, 50
blood-brain barrier (BBB), 184
body art
  extreme changes, 75–80
  political viewpoints, 245–246

religious viewpoints
  Christianity, 247–248
  Hinduism, 248–249
  Islam, 248
  Judaism, 246–247
tattoos, 86–89
  common symbols, 87
  gadget-activated tattoos, 89
  invisible tattoos, 88
  LED tattoos, 89
  removable ink, 88
  white tattoos, 89
body image, 71–73
body parts, 3D printing, 149–152
body spacers, 90
body-shaping tools, 90
Bohr, Niels, 24
Borg, 216
Botox, 83–84
Botvinick, Matthew, 172
Boxtel, Amanda, 230–231
brain. See also neuroscience
  characteristics of
    evolution and, 8–10, 200–203
    neurogenesis, 184–186
    plasticity. See neuroplasticity
  cognitive enhancers, 171
    learning, 173–174
    meditation, 175–177
    nootropics, 185–187
    pharmaceuticals, 187, 187–188
    psychotherapy, 175
    rubber hand illusion (RHI), 172
    video games, 188–190
  gut and, 177–178
  mysteries of
    consciousness, 196–205, 206–208
    free will, 197–198
    list of, 199
    memory processing, 198–199
    sleep, 197

neurological disorders, 178–179
  treatment options, 179–184
robotic control, 193–194
  brain-to-brain communication, 194–196
  exoskeleton connections, 191–192
  OpenWorm, 192–193
BrainGate, 195–196, 232–233
The Brain's Way of Healing (Doidge), 171
brain-to-brain communication, 194–196
Brave New World (Huxley), 250, 252
breast augmentation, 110–117
  breastfeeding and, 117
  fat transfer, 115–116
  history of, 111
  Ideal implants, 116
  implants, 112–115
  largest-breasted woman, 80
breast reduction, 110
breastfeeding
  breast augmentation and, 117
  formula feeding versus, 27–31
breathing underwater, 218–219
Brewer, Judson, 177
Brown, Louise Joy, 42, 43
Broyde, Michael J., 254
Bruce, Lenny, 246
Bruno, Giordano, 243
Buettner, Dan, 271, 272, 272–275
Bulletproof Coffee, 186
Busbice, Timothy, 192–193
Bush, George H. W., 243, 244, 254–255, 256–257
Bütschli, Otto, 54
buttock augmentation, 74
Byrne, Patrick J., 299

C

Caenorhabditis elegans
(roundworm), 165–192
Caidin, Martin, 216
Calment, Jeanne, 267
calorie restriction, 278–281
Canavero, Sergio, 130
Cannon, Tim, 237
Caplan, Art, 65
Carr, Nicholas, 260, 261
carrier tests in prenatal
testing, 62
Case, Amber, 212–213, 239,
313, 314
Caspari, Rachel, 265
CAT (computerized axial
tomography) scans, 137–138
Catherine de' Medici, 90
CBT (cognitive behavior
therapy), 175
cell atrophy, 290
cellular garbage, 290
centenarians
predictions about, 267–268
studies of
Blue Zones, 271–272
Longevity Genes Project,
277–278
*New England
Centenarian Study*,
275–277
Power 9 lifestyle list,
272–275
cervical cap, 35
Chambers, David, 207–208
Chen, Danica, 285
Chen, Jane, 51
Cheng, Gordon, 191
Chilcott, Warner, 36
cholesterol (HDL), 278
chorionic villus sampling
(CVS), 48, 63
Chorost, Michael, 228
Christianity
body modification,
247–248
cloning, 253
genetic engineering,
250–252
stem cell research, 257–258

chromosomes
in gender selection, 54–55
mutations, 291
Churman, Sarah, 221
Cialis, 40
Cidofovir, 92
circumcision, 247, 248
Clark, Arthur C., 318
Clark, Barney, 130
Clinton, Bill, 254
Clinton, Hillary, 91, 257
clocks, as cyborg
technology, 218
cloning
hair, 96–97
political viewpoints,
252–253
religious viewpoints,
253–254
"Cloning People and Jewish
Law: A Preliminary
Analysis" (Broyde), 254
clothing
as cyborg technology,
217–218
in medicine, 154–155
clotting innovations, 152–153
Clynes, Manfred E., 213
cocaine, 127
cochlear implants, 221, 228
cognitive behavior therapy
(CBT), 175
cognitive enhancers, 171
learning, 173–174
meditation, 175–177
nootropics, 185–187
pharmaceuticals, 187,
187–188
psychotherapy, 175
rubber hand illusion
(RHI), 172
video games, 188–190
Cohen, Jonathan, 172
Cohn, Billy, 126, 131–132
Colen, Cynthia, 30
colors, hearing, 226–227
computerized axial
tomography (CAT) scans,
137–138
computers, 21
Comstock Act, 34

conception
artificial reproduction
technologies, 42–47
*artificial insemination
(AI)*, 44–45
DNA sequencing, 45–47
*in vitro fertilization
(IVF)*, 42–44
birth control technologies.
See birth control
technologies
designer babies. *See*
designer babies
erectile dysfunction (ED)
treatments, 39–41
lab-grown vaginas, 67–68
male pregnancy, 66–67
pregnancy tests, 36–37
ConceptNet, 210
condoms, 34
conductive makeup, 86
consciousness
as mystery, 196–197,
206–208
robot consciousness,
208–209
contact lens, glucose-sensing,
153–154
continuous-flow artificial
heart, 126, 131–132
contraception. *See* birth
control technologies
contraceptive implant, 35
contraceptive patch, 35
contraceptive sponge, 35
Cook, James, 86
cord blood, 47–48
Cordell, Joe, 280
cordocentesis, 63
Cornaro, Liugi, 278–279
corrective lenses, 218, 221
corsets, 90
cosmeceuticals, 81–83
cosmetic dentistry, 110
cosmetic stickers, 86
cosmetic surgery
abdominoplasty, 109
blepharoplasty, 109
breast augmentation,
110–117
breastfeeding and, 117

fat transfer, 115–116
history of, 111
Ideal implants, 116
implants, 112–115
breast reduction, 110
cosmetic dentistry, 110
dermabrasion, 109
double eyelid surgery, 119
extreme changes, 75–80
foot surgery, 120
future of, 299–301
history of, 106–107
iris surgery, 119–120
limb lengthening, 120
liposuction, 108
palm alteration, 119
plastic surgery versus,
    107–108
popularity of, 120
rhinoplasty, 108
rhytidectomy, 109
sex reassignment, 117–118
stem cells in, 121–123,
    299–300
trends in, 60, 73
vaginoplasty, 110
cosmetics, 81
    Botox, 83–84
    cosmeceuticals, 81–83
    cosmetic stickers, 86
    electro cosmetics, 85–86
    future of, 300
    for men, 84–85
Crick, Francis, 207, 250
CRISPR-Cas9 technology, 250
Cristerna, Maria Jose, 79
Cronin, Thomas, 111
Crowell, Mary, 279
Cruz, Ted, 257
cryonics, 292–293
Cryonics Institute, 293
Csikszentmihalyi, Mihaly, 273
CT (computerized axial
    tomography) scans, 137–138
The Cure in the Code (Huber),
    147, 149
CVS (chorionic villus
    sampling), 48, 63
cybernetics, origin of term,
    215–216

*Cybernetics: Or Control and
    Communication in the
    Animal and the Machine*
    (Wiener), 216
cyborg
    bio-hacking
        dangers of, 236–237
        future of, 306–307
        *night vision
            enhancement,* 236
    bionic versus, 213–215
    cochlear implants, 221
    electronic accessories,
        221–225
        fashion accessories,
            224–225
        Google Glass, 223–224
        GoPro camera, 224
        health trackers, 223,
            249–250
        *mobile phones and
            PDAs,* 221–223
        tattoos, 225
        virtual-reality eyewear,
            224
    eyesight correction, 221
    future of, 238–239
    history of technologies,
        217–219
        clothing, 217–218
        corrective lenses, 218
        time-keeping, 218
        *underwater breathing
            apparatus,* 218–219
    humans as, 211–213
        Ackland, Nigel, 235–236
        Arcadiou, Stelios, 238
        Boxtel, Amanda,
            230–231
        Chorost, Michael, 228
        Graafstra, Amal,
            231–232
        Harbisson, Neil,
            226–227
        Jalava, Jerry, 231
        Licina, Gabriel, 236
        Mann, Steve, 225–226
        Mistry, Pranav, 233
        Naumann, Jens, 227
        Pinto, Juliano, 191–192

Spence, Rob, 228–229
Sullivan, Jesse, 235
Wake, Byron, 234
Warwick, Kevin,
    195–196, 232–233
implants
    BrainGate, 232–233
    RFID chips, 234
limb replacement, 219–220
in literature and
    media, 216
organ replacement,
    220–221
Cyborg (Caidin), 216
Cylons, 216

D

D'Armate, Salvino, 42
Darth Vader, 216
Darwin, Charles, 6, 250
DBS (deep brain stimulation),
    182–183
de Grey, Aubrey, 268,
    289–292, 308–309
death, leading causes of, 264
deep brain stimulation (DBS),
    182–183
default mode network
    (DMN), 177
Democratic Party viewpoints,
    stem cell research, 257
Dendy, Walter Cooper, 175
dentistry, cosmetic, 110
depression
    deep brain stimulation
        (DBS), 183
    self-help website, 263
dermabrasion, 109
Descartes, René, 167–168
designer babies
    future of, 301–303
    gender selection, 54–58
        amniocentesis, 55–56
        chromosomes in, 54–55
        Ericsson Method, 56
        genetic chemistry and, 60
        *preimplantation genetic
            diagnosis (PGD),* 57–58

sperm spinning, 56–57
genetic engineering, 58–65
  controversy surrounding,
    58–60, 250
  genetic compatibility
    testing, 61–62
  mitochondrial DNA
    replacement, 64–65
  prenatal testing and,
    62–63
  trait selection, 60–61
Morphthing.com, 52–54
Deus Ex: Human Revolution
(video game), 229
diaphragm, 35
Diedrichs, Phillippa, 72
diet. *See* lifestyle, longevity
and
digital smart pills, 140
*Discourses on the Temperate
Life* (Cornaro), 278–279
DMN (default mode
network), 177
Dmrt1 gene, 118
DNA sequencing in artificial
reproduction, 45–47. *See also*
genome sequencing
Dobelle, William, 227
Doidge, Norman, 171
do-it-yourself bio-hacking
  dangers of, 236–237
  night vision enhancement,
    236
do-it-yourself healthcare,
160–161
double eyelid surgery, 119
drugs. *See* pharmaceuticals
dualism, 167
dystopia, 299

E

economics in future society,
314–317
ECT (electroconvulsive
therapy), 180–181
ED (erectile dysfunction)
treatments, 39–41
Edwards, Robert, 43
Edwin Smith Papyrus, 166
electro cosmetics, 85–86

electroconvulsive therapy
(ECT), 180–181
electrolysis, 91–92
electronics
  cyborg technologies,
    221–225
    *fashion accessories,
      224–225*
    Google Glass, 223–224
    GoPro camera, 224
    *health trackers, 223,
      249–250*
    *mobile phones and
      PDAs, 221–223*
    tattoos, 225
    *virtual-reality eyewear,
      224*
  in medicine, 141–142
    Google Glass, 143–144
    *high-intensity
      focused ultrasound
      (HIFU), 143*
    iPads, 142–143
    *organ-on-a-chip
      simulations, 144–146*
    sensors, 153–155
    video games, 142
Embrace Infant Warmer, 51
emerging technologies,
  political viewpoints on,
  260–262
employment
  beauty and, 72–73
  robot job replacement
    *Stage 2 (repetitive job
      replacement), 311–312*
    *Stage 3
      (information-based
      job replacement), 312*
    *Stage 4 (total job
      replacement), 312–314*
enhancing brain function. *See*
cognitive enhancers
Enovid-10, 34
EPA (Environmental
Protection Agency), 243
erectile dysfunction (ED)
treatments, 39–41
Ericsson, Ronald J., 56
Ericsson Method, 56
Ettinger, Robert, 293
evolution

brain characteristics and,
  8–10, 200–203
current state of, 11–13
mutations, 12–13
natural selection, 5–6
opposable thumbs, 10–11
technology and, 13–14
  speed of change, 14–20
  *technology singularity,
    20–24, 297–299*
tool usage, 11
exorcism, 181
exoskeletons, 191–192,
230–231
eye color, changing, 119–120
Eyeborg, 229
eyelid surgery, 109
eyesight correction
  as cyborg technology,
    218, 221
  night vision enhancement,
    236
  vision replacement
    technology, 170–171, 227,
    228–229

F

facelift, 109
*Fantastic Voyage: Live Long
Enough to Live Forever*
(Kurzweil and Grossman),
294
Farra, Robert, 38
fashion accessories, as cyborg
technology, 224–225
fat grafting, 74
fat transfer breast
augmentation, 115–116
fatal familial insomnia
(FFI), 197
feeding babies, 27–31
female circumcision, 248
female sterilization, 36
Feng, Guipong, 251–252
Ferriss, Tim, 186
fertility tourism, 43–44
fever, as neurological
treatment, 181
FFI (fatal familial insomnia),
197
finasteride, 95

Finkel, Toren, 286–287
Fitbit, 223
fitness trackers. *See* health trackers
Flatline (film), 132
Fleming, Alexander, 128
Fleuss, Henry, 219
Flourens, Pierre, 168
"flow", 273
food. *See* weight loss
foot binding, 90
foot surgery, 120
Forever Young (film), 292
formula feeding versus breastfeeding, 27–31
Fox, Kate, 71–72
Foxl2 gene, 118
Frankl, Viktor, 273
Fraunhofer Institute for Medical Image Computing, 142–143
Frazier, Bud, 126, 131–132
free will, 197–198
Freedom-2-Ink, 88
Freud, Sigmund, 175
Frey, Carl Benedikt, 311, 312
Friedman, David, 287
frontal lobe, 8
Fussenegger, Martin, 103–105
future
    of bio-hacking, 306–307
    of cosmetic surgery, 299–301
    of cosmetics, 300
    of designer babies, 301–303
    of genetic engineering, 303–306
    of immortality, 308–309
    of medicine, 303–306
    of neuroscience, 200
        artificial intelligence (AI), 203–205
        human intelligence expansion, 200–203
        relationship between humans and AI, 205–206
        robot consciousness, 208–209
        technology singularity, 209–210
    of robotics, 310

    Stage 1 (current technology), 311
    Stage 2 (repetitive job replacement), 311–312
    Stage 3 (information-based job replacement), 312
    Stage 4 (total job replacement), 312–314
    of society, 314–317
"The Future of Employment: How Susceptible are Jobs to Computerisation?" (Frey and Osborne), 311

G

gadget-activated tattoos, 89
Gage, Greg, 179
Gage, Phineas, 178
Galen of Pergamon, 166–167
Galileo Galilei, 242
Galvani, Luigi, 168
Gao, Wen-Jun, 187
gastric band surgery, 101–102
gastric bypass surgery, 102–103
Gates, Bill, 37, 298
Gates, Melinda, 37
Gautama Buddha, 175
Gazzaley, Adam, 164, 189–190
gender dysphoria (GD), 117–118
gender selection for babies, 54–58
    amniocentesis, 55–56
    chromosomes in, 54–55
    Ericsson Method, 56
    genetic chemistry and, 60
    preimplantation genetic diagnosis (PGD), 57–58
    sperm spinning, 56–57
gene therapy for erectile dysfunction, 41
GenePeeks, 61
genetic compatibility testing, 61–62
genetic engineering, 58–65
    controversy surrounding, 58–60, 250
    future of, 303–306

gender selection. *See* gender selection for babies
genetic compatibility testing, 61–62
    mitochondrial DNA replacement, 64–65
    prenatal testing and, 62–63
    sex change genes, 118
    trait selection, 60–61
genetics, 21
    longevity and, 268–270
        AMPK pathway, 289
        insulin signaling pathway, 287–288
        Longevity Genes Project, 277–278
        mTOR protein, 286–287
        New England Centenarian Study, 276–277
        SENS Foundation research, 289–292
        sirtuin studies, 283–286
"Genetics and Faith: Power, Choice and Responsibility" (Evangelical Lutheran Church in America), 253, 258
genome sequencing
    23andMe, 63
    for babies, 49–50
    ease of, 146–148
    pharmaceuticals and, 148–149
geocentrism, 242–243
Gerow, Frank, 111
Gibson, Mel, 292
Gilbert, Dan, 308
Gillies, Harold, 108
Giurgea, Corneliu E., 186
The Glass Cage (Carr), 260–261
glasses, 218
glucose-sensing contact lens, 153–154
Gomez, Andres, 94
Gomez, Max, 258
Google Glass, 143–144, 223–224
Goostman, Eugene, 204
GoPro camera, 224
Graafstra, Amal, 231–232
Grant, Dionne, 67

Graziano, Michael, 208–209
Green, Glenn, 51, 52
Greenberg, Stephen, 117
grinders. *See* do-it-yourself
   bio-hacking
Grossman, Terry, 294
Grossmann, Rafael, 144
Guarente, Leonard P., 283
Guevara-Aguirre, Jaime, 288
Guha, Sujoy K., 38
The Guide to the Future of
   Medicine (Meskó), 139–140
gut, brain and, 177–178
gut sleeves, 103

H

hair, 91
   loss and restoration, 93–97
      cloning, 96–97
      history of, 93–94
      Propecia, 95
      robotic hair
        restoration, 96
      Rogaine, 94–95
      stem cells, 97
      transplant surgery,
        95–96
   removing, 91–93
Halperin, Levi Yitzschak, 259
Hamermesh, Daniel, 72
Hamilton, Geraldine, 145,
   145–146
hand gestures, in cyborg
   technology, 233
Hansen, James E., 243
"The Happy Movie"
   (Belic), 274
Harbisson, Neil, 226–227
Harper, Stephen, 244
Hart, Mickey, 190
Hawking, Stephen, 298
HDL (good cholesterol), 278
The Healing Cell: How the
   Greatest Revolution in
   Medical History Is Changing
   Your Life (Smith, et al), 258
health trackers
   as cyborg technology, 223
   tattoos and, 249–250

healthcare. *See* medicine
healthy eating. *See* weight loss
hearing
   with cochlear implants,
      221, 228
   colors, 226–227
heart disease
   artificial heart
      continuous-flow heart,
        126, 131–132
      as cyborg technology,
        220–221
      history of, 130–131
   pacemakers, 128–129
HeartMate II ventricular assist
   device, 131
heliocentrism, 242–243
Hensch, Takao, 187–188
Hensley, Samuel, 302
Herculano-Houzel, Suzana, 10
heroin, 128
Heskett, Jame, 105
Heston, Charlton, 98
HIF (hypoxia inducible
   transcription factor), 285
high-intensity focused
   ultrasound (HIFU), 33, 143
Hinduism
   body modification,
      248–249
   cloning, 253
   stem cell research, 259
Hippocrates, 127, 166
Hippocratic Oath, 127
history
   of artificial heart, 130–131
   of birth control
      technologies, 33–34
   of breast augmentation,
      111
   of cosmetic surgery,
      106–107
   of cyborg technologies,
      217–219
      clothing, 217–218
      corrective lenses, 218
      time-keeping, 218
      underwater breathing
        apparatus, 218–219
   of hair loss and restoration,
      93–94

   of imaging technologies,
      137–138
   of life expectancy rates,
      264–266
   of medicine, 126–128
   of microscopes, 42
   of neurological treatment
      options, 181–182
   of neuroscience, 165–169
   of organ transplants,
      129–130
   of pregnancy tests, 36–37
   of smartphones, 221–223
HMOs (human milk
   oligosaccharides), 30
Hollister, Scott, 51
HopeLab, 142
Hoppenstedt, Max, 237
hormone treatment methods
   (birth control), 35
Houghton, Peter, 220–221
Huber, Peter W., 149
HULC (Human Universal
   Load Carrier), 230
Hull, Chuck, 51
human intelligence expansion,
   200–203
human milk oligosaccharides
   (HMOs), 30
Human Universal Load
   Carrier (HULC), 230
humans, relationship with AI,
   205–206
humorism, 166
Hunter, John, 44
Huxley, Aldous, 250, 252
hypoxia inducible
   transcription factor
   (HIF), 285

I

Icaria, Greece, 271
ice water, as neurological
   treatment, 181
Ideal implants, 116
IGF-1 hormone, 288
IIT (integrated information
   theory), 208

imaging technologies in
medicine, 137–138

immaterialists, 198

immortality, 294–295
future of, 308–309
planning for, 309

implants
bio-hacking
dangers of, 236–237
future of, 306–307
BrainGate, 195–196,
232–233
for breast augmentation,
112–115
breastfeeding and, 117
RFID chips, 231–232, 234

income inequality, 315–316

incubation technologies,
50–51

induced pluripotent stem cells
(iPSCs), 97

infant formula versus
breastfeeding, 27–31

infant mortality rates, 6–7

infection control, history
of, 128

inframammary (IMF)
insertion point, 113

insertion points for breast
implants, 113–115

insulin, 104

insulin coma as neurological
treatment, 181

insulin signaling pathway
research studies, 287–288

integrated information theory
(IIT), 208

intelligence
artificial intelligence (AI),
203–205
future of. See artificial
intelligence (AI),
future of
relationship with
humans, 205–206
robot consciousness,
208–209
technology singularity,
209–210
human intelligence
expansion, 200–203

intracervical insemination, 44

intrauterine device (IUD), 35

intrauterine insemination, 44–45

intrauterine system (IUS), 35

intrauterine tuboperitoneal
insemination, 45

invisible tattoos, 88

in vitro fertilization (IVF),
42–44, 45–47

iPads in medicine, 142–143

iPSCs (induced pluripotent
stem cells), 97

iris surgery, 119–120

"Is Google Making Us
Stupid?" (Carr), 260

Islam
body modification, 248
cloning, 254
stem cell research, 258–259

Istvan, Zoltan, 206, 210, 239,
251, 252, 255, 260, 261–262,
298, 302, 303, 304, 311,
312–313, 314, 315, 316

IUD (intrauterine device), 35

IUS (intrauterine system), 35

IVF (in vitro fertilization),
42–44, 45–47

J

Jacobstein, Neil, 206

Jahoda, Colin, 96

Jalava, Jerry, 231

Jarvik 7 artificial heart, 130

Jarvik 2000 ventricular assist
device, 220–221

Jehovah's Witness, 249

Jeopardy! (television series),
204–205

Jesus Christ, tattoos of, 247

The Jetsons (television series),
203

John Paul II (pope), 242, 254,
257–258

Johnson, Tom, 287

Jorgensen, Christine, 118

Judaism
body modification, 246–247
cloning, 254
stem cell research, 259

Judlica, Justin, 79

K

Kahn, Irving, 278

Ken doll look-alike, 79

Kendall, Mark, 139

Khan, Razib, 48–49

Kimura, Jiroemon, 267

Kline, Nathan S., 213

Knauss, Sarah, 267

knowledge transference,
longevity and, 265

Koch, Christof, 208

Kocher, Theodor, 129

Kotler, Steven, 219

Krentcil, Patricia, 79–80

Kughen, Rick, 214

Kurzweil, Ray, 22–24, 50, 140,
164, 166, 201–202, 205,
205–206, 294–295, 297,
297–298, 307, 310, 318

L

lab-grown vaginas, 67–68

labiaplasty, 74

LaGondino Tracy Lehuanani,
66

Landolina, Joe, 153

Lanphier, Edward, 251

largest-breasted woman, 80

Laron's syndrome, 288

Larsson, Arne, 128, 132

laser hair removal, 92

Last Ape Standing: The
Seven-Million-Year Story of
How and Why We Survived
(Walter), 14

Lawler, Peter, 302

Lazar, Sara, 177

learning, training brain via,
173–174

LED tattoos, 89

Lee, Sang-Hee, 265

Leeuwenhoek, Antonie van, 42

Levitra, 40

levonorgestrel, 37

Lewis, Craig, 131

libido drugs, 41

Licina, Gabriel, 236

Liebig, Justus von, 28

life expectancy, 138
centenarians, predictions
  about, 267–268
current, 264
history of rates, 264–266
life-saving technologies for
  babies, 47–52
  3D printing, 51–52
  cord blood, 47–48
  genome sequencing, 48–49
  low-cost incubation, 50–51
  nanotechnology, 49–50
lifestyle, longevity and,
  270–271
  calorie restriction, 278–281
  New England
    Centenarian Study, 276
  Power 9 lifestyle list,
    272–275
  red wine intake, 281–283
  spouse age differences, 277
limb lengthening surgery, 120
limb regeneration in starfish,
  122
limb replacement, 191–192,
  219–220, 235–236
Lindsey, Timmie Jean, 111
liposonix, 106
liposuction, 74–75, 108
Litt, Brian, 89
Lizardman, 75–78
lobes (of brain), 8
lobotomy, 179, 181
logarithmic improvements to
  technology, 14–20, 42
logotherapy, 273
Loma Linda, California, 271
longevity
  causes of aging, 290–291
  centenarians
    Blue Zones study,
      271–272
    Longevity Genes Project,
      277–278
    New England
      Centenarian Study,
      275–277
    Power 9 lifestyle list,
      272–275
    predictions about,
      267–268
  cryonics, 292–293
  genetics and, 268–270

AMPK pathway, 289
insulin signaling
  pathway, 287–288
Longevity Genes Project,
  277–278
mTOR protein, 286–287
New England
  Centenarian Study,
  276–277
SENS Foundation
  research, 289–292
sirtuin studies, 283–286
history of life expectancy
  rates, 264–266
immortality, 294–295
  future of, 308–309
  planning for, 309
lifestyle and, 270–271
  calorie restriction,
    278–281
  New England
    Centenarian Study, 276
  Power 9 lifestyle list,
    272–275
  red wine intake, 281–283
  spouse age differences,
    277
steps for, 263–264
supercentenarians,
  examples of, 266–267
Longevity Genes Project,
  277–278
Longo, Valter, 274, 288
losing
  hair. See hair, loss and
    restoration
  weight. See weight loss
Louis XIII, King of France, 93
low-cost incubation
  technologies, 50–51
Lukyanova, Valeria, 78
luxury communism, 316–317
Lysenko, Trofim, 243
Lysenkoism, 243

M

MacWilliam, John Alexander,
  128
magnetic resonance imaging
  (MRI) scans, 138, 177
magneto-acoustic imaging, 33
Maguire, Eleanor, 171

makeup. See cosmetics
male circumcision, 247, 248
male contraception, 38–39
male cosmetics, 84–85
male pregnancy, 66–67
male sterilization, 36
Malepregnancy.com, 66
mammaplasty. See breast
  augmentation
"The Man That Was Used
  Up" (Poe), 216
Mann, Steve, 225–226
materialists, 198
Mayberg, Helen, 183
Mayer-Rokitansky-Kuster-
  Hauser syndrome
  (MRKH), 67
MBSR (mindfulness-based
  stress reduction), 177
McCarthy, Leon, 219–220
McCay, Clive, 279
medicine
  3D printing in, 149–152
  blood-clotting innovations,
    152–153
  cochlear implants, 221
  do-it-yourself healthcare,
    160–161
  electronics in, 141–142
    Google Glass, 143–144
    high-intensity focused
      ultrasound (HIFU), 143
    iPads, 142–143
    organ-on-a-chip
      simulations, 144–146
    sensors, 153–155
    video games, 142
  future of, 303–306
  genome sequencing
    23andMe, 63
    for babies, 49–50
    ease of, 146–148
    pharmaceuticals and,
      148–149
  heart disease
    artificial heart,
      220–221
    continuous-flow
      artificial heart, 126,
      131–132
    history of artificial
      heart, 130–131
    pacemakers, 128–129

history of, 126–128
imaging technologies,
137–138
nanotechnology in,
138–141
neuroscience. *See*
neuroscience
organ transplants
history of, 129–130
nanotechnology and,
49–50
via 3D printing,
150–151
pharmaceuticals. *See also*
cosmeceuticals
as cognitive enhancers,
187
digital smart pills,
140–144
genome sequencing and,
148–149
for neurological
disorders, 183–184
for neuroplasticity
improvement,
187–188
organ-on-a-chip
simulations, 145–146
plastics in, 136–137
robotics in
as assistants,
158–159
surgery, 156–158
surgery. *See* surgery
meditation, 175–177
melanocortin receptor
agonists, 41
Mellinger, Frederick, 90
Melton, Doug, 256
memory processing,
198–199
Mendel, Gregor, 250
mental illness. *See*
neurological disorders
Meskó, Bertalan, 48, 139–140,
144, 146, 147, 148, 158, 304,
305, 311, 316
metabolism-regulating weight
loss implant, 103–105
metformin, 289
Metrazol, 182
Michelangelo, 211–212
Micra pacemaker, 129

microCHIPS, 37–38
microchips. *See* RFID
(radio-frequency
identification) implants
microdermabrasion, 109
microscopes, history of, 42
MicroSort, 57
Miller, Michael, 30
mind file, 205
mindfulness-based stress
reduction (MBSR), 177
MinION, 147
minoxidil, 94–95
Mishra, Pankaj, 259
Mistry, Pranav, 233
mitochondrial breakdown,
284–285, 290
mitochondrial DNA
replacement, 64–65
mobile phones, history of,
221–223
MobiUS SP1 Ultrasound
Imaging System, 32
Moore, Gordon, 14
moral discipline, as
neurological treatment,
181
More, Thomas, 244–245
morphine, 128
Morphthing.com, 52–54
Morriss, Anne, 61–62
Morse, Sam, 195
Mosso, Angelo, 168
MRI (magnetic resonance
imaging) scans, 138, 177
MRKH (Mayer-Rokitansky-
Kuster-Hauser syndrome),
67
mTOR protein research
studies, 286–287
Muller, Johannes Peter, 168
Murphy, Robert, 121–123,
299–300, 305
Musk, Elon, 298
mutations, 12–13, 291
myelin, 174

N

NAD (nicotinamide adenine
dinucleotide), 284–285
nanobots, 201, 202–203
Nanopatch, 139

nanotechnology, 21
for babies, 49–50
in medicine, 138–141
Nash, Adam, 57
Nash, Molly, 57
National Institute of Aging
(NIA), 281
natural selection, 5–6, 12–13
Naumann, Jens, 221, 227
neocortex, 200, 8
Nestlé boycott, 29
neurogastroenterology, 177
neurogenesis, 184–186
neurological disorders,
178–179
treatment options, 179
deep brain stimulation
(DBS), 182–183
electroconvulsive
therapy, 180–181
history of, 181–182
pharmaceuticals,
183–184
neurons, 10
neuroplasticity, 169–172
improving
rubber hand illusion
(RHI), 172
via learning, 173–174
via meditation, 175–177
via pharmaceuticals,
187–188
via psychotherapy, 175
neuropreservation, 293
Neuroracer (video game), 190
neuroscience. *See also* brain
current state of, 163–165
future of, 200
artificial intelligence
(AI), 203–205
human intelligence
expansion, 200–203
relationship between
humans and AI,
205–206
robot consciousness,
208–209
technology singularity,
209–210
history of, 165–169
mysteries of
consciousness, 196–197,
206–208

free will, 197–198
list of, 199
memory processing,
198–199
sleep, 197
neurological disorders,
178–179
treatment options,
179–184
neurotransmitters, 183–184
New England Centenarian
Study, 275–277
NIA (National Institute of
Aging), 281
Nicolelis, Miguel, 191–192,
194–195, 202
nicotinamide adenine
dinucleotide (NAD), 284–285
Nicoya, Costa Rica, 272
night vision enhancement, 236
Nixon, Richard, 243
nootropics, 185–187
Norton, Michael, 315
nose job (rhinoplasty), 108
The Nyctalope, 216

O

Obama, Barack, 253, 255
occipital lobe, 8
Okinawa, Japan, 272
Oleynikov, Dmitry, 156
Omidyar, Pam, 142
On Injuries of the Head
(Hippocrates), 166
Open (Agassi), 94
OpenWorm, 192–193
opposable thumbs, 10–11
oral contraception, 35
Orentreich, Norman, 94
organ transplants
history of, 129–130
nanotechnology and,
49–50
via 3D printing, 150–151
organ-on-a-chip simulations,
144–146
The Origin of Species by
Means of Natural Selection
(Darwin), 6, 250
ORLAN, 80
Orphan Black (television
series), 253

Osborne, Michael, 311, 312
Össur, 219
Owen, Ivan, 220
OY ratio (old to young ratio),
265
Ozcan, Aydogan, 155

P

pacemakers, 128–129
pain control, history of,
127–128
palm alteration surgery, 119
Panicker, Rahul, 51
parietal lobe, 8
Parkinson's disease, 182–183
Paul, Rand, 245
PDAs (personal digital
assistants), history of, 221–223
penile prosthesis, 40
Penrose, Roger, 207
Pepper, John, 171
peptides, 81, 82
performance art
body as, 80
with cyborg technology,
238
periareolar insertion point, 114
personal digital assistants
(PDAs), history of, 221–223
Peterson, Garrett, 51
pets, cryopreservation of, 293
PGD (preimplantation genetic
diagnosis), 57–58
phantom limb syndrome, 170
pharmaceuticals. *See also*
cosmeceuticals
as cognitive enhancers, 187
digital smart pills, 140–144
genome sequencing and,
148–149
for neurological disorders,
183–184
for neuroplasticity
improvement, 187–188
organ-on-a-chip
simulations, 145–146
phrenology, 168
physical restraints, as
neurological treatment, 181
"the pill", 35
pills. *See* pharmaceuticals
Pinto, Juliano, 191–192

plastic surgery
cosmetic surgery versus,
107–108
future of, 299–301
plasticity of brain. *See*
neuroplasticity
plastics in medicine, 136–137
Poe, Edgar Allen, 216
political viewpoints
body modification,
245–246
cloning, 252–253
emerging technologies,
260–262
future of society, 314–317
genetic engineering, 252
stem cell research,
254–257
Power 9 lifestyle list, 272–275
prayer, 181
pregnancy. *See* conception
pregnancy tests, 36–37
preimplantation genetic
diagnosis (PGD), 57–58
prenatal testing
amniocentesis, 63
chorionic villus sampling
(CVS), 48, 63
cordocentesis, 63
gender selection, 54–58
amniocentesis, 55–56
chromosomes in, 54–55
Ericsson Method, 56
preimplantation genetic
diagnosis (PGD),
57–58
sperm spinning, 56–57
genetic engineering and,
62–63
genome sequencing,
48–49
ultrasound technology,
31–33
Propecia, 95
prosthetics, 219–220
protein cross links, 290
Proteus Digital Health, 140
psychotherapy, 175
push-up bras, 90

Q

QC Bot, 158–159

R

radio-frequency identification (RFID) implants, 231–232, 234
Ramm, Olaf von, 31
rapamycin, 286
rapid eye movement (REM), 197
Ravnan, Britt, 48
Razi, 243
Rebuilt: How Becoming Part Computer Made Me More Human (Chorost), 228
red wine intake, 281–283
religious viewpoints
  body modification
    Christianity, 247–248
    Hinduism, 248–249
    Islam, 248
    Judaism, 246–247
  cloning, 253–254
  future of society, 314–317
  genetic engineering, 250–252
  organ transplants, 50
  stem cell research, 257
    Christianity, 257–258
    Hinduism, 259
    Islam, 258–259
    Judaism, 259
REM (rapid eye movement), 197
Re-Mission (video game), 142
remote control birth control, 37–38
removing
  hair, 91–93
  tattoos, 87, 88
reproduction. See conception
Republican Party viewpoints, stem cell research, 256–257
resistance to scientific progress, 241–244
  body modification
    political viewpoints, 245–246
    religious viewpoints, 246–249
  cloning, 252–254
  emerging technologies, 260–262
  genetic engineering, 250–252

stem cell research
  political viewpoints, 254–257
  religious viewpoints, 257–259
restoring hair, 93–97
  cloning, 96–97
  history of, 93–94
  Propecia, 95
  robotic hair restoration, 96
  Rogaine, 94–95
  stem cells, 97
  transplant surgery, 95–96
restructuring of society, 314–317
resveratrol, 281–283
retinoids, 83
retinol, 81
Reversible Inhibition of Sperm Under Guidance (RISUG), 38–39
RFID (radio-frequency identification) implants, 231–232, 234
RHI (rubber hand illusion), 172
Rhinehart, Rob, 98–101
rhinoplasty, 108
Rhythmicity (video game), 190
rhytidectomy, 109
RISUG (Reversible Inhibition of Sperm Under Guidance), 38–39
robotic hair restoration, 96
robotics, 21
  artificial intelligence (AI), 203–205
    relationship with humans, 205–206
    robot consciousness, 208–209
    technology singularity, 209–210
  artificial limbs, 235–236
  brain control, 193–194
    brain-to-brain communication, 194–196
    exoskeleton connections, 191–192
    OpenWorm, 192–193
  future of, 310
    Stage 1 (current technology), 311

    Stage 2 (repetitive job replacement), 311–312
    Stage 3 (information-based job replacement), 312
    Stage 4 (total job replacement), 312–314
  in medicine
    as assistants, 158–159
    surgery, 156–158
  nanobots, 201, 202–203
  in nanotechnology, 139–141
  political viewpoints, 260–262
Rogaine, 94–95
Rogers, John, 86
Röntgen, William, 137
Rosenblueth, Arturo, 215
Rosenthal, Kenneth, 119
Rotch, Thomas Morgan, 28
Rothblatt, Bina, 205
Rothblatt, Martine, 205
roundworm (*Caenorhabditis elegans*), 165, 192–193
Rozelle, David, 219
rubber hand illusion (RHI), 172

S

saline breast implants, 113
Sanders, Bernie, 257
Sanger, Margaret, 34
Sardinia, Italy, 271
Schroeder, William, 130
scientific progress, resistance to, 241–244
  body modification, 245–249
  cloning, 252–254
  emerging technologies, 260–262
  genetic engineering, 250–252
  stem cell research, 254–259
scuba (self-contained breathing apparatus), 218–219
Search for Paradise: A Patient's Account of the Artificial Vision Experiment (Naumann), 227
SENS Foundation longevity research, 289–292
sensors in medicine, 153–155
serotonin, 178

Seventh-Day Adventists, 271
sex reassignment surgery,
    117–118
Sherman, Jerome K., 44
Shields, Brooke, 94
Shuler, Michael, 145
silicone breast implants, 113
Silver, Lee, 61–62
The Simpsons (television
    series), 222
Sinclair, David, 282, 283,
    284–285
singularity, 20–24, 209–210,
    297–299
The Singularity is Near
    (Kurzweil), 295
Sir2 genes, 283–284
SIRT1 genes, 284–285
SIRT3 genes, 285
SIRT6 genes, 285–286
sirtuins
    red wine intake and, 282
    research studies, 283–286
Sitti, Metin, 141
The Six Million Dollar Man
    (television series), 214, 216
skin, synthetic, 151
skin whitening, 249
sleep, 197
sleeve gastrectomy, 103
smart fabrics, 154–155
smartphones
    history of, 221–223
    sensors in, 155
    ultrasound technology, 32
Smith, Robin, 258
Smith, Stephen, 31
Snyder, Michael, 147
society, restructuring of,
    314–317
sonography. *See* ultrasound
    technology
Soylent, 98–101
Soylent Green (film), 98
spatial resolution in MRI
    scans, 177
Spence, Rob, 228–229
sperm
    in artificial insemination,
        44–45
    genetic compatibility
        testing, 61–62
    separating via spinning,
        56–57

Spletter, Kimberly, 143
spouse age differences, effect
    on longevity, 277
Sprague, Erik, 75–78
Stapleton, John, 84
*Star Trek: The Next
    Generation* (television
    series), 216
starfish, limb regeneration, 122
Staxyn, 40
Steinberg, Jeffrey, 58–61, 252,
    301, 302
Stelarc, 238
stem cell research
    political viewpoints,
        254–257
    religious viewpoints, 257
        Christianity, 257–258
        Hinduism, 259
        Islam, 258–259
        Judaism, 259
stem cells
    in cord blood, 47–48
    in cosmetic surgery,
        121–123, 299–300
    for hair restoration, 97
Stendra, 40
Steptoe, Patrick, 43
sterilization (birth control), 36
stimulants, 187
Stock, Jay T., 13
stress, avoiding, 273
Stumbling on Happiness
    (Gilbert), 308
suicide gene therapy, 290
Sullivan, Jesse, 235
sunscreens, 83
supercentenarians, examples
    of, 266–267
surgery
    birth control, 36
    cosmetic surgery. *See*
        cosmetic surgery
    with Google Glass, 143–144
    hair transplant, 95–96
    for neurological disorders,
        179
    robotics in, 156–158
    sex reassignment, 117–118
    for weight loss
        gastric band, 101–102
        gastric bypass, 102–103
        gut sleeves, 103
        sleeve gastrectomy, 103

T

Tagliacozzi, Gasparo, 107
Takaoka, Michio, 282
talk therapy, 175
Tanning Mom, 79–80
tattoos, 86–89
    common symbols, 87
    as cyborg technology, 225
    gadget-activated tattoos, 89
    health trackers and, 249–250
    invisible tattoos, 88
    LED tattoos, 89
    political viewpoints,
        245–246
    religious viewpoints
        Christianity, 247–248
        Hinduism, 248–249
        Islam, 248
        Judaism, 246–247
    removable ink, 88
    white tattoos, 89
tDBS (transcranial deep brain
    stimulation), 181
technology
    evolution and, 13–14
        speed of change, 14–20
        *technology singularity,
            20–24, 297–299*
    infant mortality rates and,
        6–7
    resistance to, 241–244
        *body modification,
            245–249*
        cloning, 252–254
        *emerging technologies,
            260–262*
        *genetic engineering,
            250–252*
        *stem cell research,
            254–259*
    tool usage, 11
technology singularity, 20–24,
    209–210, 297–299
temporal lobe, 8
Terasem Hypothesis, 205
Theobold, Daniel, 159
third ear, 238
thought communication,
    195–196, 232–233, 307
three-parent babies, 64–65
Thuret, Sandrine, 184
Tijo, Joe Hin, 54

time-keeping technologies, 218
Timms, Daniel, 132
TMS (transcranial magnetic
    stimulation), 181
Tomorrowland (Kotler), 219
Tononi, Giulio, 208
tool usage, 11
topical hair removal gel, 92–93
To-Robo, 210
Trafny, Tomasz, 258
trait selection in babies, 60–61
transabdominal insertion
    point, 114
transaxillary insertion
    point, 114
*Transcend: Nine Steps to Living
    Well Forever* (Kurzweil and
    Grossman), 294
transcranial deep brain
    stimulation (tDBS), 181
transcranial magnetic
    stimulation (TMS), 181
transgendered people,
    117–118
transistors, 15
transplant surgery
    for hair, 95–96
    organ transplants.
        See organ transplants
Transpolitica (Wood), 262
transumbilical insertion
    point, 114
trepanation, 126–127, 181
Trump, Donald, 257
tubal ligation, 36
tummy tuck, 109
Turing, Alan, 204
Turing test, 203–204
Tylenol, 184

U

ultrasound technology, 31–33
    high-intensity focused
        ultrasound (HIFU),
        33, 143
universal income, 313, 316, 317
unwashed sperm, 45
Uprima, 41
Urban, Kimberly R., 187
USB drives, 231
utopia, 244–245, 299
Utopia (More), 244–245

V

vaginal ring, 35
vaginas, lab-grown, 67–68
vaginoplasty, 74, 110
vagus nerve stimulation
    (VNS), 181
Valporate, 188
Vampire Mom, 79
Van Roekel, Eeske, 198
VasalGel, 38–39
vasectomy, 36
Vega, Katia, 85–86
ventricular assist devices, 130,
    131, 220–221
Veti-gel, 153
Viagra, 40
video games
    as cognitive enhancers,
        188–190
    in medicine, 142
Vinge, Vernor, 23
Virtual Incision, 156–157
virtual-reality eyewear, 224,
    225–226
viruses as nanobots, 203
vision replacement
    technology, 170–171, 227,
    228–229
Vitruvian Man, 211–212
VNS (vagus nerve
    stimulation), 181
Von As, Richard, 220

W

Wagner, John, 57
Wake, Byron, 234
Walford, Roy, 279
Walsh, John, 168
Walter, Chip, 14
Wampler, Richard, 131
Warwick, Kevin, 163,
    193–194, 195–196, 203–204,
    232–233, 237, 297, 306–307,
    312
washed sperm, 45
watches, as cyborg technology,
    218
Watson (IBM supercomputer),
    204–205, 311
Watson, James, 250

wearable computing. *See*
    cyborg
Webber, David, 171
weight loss, 97
    non-surgical techniques,
        103–106
        liposonix, 106
        *metabolism-regulating
            implant, 103–105*
        Zerona laser, 105–106
    Soylent, 98–101
    surgery for
        gastric band, 101–102
        gastric bypass, 102–103
        gut sleeves, 103
        sleeve gastrectomy, 103
Weindruch, Richard, 279
Wells, Dagan, 45–47
White, Robert, 129
white tattoos, 89
Wiener, Norbert, 215–216,
    238
Wildd, Lacey, 80
Williams, Ted, 293
Willis, Thomas, 167
Wisconsin National Primate
    Research Centre (WNPRC),
    280–281
Wood, David, 262
*World Wide Mind: The
    Coming Integration of
    Humanity, Machines and the
    Internet* (Chorost), 228
worms. *See* roundworm
    (*Caenorhabditis elegans*)

X

X-ray imaging, 137
XStat, 152
Xu, George, 97

Z

Zarkower, David, 118
Zerona laser treatments,
    105–106
Zombie Smokeout (video
    game), 142
Zong, Oliver, 120

REGISTER THIS PRODUCT
SAVE 35%*
ON YOUR NEXT PURCHASE!

How to Register Your Product

- Go to quepublishing.com/register
- Sign in or create an account
- Enter the 10- or 13-digit ISBN that appears on the back cover of your book or on the copyright page of your eBook

Benefits of Registering

- Ability to download product updates
- Access to bonus chapters and workshop files
- A 35% coupon to be used on your next purchase – valid for 30 days
  - To obtain your coupon, click on "Manage Codes" in the right column of your Account page
- Receive special offers on new editions and related Que products

Please note that the benefits for registering may vary by product. Benefits will be listed on your Account page under Registered Products.

We value and respect your privacy. Your email address will not be sold to any third party company.

* 35% discount code presented after product registration is valid on most print books, eBooks, and full-course videos sold on QuePublishing.com. Discount may not be combined with any other offer and is not redeemable for cash. Discount code expires after 30 days from the time of product registration. Offer subject to change.

quepublishing.com